£17.95

[illegible] on the op amp, [illegible] enable the reader [illegible]

[illegible] me rather than a teaching text. [illegible] er in its pages an extensive and invaluable source of functional [illegible] analog circuits, from integrators and differentiators to logarithmic amplifiers, from instrumentation amplifiers to filters. The circuits are conveniently grouped according to function, and the approach followed is to build up slowly from the basic textbook examples towards a series of practical, workable circuits.

Students who need to build and test particular types of analog circuitry as part of an assignment or project based activities will find this book invaluable. Professional engineers will also find the book useful for design and development work. The coverage is extensive and up-to-date, and provides a wealth of expert, technical advice on the selected circuits.

Analog Electronics with Op Amps

Analog Electronics with Op Amps

A Source Book of Practical Circuits

A. J. PEYTON

University of Manchester Institute of Science and Technology

V. WALSH

British Aerospace Ltd

Published by the Press Syndicate of the University of Cambridge
The Pitt Building, Trumpington Street, Cambridge CB2 1RP
40 West 20th Street, New York, NY 10011-4211, USA
10 Stamford Road, Oakleigh, Melbourne 3166, Australia

First published 1993

Printed in Great Britain at the University Press, Cambridge

A catalogue record for this book is available from the British Library

Library of Congress cataloguing in publication data
Peyton, A. J.
Analog electronics with Op Amps: a source book of practical
circuits / A. J. Peyton, V. Walsh.
p. cm.
Includes bibliographical references and index.
ISBN 0 521 33305 9. ISBN 0 521 33604 X (pbk.)
1. Linear integrated circuits. 2. Operational amplifiers.
I. Walsh, V. II. Title.
TK874.P459 1993 621.3815—dc20 92-22691 CIP

ISBN 0 521 33305 9 hardback
ISBN 0 521 33604 X paperback

KW

Contents

Preface

In recent years we have seen the emergence of a new subject in electronics, that of digital signal processing (DSP). In DSP, which is based on the computational power of the microprocessor, many new application areas have been pioneered and at the same time old ones have been given a fresh impetus. Results have been produced, through various software techniques, which would previously have been possible, if at all, only through the prohibitively extensive use of hardware. Over this same period, a new technology has also come to maturity centred around the creation of monolithic integrated circuits which combine both analog and digital operations on a single silicon substrate. These hybrid ics and powerful DSP systems have produced enormous benefits for the design engineer in terms of reduced costs, increased performance and greater flexibility. Analog electronics, however, tends to have been overshadowed, based as it is on the more mature technology of the op amp. Yet a sound grasp of analog electronics is probably more important now than ever since DSP has opened up so many new applications, all of which require an analog front-end for their operation. There will also continue to be a need for prototyping new designs in hardware in the early stages of development work, whether this prototyping is done in the industrial laboratory by the experienced design engineer or in the college or university laboratory by students who are just setting out on the engineering path. Useful as software simulations of analog circuits are to the engineer and student, there is still no substitute for the 'real-world' experience provided by a hands-on approach.

We are conscious, owing to limitations of space, of the absence of conversion electronics in these pages, especially in the use of digital to analog converters and analog to digital converters. Apart from this omission, readers will find that many practical circuits from analog electronics are usefully described and outlined. However, this book is not intended to be a textbook in analog electronics, nor is it an introduction to the fundamentals of the op amp. Many other works carry out this role perfectly well. Instead, it is offered as a source of practical circuits in analog electronics so that the reader can readily and speedily obtain information and advice on the particular task which needs to be carried out using that work-horse of analog electronics, the op amp. Using this superbly flexible building block, and a suitable addition of resistors, capacitors, diodes and discrete transistors, a remarkable range of operations can be carried out. If this volume is the first book which readers consult when they begin their task of designing, building and testing a particular analog circuit, then our work will have succeeded.

I wish to thank my wife, Denise, for her constant help and encouragement, and my daughters Lucy and Anna. With thanks also to my parents. [AP]

I wish to thank my wife Mary and my children, Katherine, Sean, Nicola and Daniel for their tolerance and support during the writing of this manuscript. [VW]

1

Instrumentation amplifiers

An instrumentation amplifier is a device with two balanced differential inputs. The amplifier is configured so that it accurately amplifies the differences between the voltages applied to its two input terminals ($V_{IN2} - V_{IN1}$) without being affected by the common mode voltage on both the inputs as shown here. For many typical instrumentation amplifier ics the voltage gain A_V is set between 1 and 1000.

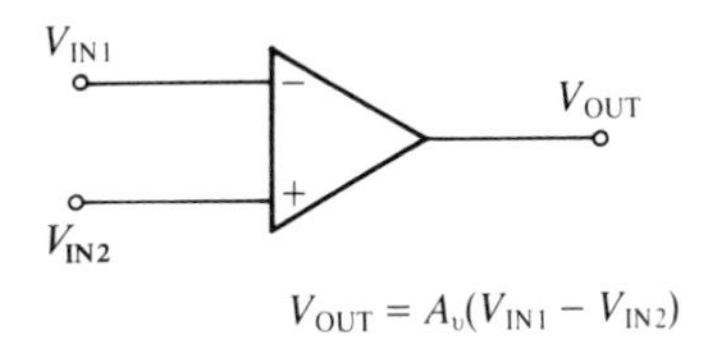

$V_{OUT} = A_v(V_{IN1} - V_{IN2})$

Engineers are always contrasting real world devices with fictional devices, this is a useful approach for providing a 'feel' for the targets being aimed at. We may not hit dead centre but at least we will have a good idea of where the shots are landing. So, in the ideal case, an instrumentation amplifier will have the following characteristics amongst others: a constant and perfectly linear gain independent of time, frequency, load, temperature or humidity; an infinite common mode rejection ratio and power supply rejection ratio; zero input and output offsets and offset drifts and zero output impedance for maximum signal delivered to the load from the amplifier. You will usually find an instrumentation amplifier in the first stage of a measurement or instrumentation system where accurate measurement is the fundamental requirement for the application. In many cases, the input signal to an instrumentation amplifier will be derived from a bridge network or some form of transducer which has converted a physically varying quantity into an analog electrical signal. The main problems for the design engineer, in amplifying this signal for further processing by later stages, are concerned with restricting unwanted noise and controlling movements in the gain of the amplifier due to environmental changes.

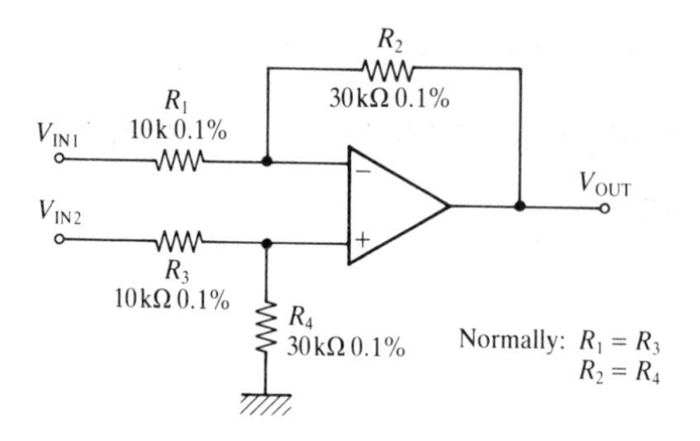

Fig. 1.1. Single op amp instrumentation amplifier circuit.

1.1 Single op amp instrumentation amplifiers

The configuration shown in Fig. 1.1 is the simplest and lowest-cost option for an instrumentation amplifier. R_3 and R_4 act as a potential divider for the non-inverting input of the op amp. The inverting input of the op amp is forced to the same voltage as the non-inverting input due to feedback through R_1 and R_2 and the very large gain of the op amp. The ratio R_2/R_1 sets the gain of the amplifier. When $R_1/R_2 = R_3/R_4$ the differential voltage gain will be much greater than the common mode voltage gain and the common mode rejection ratio (CMRR) will be maximized. For the resistor values shown here, you will get a differential gain of $\times 3$ and a CMRR of 1000, i.e. 60 dB.

Differential gain $= A_D = V_{OUT}/(V_{IN2} - V_{IN1}) = \dfrac{R_2}{R_1} \cdot \dfrac{1}{\left(1 + \dfrac{R_2}{R_1} \cdot \dfrac{1}{A_V}\right)}$
(A_V is the op amp gain)

$$\simeq R_2/R_1 \ (A_V \text{ very large})$$

Common mode gain $= A_{CM1} = \dfrac{R_1 R_4 - R_2 R_3}{R_1(R_3 + R_4)}$
(due to resistor mismatch)

Common mode gain $= A_{CM2} = \dfrac{R_2}{R_1 \cdot \text{CMRR}}$

(due to finite CMRR of op amp.)
(Note that CMRR is expressed as a ratio and not in dB)

$$\text{Common mode rejection ratio} = \text{CMRR} = \frac{A_D}{A_{CM1} + A_{CM2}}$$

Differential input resistance $= R_1 + R_3$

Common mode input resistance $= (R_1 + R_2)//(R_3 + R_4)$

(assuming CMRR $= \infty$)

Output offset voltage $= \left(1 + \frac{R_2}{R_1}\right) V_{IO} + I_{OS} R_2$ if $R_1 = R_3$ and
(worse case) $R_1 = R_4$

where

V_{IO} = input offset voltage of op amp

I_{OS} = input offset bias current.

The circuit in Fig. 1.1 has a low input impedance (around 20 kΩ in this example) and is really only of use in applications which have a low source impedance. Driving the circuit from a high impedance signal source will result in attenuation of the input signal due to loading and a poor common mode performance. Increasing the values of the input resistors (R_1, R_3) will increase the input resistance but at the cost of increasing offset drift due to finite input offset current, reduction of the bandwidth due to stray capacitance and an increase in noise. The values of R_1 and R_3 need to be selected for a trade-off between input resistance and input offset current, input noise current and bandwidth.

A high CMRR is achieved by making $R_1/R_2 = R_3/R_4$ but do not forget that R_1 and R_3 must include the inverting and non-inverting input source impedances otherwise gain errors and common mode errors will be introduced. Similarly, changes in resistor values can cause problems: high-precision, high-quality resistors may be needed.

The bandwidth of this amplifier is going to be limited by either the finite bandwidth of the op amp or by stray capacitances. If the bandwidth is limited by the op amp, assuming that the op amp is fully compensated, then the bandwidth will be approximately $f_A \cdot R_1/(R_1 + R_2)$ where f_A is the gain–bandwidth product (equal to the unity gain crossover frequency for a fully compensated op amp) of the op amp. If the bandwidth is limited by stray capacitances, in particular the strays across R_2 and R_4, and assuming that $R_1 = R_3$, $R_2 = R_4$, then the 3 dB frequency is given approximately by $1/2\pi R_2 C_{STR}$ where C_{STR} is the capacitance across R_2 which is typically 10 pF or less (e.g. if $R_2 = 1$ MΩ, $C_{STR} = 10$ pF, then the bandwidth will be limited by this effect to only 16 kHz). So, if you want a wide bandwidth you must use a fast op amp and resistor values must be kept low.

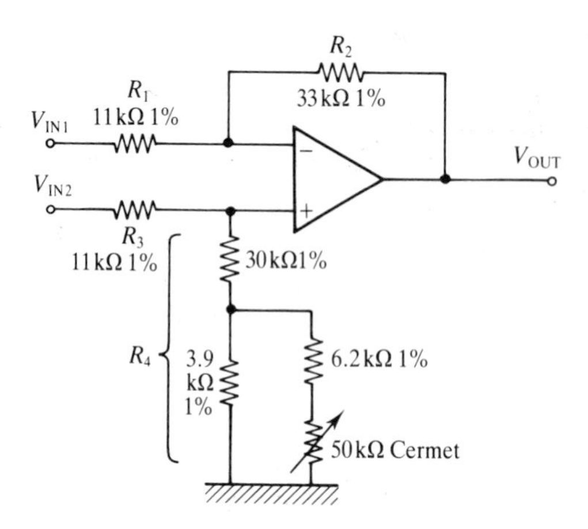

Fig. 1.2. Using a trimmer to maximize CMRR.

Note that the CMRR of the differential amplifier becomes very poor at higher frequencies due to the limitations of the op amp and mismatch of the impedance of R_1 and R_2, R_3 and R_4. In one-off applications, you can use a trimming pot to maximize CMRR. A very high CMRR at dc is possible by trimming since the CMRR of the op amp is cancelled by the common mode contribution of the resistor mismatch. The trimmed value will vary, however, due to drift and ageing of the circuit. The configuration shown in Fig. 1.2. allows you to use a higher value trimmer since low value trimmers tend to be less stable.

A good practice in instrumentation amplifier construction is to lay out the discrete components in a mirror-image fashion onto the board so that any stray capacitances are operating equally on both inputs, this practice is effective in maximizing the common mode frequency response. In some cases you may want to limit the bandwidth of the amplifier; to do this, identical capacitors must be added across both R_2 and R_4; be careful with the tolerances (use 1% if possible) otherwise the amplifier may have a poor high frequency performance.

If you want to increase the gain of the amplifier, the circuit in Fig. 1.3 could be used. This approach achieves high gains without the use of high value resistors. The buffer shown in this circuit will not be needed if R_5 and R_6 are sufficiently low in value compared to R_2 so that loading does not take place.

An increase in amplifier gain can also be obtained by using a T-network for the feedback resistors as shown in Fig. 1.4. This circuit allows you continuously to vary the gain without significantly affecting the CMRR of the circuit. This configuration also allows higher-value resistors to be used in other parts of the circuit to increase input resistance. Notice that the gain is not a linear function of K and the circuit requires three pairs of matched resistors; the four R_2 resistors,

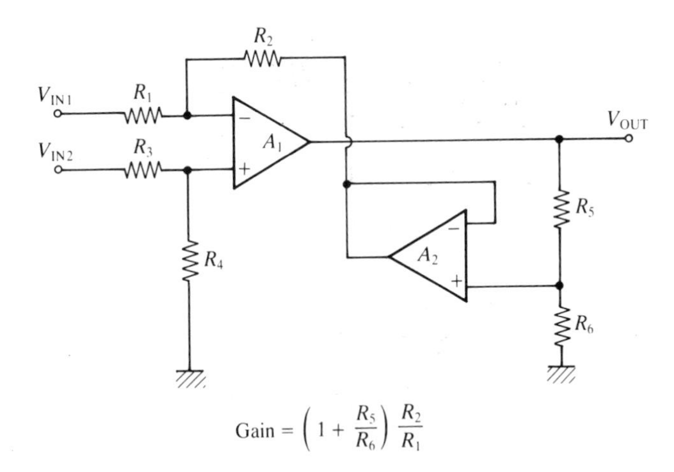

Fig. 1.3. Increasing amplifier gain without using high value resistors.

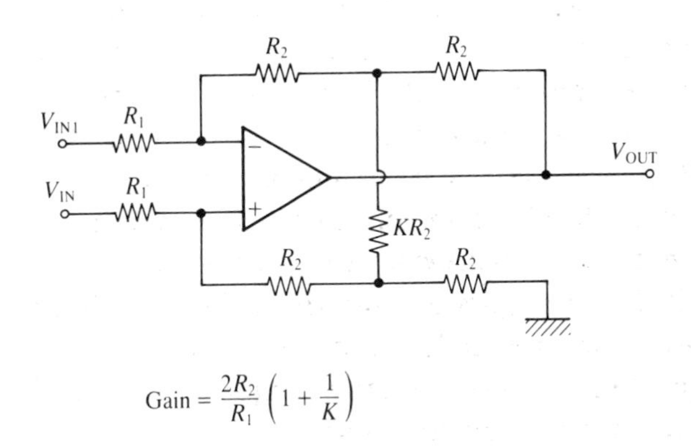

Fig. 1.4. Using a T-network to increase circuit gain.

however, can be on the same package to increase temperature tracking.

One point to watch out for with higher gain (e.g. 1000 or more) differential amplifiers is the finite open loop gain of the op amp. If the open loop gain is not sufficiently high, an excessive differential gain error may result.

If you are expecting high common mode voltages beyond the supply rails, the circuit shown in Fig. 1.5 can be used. Note that A_2 nulls the common mode voltage at the input of A_1 through the action of R_5 and R_6. Since no common mode voltage is applied to either A_1 or A_2, this circuit gives good CMRR. The limits of the CMRR depend on how well R_1/R_5 can be matched to R_3/R_6. Usually $R_5 = R_6$, $R_7 = R_8$, $R_1 = R_3$ and $R_2 = R_4$.

Standard offset nulling techniques can be used for instrumentation amplifiers. The circuit shown in Fig. 1.6 is useful for instrumentation amplifiers since varying the reference level at the non-inverting input of the op amp offers quite effective offset nulling. This technique has the

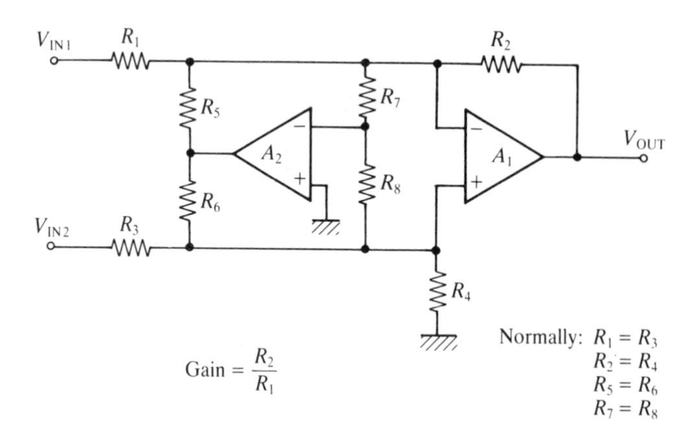

Fig. 1.5. High common mode voltage circuit.

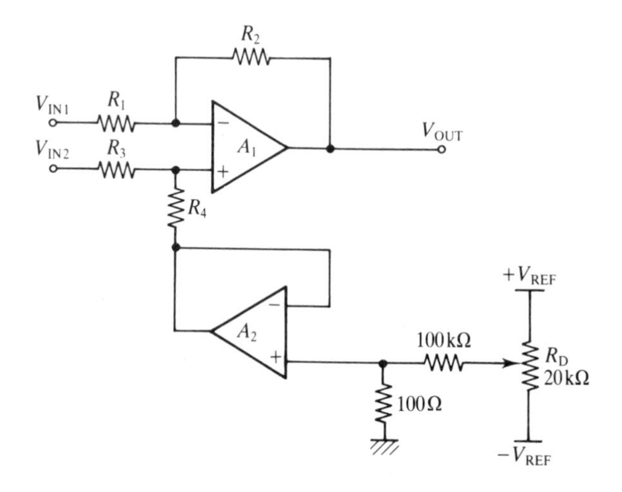

Fig. 1.6. Reference level nulling.

disadvantage of requiring a low output impedance reference, hence the use of a second op amp A_2. A lower value of divider R_D could be used, with A_2 omitted, but this may degrade the performance of the amplifier. Note that this circuit effectively nulls the offset voltage at the output. So, only small output voltage offsets well within the range of output swing of the op amp can be nulled.

1.2 Two op amp instrumentation amplifiers

In the configuration shown in Fig. 1.7 both op amps are connected together as non-inverting amplifiers where the first non-inverting amplifier (A_1) varies the reference level of the second, (A_2). The output of A_1 is fed to the inverting input so that A_1 amplifies the differential input signal ($V_{IN2} - V_{IN1}$). This circuit results in a much higher input impedance than is possible with the single op amp configuration.

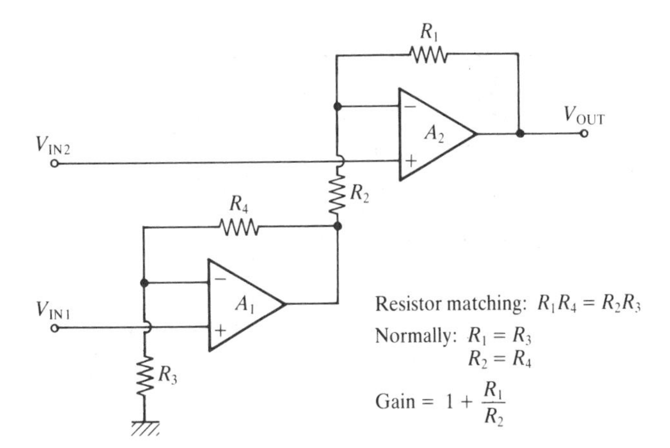

Fig. 1.7. Two op amp instrumentation amplifier circuit.

$$\text{Gain} = 1 + R_1/R_2$$

$$\text{Common mode gain} = (R_1R_4 - R_2R_3)/R_3R_2$$

$$= 1 - \frac{R_1}{R_2}\cdot\frac{R_4}{R_3}$$

$$\text{Therefore if } \frac{R_1}{R_2} = \frac{R_4}{R_3}, \quad \text{then the common mode gain} = 0$$

$$\text{Output offset} = 2(1 + R_1/R_2)V_{IO}$$

where

$$R_1/R_2 = R_3/R_4$$

and

$$V_{IO} = \text{input offset voltage of the op amp}$$

$$\begin{pmatrix}\text{Differential input impedance}\\ \text{Common mode input impedance}\end{pmatrix} \Rightarrow \begin{pmatrix}\text{dependent on the type}\\ \text{of op amp used in the circuit.}\end{pmatrix}$$

As before, the close matching of equal value resistors is essential for high common mode rejection.

Gain can be varied, if required, by adding an extra variable resistor as shown in Fig. 1.8. The gain in this circuit does not, however, vary linearly with R_3. Not shown on this circuit are the extra components needed to provide a path for input bias currents to flow (see later).

A further development of the previous configuration applies the input signal between the two inverting terminals of the op amp as shown in Fig. 1.9. Note, though, that this circuit has the disadvantage of a low input impedance (approximately equal to R_1); the gain, however, can be varied proportionally by R_2.

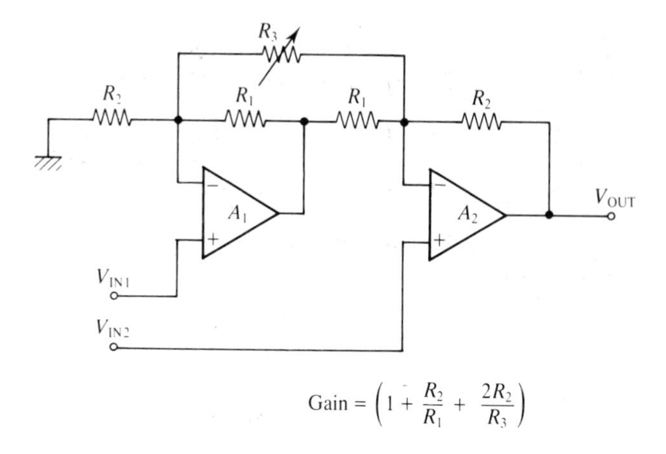

Fig. 1.8. Variable gain modification to the two op amp instrumentation amplifier.

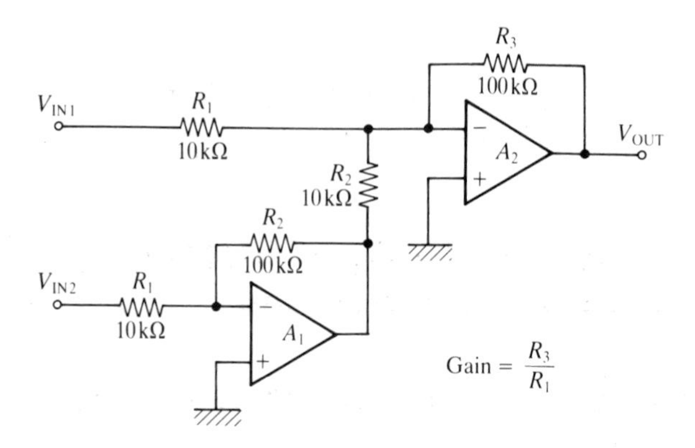

Fig. 1.9. Two op amp instrumentation amplifier using the inverting inputs.

1.3 Three op amp instrumentation amplifiers

Fig. 1.10 shows a standard, op amp based, two stage instrumentation amplifier circuit. The first stage, comprising amplifiers A_1 and A_2, amplifies the differential signal by a factor of $(R_1 + R_2 + R_3)/R_1$ and yet restricts the common mode gain to unity. The differential signal is consequently increased at the output of A_1 and A_2 without any increase in the common mode signal. The next stage, A_3, is a single op amp differential amplifier which amplifies the increased differential signal by a factor of R_5/R_4. This circuit has a higher input impedance and can give much larger gains with improved CMRR than is possible with the single op amp approach. The CMRR is also less sensitive to resistor ratios.

$$\text{Differential gain} = A_D = \frac{(R_1 + R_2 + R_3)}{R_1} \cdot \frac{R_5}{R_4}$$

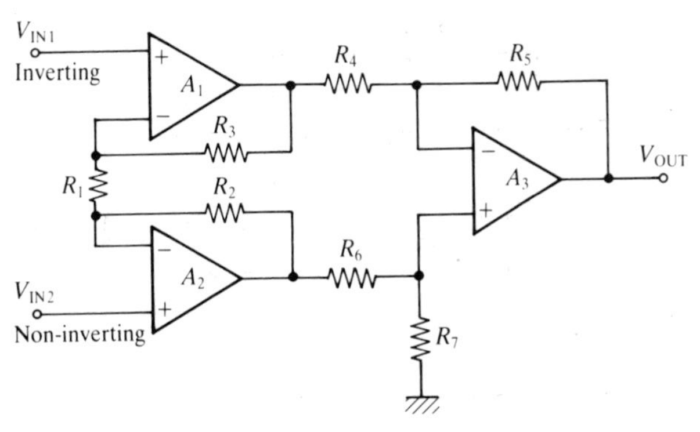

Fig. 1.10. Three op amp instrumentation amplifier circuit.

For maximum CMRR : $R_4R_7 = R_5R_6$
Normally : $R_2 = R_3 \quad R_4 = R_6 \quad R_5 = R_7$

Common mode gain $= A_{CM1} = (R_7R_4 - R_5R_6)/R_4(R_6 + R_7)$
(due to resistor imbalance)

Common mode gain $= A_{CM2} = R_5/R_4 \cdot \text{CMRR}$
(due to finite CMRR of op amp $A3$)

CMRR of instrumentation amplifier = CMRR $= \dfrac{A_D}{A_{CM1} + A_{CM2}}$
(for worst case of A_{CM1} adding to A_{CM2})

Output offset, V_{OO2}, due to offset voltage of op amps
(worse case)

$$V_{OO1} = \frac{R_1 + R_2 + R_3}{R_1} \cdot \frac{R_5}{R_4} (V_{IO1} + V_{IO2}) + \left(1 + \frac{R_5}{R_4}\right) V_{IO3}.$$

where

V_{IO1} = input offset of op amp 1

V_{IO2} = input offset of op amp 2

V_{IO3} = input offset of op amp 3

Output offset, V_{OO2}, due to op amp bias currents
(worse case with all offset errors adding)

$$V_{OO2} = [R_3I_{B2}^- - R_{S2}I_{B2}^+ - R_2I_{B1}^- + R_{S1}I_{B1}^+] \frac{R_5}{R_4} + I_{OS}R_5$$

where

$R_4 = R_6$ and $R_5 = R_7$

I_{B1}^-, I_{B1}^+ = input bias currents op amp 1

I_{B2}^-, I_{B2}^+ = input bias currents op amp 2

I_{OS3} = input offset current op amp 3

R_{S1} = inverting input source impedance

R_{S2} = non-inverting input source impedance

Total output offset $= V_{OO} = V_{OO1} + V_{OO2}$
(worse case)

It is extremely important that a path is available for bias currents to flow from ground into the inputs of the instrumentation amplifier, otherwise the amplifier will saturate. If the input signal source does not provide such a path, for example when the inputs are AC-coupled, then resistors must be added to ground as shown in Fig. 1.11. R is typically chosen 1 MΩ or above; note that R and C constitute a high pass filter which must pass all signal frequencies.

The output offset and output offset drift are usually higher with this circuit than with previous configurations since more op amps are being used. To counteract this, the input op amps can be chosen so that they have offsets which drift together. Feedback resistors can also be kept low to reduce drift due to the input bias currents of the op amps. In addition, FET input op amps are often used for A_1 and A_2, which have extremely small bias currents. It is worth noting that the bias current effects of A_1 and A_2 tend to cancel.

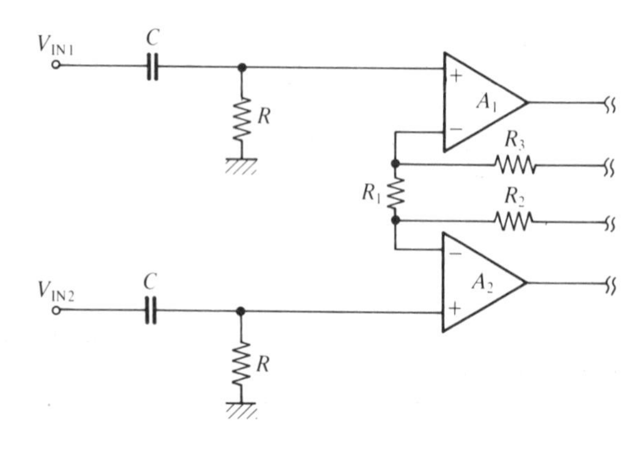

Fig. 1.11. Providing a path for input bias currents.

The gain of the circuit can be varied by changing the value of R_1 without affecting either the input impedance or the CMRR of the circuit; R_1 does not, however, give proportional control over the gain. When choosing resistor values, it is convenient to place all of the circuit gain in the first stage of the amplifier, around A_1 and A_2, as this stage preferentially amplifies the differential input voltage over the common mode voltage. The gain of A_3 can be set at unity by making $R_4 = R_5 = R_6 = R_7$. Resistors in the same package can then be used to allow very close matching of these values along with common drifts of resistor values due to environmental changes. If you put too much gain in the first stage, however, A_1 or A_2 may saturate and you may restrict the common mode input range of the amplifier.

A guard driver can be used to reduce common mode input capacitance and input leakage from connecting cables (over circuit boards and so on) by raising the surrounding area to the same common mode potential. Fig. 1.12 shows such an arrangement.

For applications with a high source impedance requiring a relatively wide bandwidth, a bootstrap technique can be used which compensates for the input capacitances to each input. The circuit shown in Fig. 1.13 produces very small input capacitances and this is important when amplifying from sources which have a high resistance (e.g. MΩs) since the source resistance and input capacitance of the amplifier form a low pass filter. For example, with a 100 MΩ source impedance and 20 pF input capacitance, there is a cut-off frequency of $1/(2\pi R_S C_{IN})$ i.e. only 80 Hz. Note that the two inputs may need individual guarding as shown

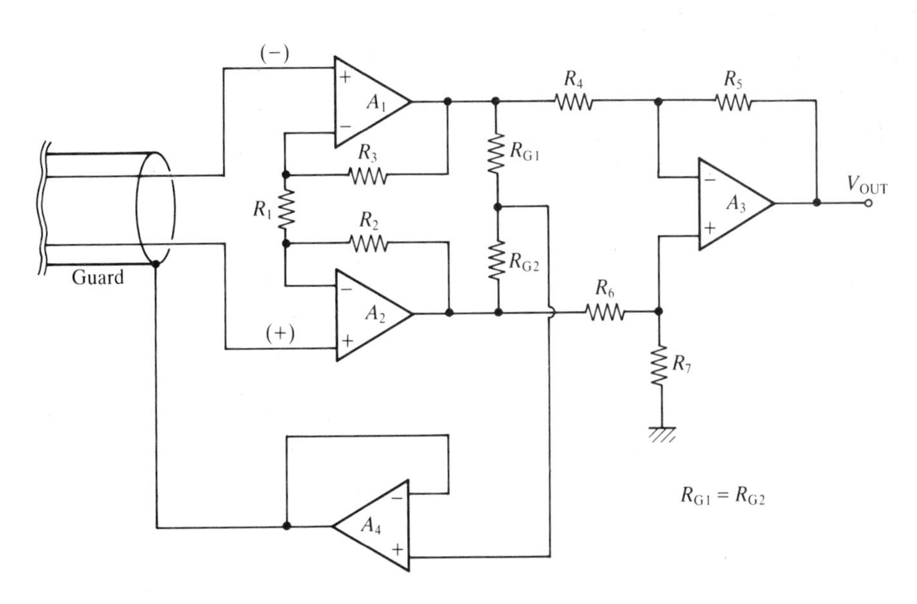

Fig. 1.12. Using a guard driver to increase the input impedance.

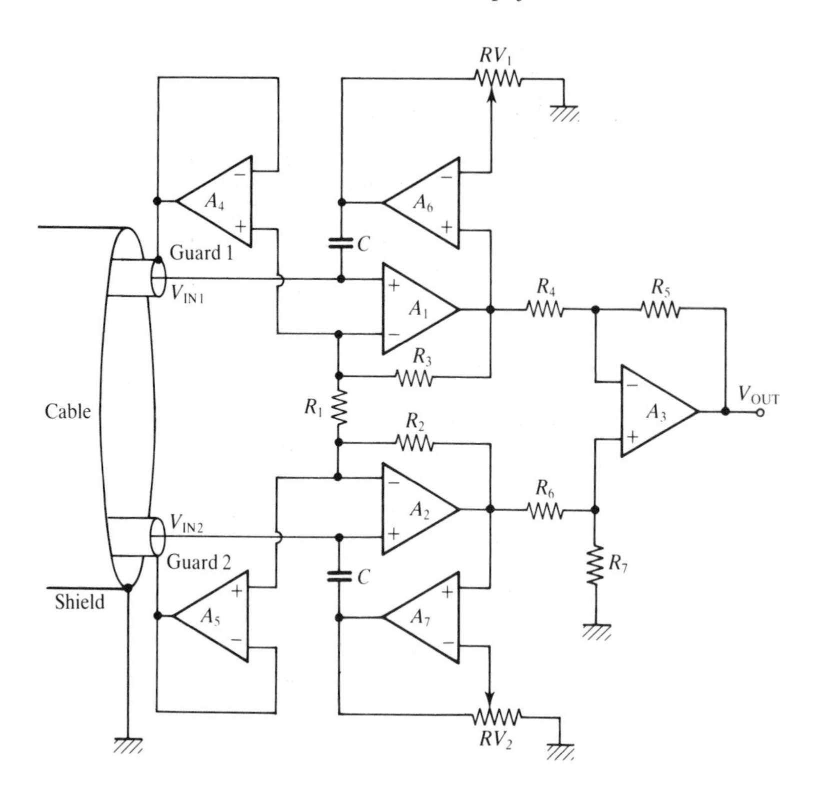

Fig. 1.13. Compensating for the input capacitances by bootstrapping.

in the circuit. Also, the input op amps (A_1 and A_2) must be FET input types. The value of C should be small, in the region of 10 pF. Output offsets can be nulled using any of the standard techniques but note that you can null a larger range of offsets on A_1 and A_2 than on A_3 alone. RV_1 and RV_2 are adjusted to compensate for cable input capacitances (the circuit will oscillate if you over-compensate!).

1.4 Matched transistor instrumentation amplifiers

The availability of highly-matched transistor pairs has enabled the designer to replace the input stage of the op amp with a circuit which is custom-designed for the application. The matched bipolar transistor circuit shown in Fig. 1.14 uses current feedback and may be especially useful for applications which require very low noise, very low drift or low supply currents. The circuit consists of an input emitter follower pair, Q_1 and Q_2, biased from the matched current source from Q_3 and Q_4. Transistors Q_3 and Q_4, with resistors R_{41}, R_{42}, R_5 and diodes D_1 and D_2, form a constant current source which forces equal current I to flow down each side of the input stage. The differential input voltage ($V_{IN2} - V_{IN1}$) is impressed across R_3 by the emitter follower action of

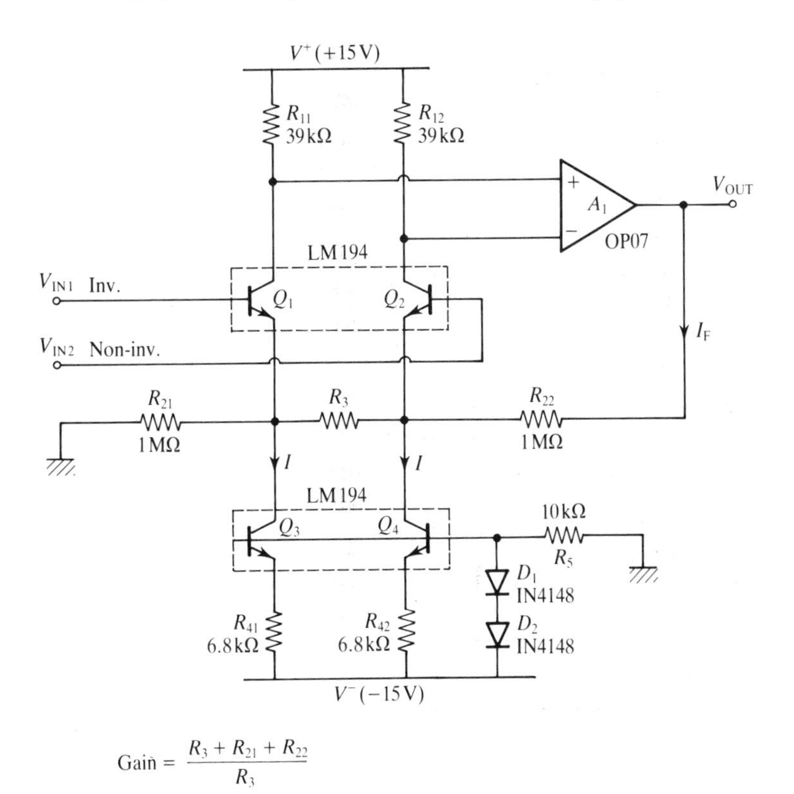

Fig. 1.14. Matched transistor instrumentation amplifier.

Normally: $R_{11} = R_{12} = R_1 \quad R_{21} = R_{22} = R_2 \quad R_{41} = R_{42} = R_4$

input transistors Q_1 and Q_2. Current flowing through R_3 causes an initial imbalance in the currents in each side of the input stage. This current imbalance creates a differential voltage at the input terminals of A_1 due to different voltage drops across equal resistors R_{11} and R_{12}. A_1 corrects this initial imbalance by injecting current I_{12} back into the input stage, through R_{21}, R_{22} and R_3, thus producing output voltage V_{OUT}.

$$\text{Differential gain} = A_{\mathrm{D}} = 1 + (2R_2/R_3)$$

Common mode gain:

(i) due to mismatch in resistors R_{21} and R_{22} of $\Delta R_2 = A_{\mathrm{CM1}} = \Delta R_2/R_2$

(ii) due to mismatch in resistors R_{11} and R_{12} of $\Delta R_1 = A_{\mathrm{CM2}} = \Delta R_1/R_1$

(iii) due to mismatch in the input resistors' dynamic emitter impedance

$$\Delta r_{\mathrm{e}} = A_{\mathrm{CM3}} = \frac{\Delta r_{\mathrm{e}} \cdot (R_3 + 2R_2)}{R_2 \cdot R_3}$$

(iv) due to CMRR of op amp A_1 of $\mathrm{CMRR} = A_{\mathrm{CM4}} = \dfrac{1}{\mathrm{CMRR}}$

Common mode rejection ratio = CMRR = $\dfrac{A_D}{A_{CM1} + A_{CM2} + A_{CM3} + A_{CM4}}$

Differential input resistance = $R_{IND} = \beta(R_3 // 2R_2)$

where

$$\beta = \text{current gain of } Q_1 \text{ and } Q_2$$

Common mode input resistance = $R_{INCM} = \beta R_2/2$

Input offset voltage:

(i) due to imperfect matching of Q_1 and Q_2 of $V_{IO2} = V_{IO2}$

(ii) due to imperfect matching of Q_3 and Q_4 of $V_{IO3} = \dfrac{V_{IO3} R_3}{R_4}$

(iii) due to mismatch of R_{41} and R_{42} by $\Delta R_4 = \dfrac{\Delta R_4}{R_4} \cdot I R_3$

(iv) due to mismatch of R_{11} and R_{12} by $\Delta R_1 = \dfrac{\Delta R_1}{R_1} \cdot I R_3$

(v) due to the input offset voltage in A_1 of $V_{IO1} = \dfrac{R_3}{R_1} V_{IO1}$

(vi) due to the input offset bias current of A_1 of $I_{IOS} = I_{OS} \cdot R_3$

$$\text{Total input offset voltage} = V_{IO2} + \frac{R_3}{R_4} V_{IO3} + \frac{\Delta R}{R_4} I R_3 + \frac{\Delta R_1}{R_1} I R_3 + \frac{R_3}{R_1} V_{IO1} + I_{OS} R_3$$

Current source $I = \dfrac{0.7}{R_4}$ for a silicon transistor and diodes.

There are several points to note when choosing components and component values:

The input voltage range of the instrumentation amplifier is limited at the positive end by the voltage range of the op amp inputs and at the negative end by the voltage compliance of the current source.

The inputs to the op amp A_1 must be within the working input voltage range of this op amp. Choose an op amp which can tolerate its inputs being close to its positive supply line.

Resistors R_{21} and R_{22} draw current from the emitters of Q_1 and Q_2 when common mode voltages are applied. So, R_{21} and R_{22} are generally made large so that the current they draw is insignificant to the normal emitter currents of Q_1 and Q_2. As a rule of thumb, make these resistors at least ten times the value of R_{11} and R_{12}.

As a simple example, assume that an OP07 is used for A_1 (the OP07 is a widely used, low drift precision op amp) and that the popular LM 194 supermatched transistor pairs are used for Q_1, Q_2, Q_3 and Q_4, with power supplies of ± 15 V.

The data sheet shows us that the minimum input voltage range of the OP07 = ± 13 V therefore R_{11} and R_{12} must have at least (15 V − 13 V) dropped across it, i.e. 2 V, therefore choose a figure of 4 V to drop across R_1. A reasonable collector current for the Q_1 and Q_2 transistors is around 100 μA so design the current source around Q_3 and Q_4 to supply 100 μA, this determines the values of R_{41} and R_{42}. The values of R_{11} and R_{12} are given by 4 V/100 μA $\simeq$ 40 kΩ so choose a value of around 47 kΩ. The value of R_2 should be at least 10 R_{11}, around 400 kΩ, so choose R_{21} and R_{22} to be 1 MΩ, this value would only draw a current of 10 μA for a 10 V common mode input. Choose R_3 to give the required gain from the gain equation.

A poor Common Mode Rejection Ratio can occur due to a number of reasons such as, mismatching of R_{11}, R_{12}, mismatching of R_{21} and R_{22}, mismatch in transistor dynamic emitter resistance, r_e, and the finite CMRR of op amp A_1. You can optimize CMRR by adding a trimming pot in series with R_{21} or R_{22}.

Output offset error can occur due to a number of different sources: mismatch of Q_1 and Q_2; mismatch of Q_3 and Q_4; mismatch of R_{41} and R_{42} and the input offset from A_1. Offset drift can be minimized by increasing the gain of the matched transistor input stage, making R_3 as small as possible compared to R_1 and R_4. This will reduce the offset contribution from the rest of the circuit except Q_1 and Q_2. The offset can be nulled by inserting a small-value trimming pot between either R_{11} and R_{12} or R_{41} and R_{42}. This technique could increase offset drift, though, since matched transistors Q_1 and Q_2 will be operating with different emitter currents.

Again, do not forget that a path must be provided for bias currents to flow into the inputs of transistors Q_1 and Q_2, otherwise the amplifier will saturate. Inputs can be given some degree of protection by placing current limiting resistors in series with the inputs. The limiting resistors will, however, generate noise which will add to the input noise and may also increase the offset voltage due to the flow of the transistors Q_1 and Q_2 base bias currents. Input noise can be minimized by increasing the gain of the input stage, where R_3 is made much smaller than both R_1 and R_4. The increase in gain will have the effect of reducing the noise contribution from all parts of the circuit apart from the Q_1 and Q_2 noise contributions. For low noise applications, the matched transistors Q_1 and Q_2 can be selected to give the optimum noise performance for the particular source resistance and signal frequency range of the application and then they can be biased at their optimum value of emitter current.

With the circuit of Fig. 1.14, the op amp has rather a complicated feedback loop containing the two transistors Q_1 and Q_2, which are configured, as far as the feedback considerations are concerned, in the common base mode (since the instrumentation amp inputs are taken as a.c. ground). Fortunately, when using fully compensated, low-to-medium frequency op amps, stability is not usually a problem since the gain of the feedback network (i.e. the common base transistors) is less than 1.0, as $R_{22} > R_{12}$. With high speed op amps, or where large values are used for R_{11}, R_{12}, R_{21}, R_{22} and R_3, you may find that stability is a problem due to the effects of stray and parasitic capacitances.

You can improve the gain linearity and CMRR by inserting an op amp in each side of the input stage before A_1 as shown in Fig. 1.15. This circuit has a similar configuration to that of the standard 3 op amp circuit with the difference that matched transistor inputs are used. The technique shown in this circuit is commonly found in high performance commercial instrumentation amplifier ics. There are three main advantages of this configuration over the one shown in Fig. 1.14.

(i) V_{REF}, R_1, A_1 and A_2 set the emitter currents of Q_1 and Q_2 at a constant value over the entire range of input voltages. This allows Q_1 and Q_2 to operate at their optimum value of emitter current for low noise or low drift over their entire working range of input voltages.

(ii) Gain non-linearity is considerably reduced since the open loop gain around Q_1 and Q_2 has now been significantly increased by A_1 and A_2 respectively.

(iii) Common mode voltages are no longer impressed upon the feedback resistors R_{21} and R_{22} and so close balancing of resistor values is not needed. Consequently, the CMRR is significantly increased.

Matched FETs are superior to bipolar transistors for applications with a very high source impedance. A FET input stage offers very high input impedance, very low input bias current and very low input noise current. FETs do have, however, higher input offset voltages, offset voltage drifts and input equivalent noise voltages. So, bipolar transistors are usually better for low source impedance applications (less than 10 kΩ) and are more commonly used. The circuit shown in Fig. 1.16 is similar to the circuit in Fig. 1.15 (apart from the use of a FET input stage) and can be improved using the same techniques. Note the use of the cascade-connected differential input stage which ensures that the two input transistors *Q*1 and *Q*2 operate under constant drain-source voltage conditions. This connection helps to improve the frequency response and reduces input bias currents. Suitable dual FET packages include U401-6 series, 2N6905-7 series and the 2N5196-9 series amongst others.

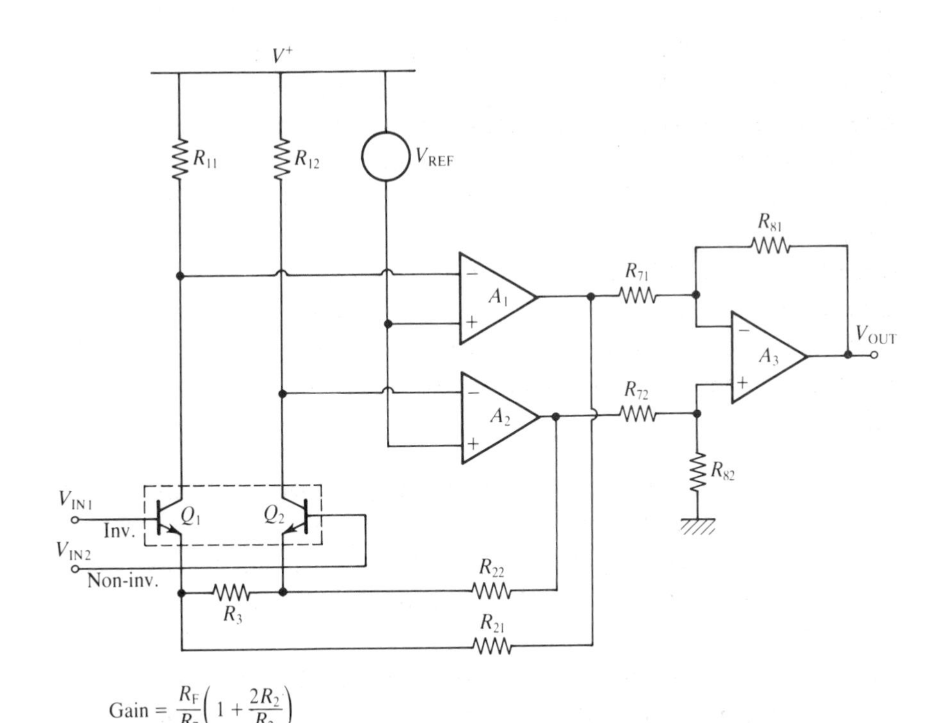

$$\text{Gain} = \frac{R_F}{R_7}\left(1 + \frac{2R_2}{R_2}\right)$$

Where: $R_{11} = R_{12} = R_1$
$R_{21} = R_{22} = R_2$
$R_{71} = R_{72} = R_7$
$R_{81} = R_{82} = R_8$

Fig. 1.15. Improving gain linearity and CMRR.

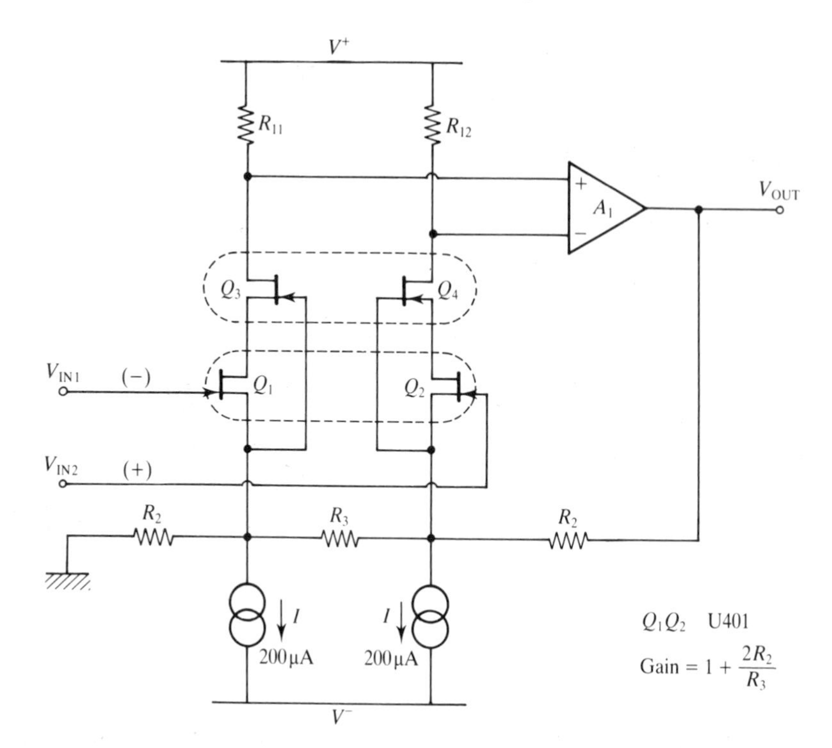

Fig. 1.16. Matched FET input instrumentation amplifier.

1.5 Using instrumentation amplifiers with transducers

A common application for instrumentation amplifiers is in amplifying the output of transducers. Examples of two such applications are shown in Fig. 1.17 and Fig. 1.18. Fig. 1.17 shows a commercial instrumentation amplifier, AD624C, in a strain gauge bridge arrangement. Fig. 1.18 shows a simple differential amplifier used to amplify the signal from a Hall probe monitoring a magnetic field.

1.6 Commercial single ic instrumentation amplifiers

Table 1.1 gives the typical performance parameters of a cross-section of commercially available single ic instrumentation amplifiers. Clearly, the particular application will determine the selection of a single ic commercial instrumentation amplifier along with such factors as price and availability. There are several instrumentation amp ics available which are digitally controllable. These are particularly suitable for microprocessor based systems.

From the technical viewpoint, the selection of any commercially

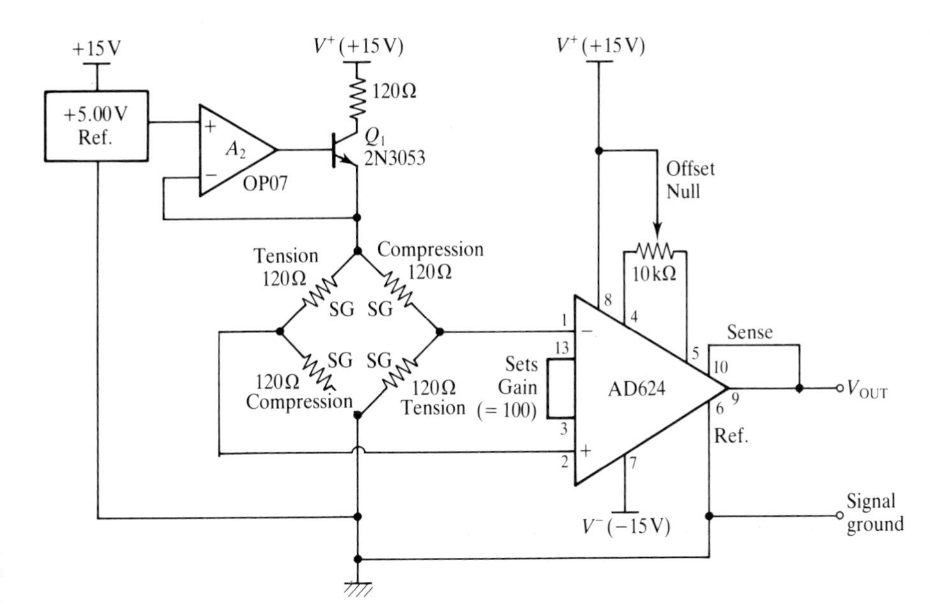

Fig. 1.17. Using an AD624C in a strain gauge (SG) bridge application.

available instrumentation amplifier is usually based on the calculation of an error budget. If you do draw up an error budget you must take all the relevant parameters for the instrumentation amplifier into account and calculate the worst-case error over the temperature range to determine whether the amplifier has satisfied the target accuracy or resolution. An error budget analysis is detailed in Table 1.2 for the Analog Devices AD624C used in the strain gauge bridge application shown in Fig. 1.17.

A modest operating temperature range of 20°C ± 20°C is assumed for this example which is well within the operating range specified by the manufacturers. The following conditions are also assumed: a maximum input of 10 mV, a maximum output of 1 V and a bandwidth of 10 Hz. Other conditions which are assumed are that the bridge is balanced and that the bridge and the amplifier have had their offsets zeroed at a temperature of 20°C. In this example, the difference between resolution and accuracy is clearly highlighted and the effect of the particular application is shown. Four different cases are taken.

Case A. The amplifier is used directly, without trimming its gain or offset errors.

Case B. The amplifier is used with trimming pots to remove gain and offset errors. However, gain and offset drifts due to temperature cannot be avoided in this case.

Table 1.1. *Performance parameters for single ic instrumentation amplifiers*

		AMP-01	AMP-05	PGA200	LH0084C
Supplies					
Max.	V	±18	±18	±18	±18
Min.	V	±4.5	±4.5	±10	±8
Supply current (+ve)	mA	4.8[A]	10.0[A]	10	12
DC offset errors					
Input offset voltage (25°C)	μV	40	500	25	300
Input offset voltage drift	μV/°C	0.3	7	1	10
Output offset voltage (25°C)	mV	2	5	0.2	0.6
Output offset voltage drift	μV/°C	50	70	10	20
Input bias current (25°C)	nA	2	0.03	10	150
Input offset bias current	nA	0.5	0.01	10	50
Noise					
Input 0.1 to 10 Hz (pK·pK)	μV	0.12[C]	4[C]	0.8	7
Input density (1 kHz)	$nV/\sqrt{H_3}$	5[C]	16	13	–
Input impedance					
Differential	GΩ	1[C]		10//3 pF	100
Common mode	GΩ	20[C]	1000//8 pF	10//3 pF	100
Gain					
Range		0.1–10 000	0.1–2000	1–1000	1–100
Accuracy	%	0.5	0.4	0.02	0.03 ($G = 100$)
Drift	ppm/°C	5	8[C]	10	1
Non-linearity	ppm	7[C]	200[C]	120[C]	20
CMRR at dc					
Gain = 1	dB	90	80	95	80
Gain = 1000	dB	125	100	120	–
Dynamic					
Slew rate	V/μs	4.5 ($G = 10$)	7.5 ($G \geqslant 10$)	0.4	13
3 dB bandwidth $G = 1$	kHz	570	3000	500	3250
1% error bandwidth $G = 1$	kHz	–	–	50	300
3 dB bandwidth $G = 1000$	kHz	26	120	2.4	–
1% error bandwidth $G = 1000$	kHz	–	–	0.3	–
Settling time 0.1% $G = 1$	μs	12[D]	5	35	2.3
Settling time 0.1% $G = 1000$	μs	50[D]	5	480	–
Comments		Bipolar input. Requires two external gain set resistors.	FET input. Reqired two gain set resistors. Contains two independent guard drivers.	Bipolar input. Digitally controlled. For gains of 1, 10 and 100, with three input channels.	Bipolar input. Pin strap for gains of 100, 200, 400, 500, 1 K, 2 K. Contains a single guard driver.

A: Max.
B: Min.
C: $G = 1000$
D: to 0.01%
E: 0.01–10 Hz
F: at 10 kHz

Table 1.1 (*Continued*)

INA105	INA101	AD624	AD625	
±18	±20	±18	±18	General powered from ±15 V
±5	±5	±6	±6	supplies.
1.5	6.7	3.5	3.5	
–	125	200[A]	50	DC offsets can usually be trimmed
–	2	2[A]	1	to zero with a small pot.
0.05	0.45	5[A]	4	Temperature drifts are more difficult
5	20	50[A]	20	to eliminate.
–	15	50[A]	30	Offset and noise can be
–	15	35[A]	2	split into two components, one at the input (V_{IO}) and one at the output (V_{OO}).
2.4[E]	0.8[E]	0.2[C]	0.2[C]	Total input error $= V_{IO} = V_{OO}/G$.
60[F]	13	4	4	
50 kΩ	10//3 pF	1//10 pF	1//4 pF	FETs for high source impedances.
50 kΩ	10//3 pF	1//10 pF	1//4 pF	Input capacitance dominates about typically 100 Hz.
1	1–1000	1–1000	1–10 000	Many amplifiers require external gain set resistors.
0.005	–	0.25[A] ($G = 100$)	0.035	
1		25[AC]	5[A]	Accuracy and drift specifications do
2	22[C]	50	100[C]	not cover these resistors. Generally, non-linearity increases with gain.
90	85	70[B]	75	CMRR increases with gain.
–	105	110[B]	115	CMRR is worst at high frequencies.
1000	300	1000	650	Slew rates and settling time concern
–	20	–	–	large signal changes.
–	2.5	25	25	Bandwidth figures are for small signals.
–	0.2	–	–	Generally, larger the gain, the slower
4	30	15[D]	15[D]	the response.
–	350	75[D]	75[D]	
Fixed gain of unity. Standard single op-amp differential circuit (Fig. 1.1).	Bipolar input. Set gain by one extra resistor. Three op amp circuit.	Bipolar input. Contains internal resistors for gains of 1, 100, 200, 500 and 1000.	Bipolar input. Gain set by three external resistors. Programmable gain with a differential multiplexer.	

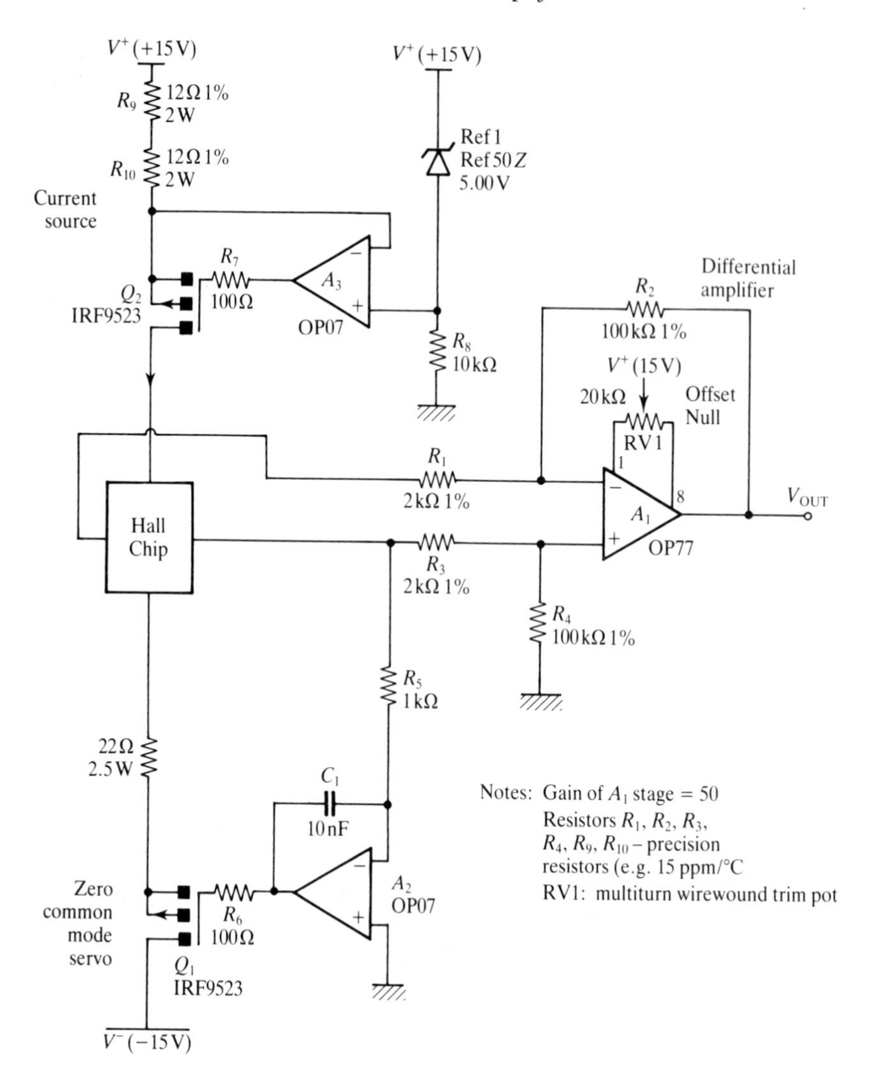

Fig. 1.18. Using a differential amplifier with a Hall probe.

Case C. The amplifier is part of a microprocessor or computer based system, which calibrates directly at the input using accurately known standards, in between taking measurements. Absolute accuracy is important. In this case, temperature variations need not be taken into account since their effects are calibrated out. The only errors which cannot be calibrated out are those due to non-linearity and noise.

Case D. The amplifier is part of a system which is only interested in detected short term changes in the strain levels (say over several seconds). The only important error source in this case is noise.

The importance of an error budget is that it allows the designer to

Table 1.2. *Error budget analysis example with the AD624C*

Error	Specification	Calculation	Untrimmed A	Trimmed B	Repeatedly re-calibrate C	Short term resolution D
Gain error	0.1%	0.1% = 1000 ppm	1000			
Gain drift	10 ppm/°C	20°C × 10 ppm/°C	200	200		
Non-linearity	0.001%	0.001% = 10 ppm	10	10	10	
Input resistance	$10^9\ \Omega$	$\frac{120\ \Omega}{10^9\ \Omega} \times 10^6$	0.1			
Input offset voltage	±25 μV	$(25\ \mu V \times 10^6)/10\ mV$	2500			
Input offset drift	±0.25 μV/°C	$(0.25\ \mu V/°C \times 10°C)/10\ mV$	500	500		
Output offset voltage	±2 mV	$(2\ mV \times 10^6)/1\ V$	2000			
Output offset drift	±10 μV/°V	$(10\ \mu V/°C \times 20°C \times 10^6)/1\ V$	200	200		
Input offset current	±10 nA	$(10\ nA \times 60\ \Omega \times 10^6)/10\ mV$	60			
Input offset current drift	100 pA/°C	$\frac{(100\ pA/°C \times 20°C \times 60\ \Omega \times 10^6)}{10\ mV}$	12	12		
Common mode rejection ratio	110 dB	$(10^{-\frac{110}{20}} \times 2.5\ V)/10\ mV$	750			
Input noise voltage	0.3 μV pk–pk (0.1–10 Hz)	$(0.3\ \mu V \times 10^6)/(2 \times 10\ mV)$	15	15	15	15
Input noise current	60 pA pk–pk (0.1–10 Hz)	$(60\ pA \times 60\ \Omega + 10^6)/(2 \times 10\ mV)$	0.2	0.2	0.2	0.2
Total			7247.3 ≈0.73%	937.2 ≈940 ppm	25.2 ≈25 ppm	15.2 ≈15 ppm

see which errors are significant and even whether automatic calibration and zeroing is required.

In addition to the large number of instrumentation amplifiers available, there are a number of manufacturers who produce specialized amplifiers designed to amplify and condition the signal from a specific transducer. These include thermocouple amplifiers, strain gauge amplifiers and conditioners, resistance thermometer amplifiers and conditioners and Hall probe amplifiers.

2
Isolation amplifiers

An isolation amplifier is characterized by an extremely high level of electrical isolation between its input stage and its output stage. You can achieve this level of isolation by using either optical coupling or transformer coupling between input and output stages. High levels of isolation are especially required where very large common mode voltages are encountered. An isolation amplifier can be designed to handle common mode voltages of several thousand volts. They are of great importance in medical applications where electrodes are attached to the body of a patient and isolation is required for safety reasons. They are also used to prevent ground loops and in applications which need a very high common mode rejection ratio (greater than 100 dB).

Fig. 2.1 shows the block diagram of a complete isolation amplifier system. The block diagram shows the main components of these amplifiers, which consist of an input section, an output section and a power section. The essential feature of these devices is that the input

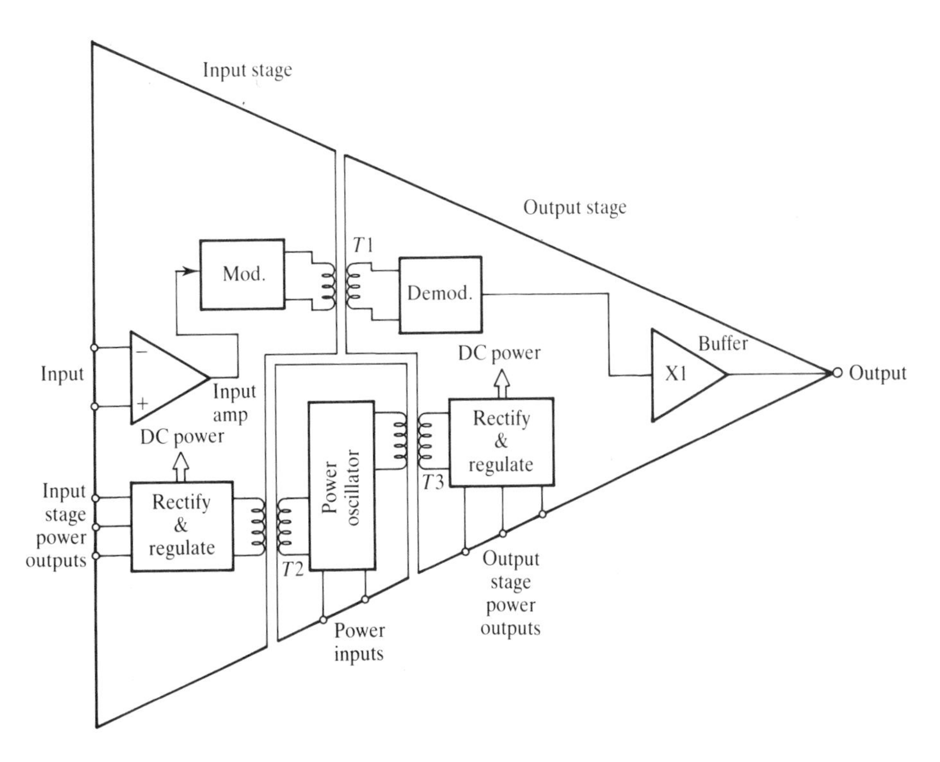

Fig. 2.1. Block diagram of an isolation amplifier.

section and the output section must have complete mutual electrical isolation of both their signal lines *and* their power supplies. The isolation amplifier shown in Fig. 2.1 has what is known as *three port* electrical isolation, which means that the input stage, the output stage and the power section are all isolated from one another. With this three port system, all the internal power is delivered by the power section. However, with some three port devices, the output stage must be powered by the signals' destination circuitry. Many isolation amplifiers are of the so-called *two port* variety, which means that their output stage and their power section are not isolated.

In all isolation amplifier systems, power is furnished to the input stage (and the output stage) by transformer coupling. Typically, a very small ferrite transformer is used. There are three approaches in popular use for coupling the signal between the input stage and the output stage: transformer coupling, optical coupling and capacitive coupling. In addition, there are two techniques commonly used for transmitting the signal through the coupling stage. These are modulation/demodulation, which is used with all three coupling methods, and linearizing feedback which is used with opto-couplers.

2.1 Isolation amplifier using modulation and demodulation

The isolation amplifier shown in Fig. 2.2 uses a dc-to-dc converter for isolated power supplies to the input circuitry. The incoming signal is amplified by a differential input amplifier before being fed to the modulator. The modulated signal is then coupled into the output section, demodulated and buffered. The modulation/demodulation stages are required because the coupling techniques either do not have a dc response or have non-linear or variable characteristics. The modulation techniques commonly used are pulse width modulation (PWM) or frequency modulation (FM). The three main coupling techniques are also shown in Fig. 2.2. The capacitive approach is a relative newcomer which involves coupling the modulation signal through a very small value capacitor (e.g. a few pF). The coupling capacitance must be as small as possible in order to limit the energy which can be transferred through it from common mode voltage transients. Burr Brown devices ISO 120/1/2 and ISO 102/6 employ this capacitive technique.

The modulation/demodulation process does give accurate performance but it also limits the bandwidth of the isolation amplifier. Consequently, the modulating carrier frequency is usually made as high as possible within the operating range of the coupler.

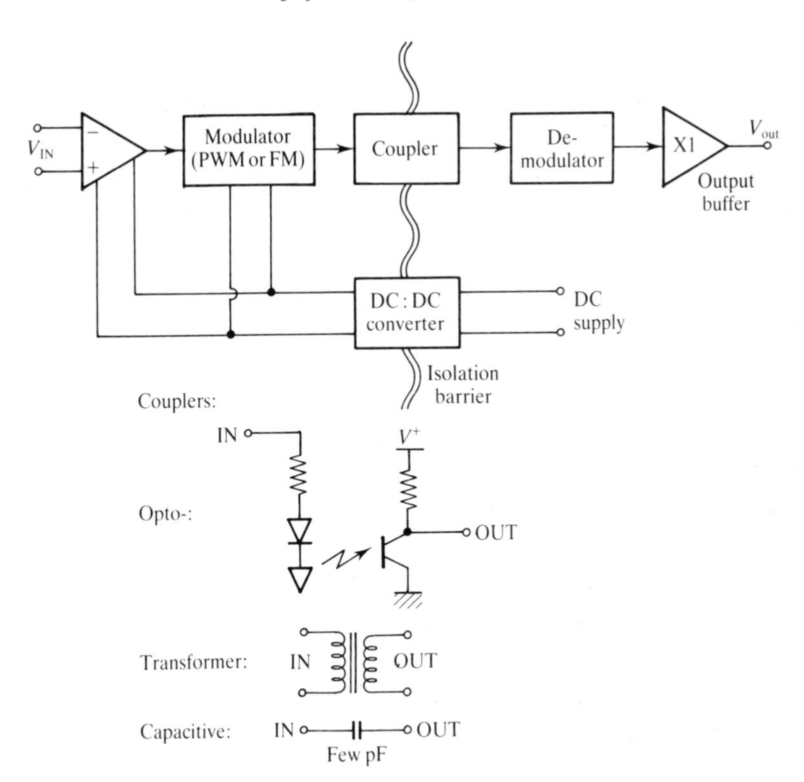

Fig. 2.2. Isolation through a modulation technique.

In applications where there are several isolation amplifiers close together, which use the modulation/demodulation techniques, it is often an advantage to operate the power oscillators of all the amplifiers at the same frequency in order to avoid intermodulation and beat frequency interference problems. Some commercial isolation amplifiers are designed with this facility.

2.2 Isolation amplifier using linearizing feedback

You can realize a low cost alternative to the previous modulation/demodulation technique by using opto-couplers. The circuit of Fig. 2.3 is designed to compensate for the non-linear response of the opto-couplers. The input signal is used to vary currents in OC_1 and OC_2. Isolation is achieved across opto-coupler OC_2. OC_1 is used to provide feedback to amplifier A_1. You should closely match OC_1 and OC_2 so that current changes in Q_2 are mirrored exactly by current changes in Q_1. Any non-linearities in the opto-isolators can then be automatically compensated by feedback through A_1. R_2 and R_4 set the operating current levels for Q_1 and Q_2. R_1 and R_5 set the current swings for the input signal. The speed of this circuit can be relatively fast since the

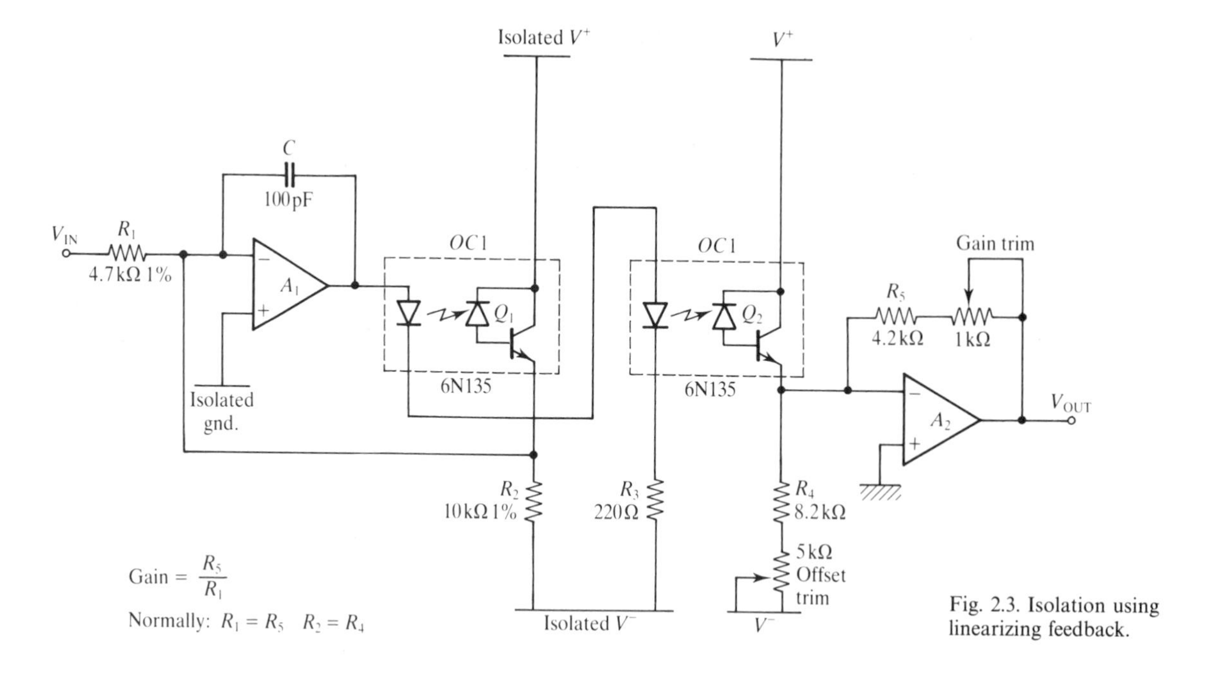

Fig. 2.3. Isolation using linearizing feedback.

voltages across Q_1 and Q_2 are constant and consequently the associated capacitances Q_1 and Q_2 are not charged and discharged as the signal changes. Although this technique offers you a wide bandwidth there is a residue of non-linearity and there are inherent drift and gain accuracy problems due to the use of opto-couplers.

To minimize distortion, gain and offset drifts, OC_1 and OC_2 should be closely matched and ideally contained in the same package. Resistors are chosen such that $R_2 = R_4$ for approximately zero output offset and $R_1 = R_5$ for unity gain. The choice of value for R_1 is generally a compromise between distortion on the one hand and noise and offset drifts on the other. For example, if R_1 is small, the signal current variations in OC_1 and OC_2 will be large and so distortion will be high but current drifts will be small. Alternatively, if R_1 is large, the situation reverses. The V^- rail and the isolated V^- rail must be stable otherwise an unstable output offset will result. You should choose a value for R_3 so that the output of A_1 is approximately 0 V when $V_{IN} = 0$. Note that capacitor C is needed for frequency compensation of the amplifier feedback loop involving A_1 and OC_1. Without this capacitor, your circuit might be unstable and start to oscillate. Values of C can range from 10s to 100s of picofarads. Offsets can be nulled by injecting current through a resistor into the inverting input of A_2. Maximum isolation voltage is limited either by the opto-isolators or by the isolated dc power supply used for A_1 and OC_1.

Table 2.1. *Selection of commercial isolation amplifiers*

A = Max.
B = Min.

Specifications	AD202K	AD210AN	ISO 122P	ISO 103	ISO 100AP
Isolation					
Max. isolation voltage (60 Hz)	1500 V rms	2500 V rms	1500 V rms	1500 V rms	750 V dc
Isolation impedance	2 GΩ//4.5 pF	5 GΩ//5 pF	10^{14} Ω//2 pF	10^{12} Ω//9 pF	10^{12} Ω//2.5 pF
CMRR: at dc ($G = 1$)	–	–			146 dB
at 60 Hz ($G = 1$)	105 dB	120 dB	140 dB	130 dB	108 dB
Gain					
Range	1–100	1–100	1	1	–
Accuracy	0.5%	2%^{A}	0.05%	0.12%	2%
Drift	20 ppm/°C	25 ppm/°C	10 ppm/°C	60 ppm/°C	300 ppm/°C
Non-linearity ($G = 1$)	250 ppm	250 ppm^{A}	80 ppm	260 ppm	1000 ppm
Offset					
Input offset voltage	5 mV^{A}	15 mV^{A}	5 mV	20 mV	–
Input offset voltage drift	10 μV/°C	10 μV/°C	200 μV/°C	300 μV/°C	–
Output offset voltage	5 mV^{A}	45 mV^{A}	–	–	–
Output offset voltage drift	10 μV/°C	30 μV/°C	–	–	–
Supplies					
Power inputs	+15 V	+15 V	±15 V	±15 V	±15 V
Quiescent current	5 mA	50 mA	–	–	–
Isolated power outputs	± 7.5 V at 400 μA	±15 V (×2) at 5 mA	– –	±15 V 15 mA	– –
Dynamic					
Bandwidth ($G = 1$)	2 kHz	20 kHz	50 kHz	20 kHz	60 kHz
Settling time (to 0.1%)	1 ms	150 μs	50 μs	75 μs	100 μs
Package type	SIL/DIL	DIL	DIL	DIP	DIP
Isolation	2 port/with power	3 port	2 port/no power	2 port/with power	2 port/no power
Technique	Transformer	Transformer	Capacitive	Capacitive	Optical
Comments	AD202K is powered with 15 V dc, common to output stage. AD204K is a similar device powered with 25 kHz external clock.	Complete 3-port isolation with separate input/output /power sections.	Requires isolated power supply for input section.	Built in isolated power supply for the input section.	Transresistance amplifier (current i/p voltage o/p). Uses linearizing feedback. Requires power supplies for inputs and outputs. Requires two sealing resistors.

2.3 Commercial isolation amplifiers

There are a large number of commercially-available isolation amplifiers which are sold in a variety of different packages and which use varying isolation techniques (both optical and transformer-based). Table 2.1 gives the important operating parameters from a cross-section of commercially-available isolation amplifiers. The values given are typical at 25°C, unless otherwise stated.

The specifications for isolation amplifiers are very similar to those of a conventional voltage amplifier. However, isolation amplifiers generally have inferior linearity, gain drift and offset drift specifications. The most important specification found on the isolation amplifier data sheet is often the maximum common mode voltage, inputs to outputs, at dc and 50 Hz/60 Hz ac. In addition, the device may be specified as meeting certain mandatory minimum isolation requirements.

Finally, many commercial isolation amplifiers have input stages which contain an uncommitted op amp for flexibility. In interfacing, some may also need the addition of an isolated power supply.

3

Charge amplifiers

A charge amplifier provides an output voltage which varies proportionally to any changes of charge stored on a device connected to its input terminals. Transducers employing a piezo-electric crystal, for example, store a charge which varies according to the mechanical forces acting upon the crystal thereby allowing such parameters as force, pressure and acceleration to be measured. Variable capacitance transducers, condenser microphones for instance, are used in series with a dc voltage so that air pressure or other variations result in changes of charge stored in the transducer. In these and other cases we use charge amplifiers to provide an output voltage. Two basic alternative approaches are the current integrating method and the high input impedance approach.

3.1 Current integrating charge amplifier

Advantages	Disadvantages
linear good frequency response connecting cables are not as critical to performance	a short circuit is, in effect, placed across the transducer which, for crystal transducers, slightly reduces the 'stiffness' of the crystal and hence its resonant frequency

The circuit of Fig. 3.1 behaves as an integrator in which capacitor C_2, connected in the feedback path of op amp A_1, integrates input current I_{IN}. Switch SW_1 may be closed periodically to discharge C_2. Resistor R_2 can be used to provide a dc path for bias currents. In some applications R_2 is omitted since it limits the low frequency response of the charge amplifier. Resistor R_2 must include circuit leakage resistances and the leakage resistance of the switch. Resistor R_1 can be included to limit the high frequency response of the amplifier, where this is desirable, and also to stabilize the closed-loop response of the op amp. R_3 and C_3 have been included in the circuit to represent the leakage resistance and leakage capacitance of the op amp, circuit board and connecting cable.

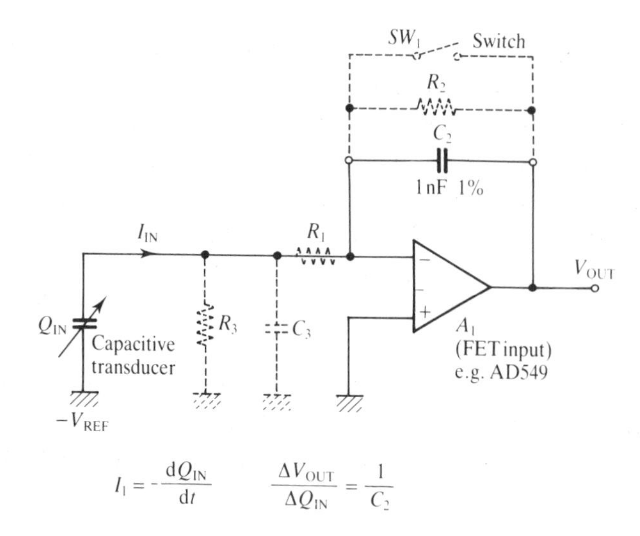

Fig. 3.1. Current integrating charge amplifier.

$$\text{Gain} = \frac{\Delta V_{\text{OUT}}}{\Delta Q_{\text{IN}}} = -\frac{1}{C_2\left(1 + \dfrac{1}{A_{\text{V}}} + \dfrac{C_3}{C_2 A_{\text{V}}}\right)} \simeq -\frac{1}{C_2}$$

(A_{V} is the voltage gain of A_1)

3 dB Bandwidth

$$\text{Low frequency } f_1 = \frac{1}{2\pi C_2 R_2} \text{ (Hz)}$$

$$\text{High frequency } f_2 = \frac{1}{2\pi C_3 R_1} \textbf{ or } \left(\frac{C_2}{C_2 + C_3}\right) \cdot f_{\text{A}} \text{ (Hz)}$$

(Whichever is smallest, where f_{A} is the unity gain crossover frequency of A_1 and assuming that A_1 is fully compensated.)

$$\text{Final output offset} = \left(1 + \frac{R_2}{R_3}\right) V_{\text{IO}} + I_{\text{B}} R_2$$

where

I_{B} is the input bias current of A_1

and

V_{IO} is the input offset voltage of A_1

$$\text{Output drift} = \frac{V_{\text{OUT}}}{C_2 R_2} + \frac{I_{\text{B}}}{C_2}$$

(output drift due to discharge of C_2 and the effect of A_1's input bias current.)

Since the amplifier operates by integrating the input current with a capacitor, the input bias current of op amp A_1 must be kept as small as possible to reduce errors and for this reason FET input op amps should generally be used. If you decide to omit R_2 from the circuit, it is essential that the capacitor is periodically discharged, usually automatically, through switch SW_1, otherwise there will be no path for dc bias currents and consequently these currents will flow through C_2 causing the output to drift in one direction. When R_2 is included, it must be small enough to ensure that the bias currents do not cause a large offset voltage at the output and yet be large enough to provide a sufficiently low frequency response. To reduce leakage current problems, the inputs to A_1 should be protected by a guard ring and care should be taken to clean the board.

Capacitor C_2 should be very stable (otherwise gain drifts will occur), have a very high insulation resistance (for a low frequency response as described above) and, for applications where the input charge can change very rapidly, then dielectric absorption could be an important factor. Suitable dielectrics include polystyrene, polypropylene and PTFE. A typical value for C_2 is between 10 pF and 10 nF.

You should use a special low noise type of cable for the input. This cable must have a high insulation resistance, otherwise excessive amounts of charge will leak from the transducer. To prevent charges building up due to flexing of the cable, the cable should be constructed with a semiconducting plastic layer in between the outer screen and the inner insulation. The length of cable used to connect the transducer to the charge amplifier may be limited by several factors. If you have too great a cable length, this can cause transmission problems at the higher frequencies due to larger cable capacitance. Ideally, the cable length should typically be less than 1/50th of the wavelength of the highest frequency of interest. Cable capacitance increases, typically, at the rate of 70 pF per metre. Increasing cable length also increases the C_3/C_2A_V term in the gain equation. Too great a cable length will therefore increase non-linearity and reduce the high frequency bandwidth.

Noise performance is limited by op amp A_1 as described below:

(i) Input noise voltage. At very low frequencies, the output noise voltage is equal to the input equivalent noise voltage multiplied by the term $(1 + R_2/R_3)$. At higher frequencies in the operating frequency range, the input noise voltage will be amplified by $(1 + C_3/C_2)$ with the result that large values of C_3 (e.g. long input cables) may cause noise problems.

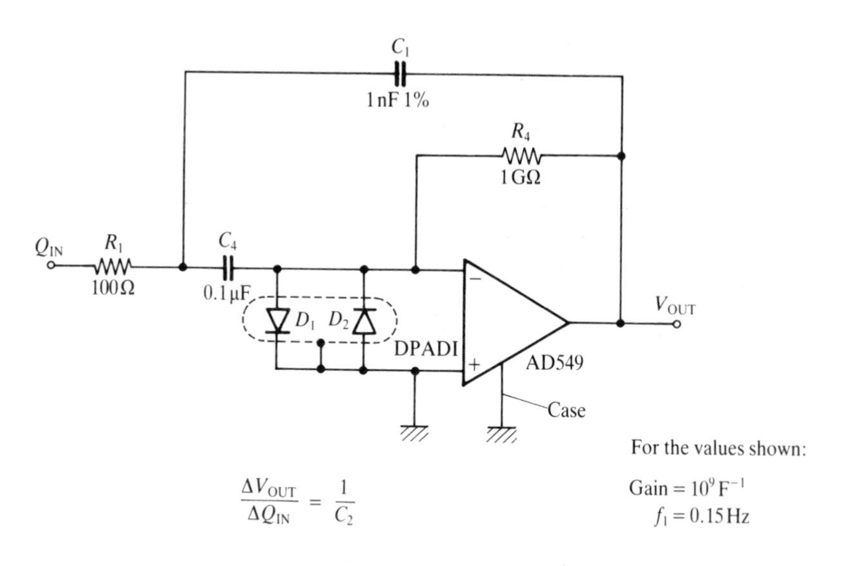

Fig. 3.2. Charge amplification with AC-coupled inputs.

(ii) Input noise current. Input noise currents flow mainly through R_2 and C_2 so that below the lower 3 dB frequency the output noise will be due to the input noise multiplied by R_2. Over the signal frequencies, input noise current will have a reducing effect.

You can use the arrangement in Fig. 3.2 to allow bias currents to flow into the inverting input of A_1. The low frequency -3 dB point of this circuit is

$$f_1 = \frac{1}{2\pi R_4} \cdot \left(\frac{1}{C_1} + \frac{1}{C_4} \right)$$

This circuit has the additional advantage that the input is ac-coupled which protects against any dc overload voltages. Also, the resistor R_1 and the back-to-back diodes D_1 and D_2 provide a degree of input protection against transient overloads.

3.2 High input impedance charge amplifier

Advantages	Disadvantages
does not load a crystal transducer	connecting cables are critical to amplifier performance
higher gain capability	

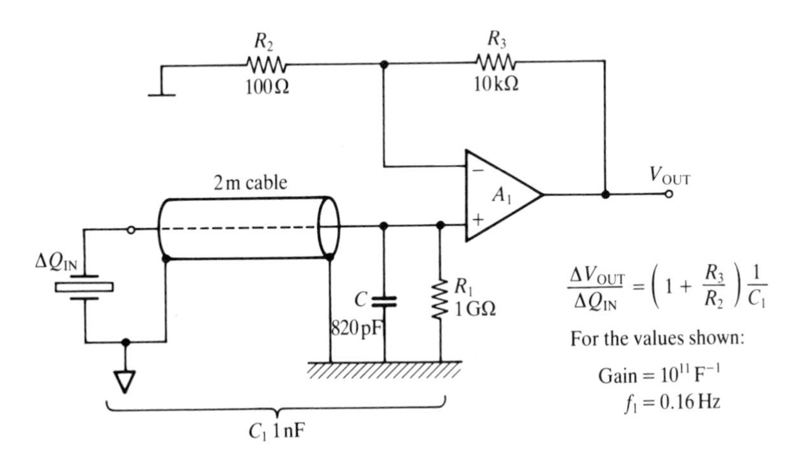

Fig. 3.3. A high input impedance charge amplifier.

This circuit shown in Fig. 3.3 is not, strictly defined, a charge amplifier in the straightforward way described previously of an amplifier which integrates its input current. It is, instead, a standard non-inverting voltage amplifier which offers a high input impedance. With capacitor C_1 at its input, any changes in charge across the input transducer will cause a change in the voltage across C_1 which will be amplified by op amp A_1. C_1 and R_1 include the capacitance and resistance of transducer, leads, board and op amp $A1$.

$$\text{Gain} = \frac{\Delta V_{OUT}}{\Delta Q_{IN}} = \left(1 + \frac{R_3}{R_2}\right) \cdot \frac{1}{C_1}$$

$$\text{Low frequency } -3 \text{ dB point } f_1 = \frac{1}{2\pi C_1 R_1} \text{ (Hz)}$$

$$\text{High frequency } -3 \text{ dB point } f_2 = \frac{R_2}{R_2 + R_3} \cdot f_A \text{ (Hz)}$$

(Where f_A is the unity gain crossover frequency of A_1, and assuming that A_1 is fully compensated.)

$$\text{Output offset} = \left(1 + \frac{R_3}{R_2}\right)(V_{IO} + I_B R_1)$$

where

V_{IO} is the input offset voltage of A_1

and

I_B is the input bias current of A_1

Since the input bias current of A_1 must be kept very small, a FET input op amp should be used. Resistor R_1 is included to provide a path for bias current to flow into the non-inverting input of A_1. Resistor value R_1 is a compromise between low frequency response and offset error, if R_1 is made too small, the low frequency breakpoint will be too high, but if R_1 is too large, the output offset will be large. The common-mode input resistance of $A1$ must be much greater than R_1.

Precautions must be taken at the input to this circuit, as described above: use a guard ring and clean boards carefully to reduce leakage currents from surface contamination and special input cable must be used as described earlier.

Choose capacitor C for very high stability with high resistance and low dielectric absorption. Remember also that C_1 includes the capacitance contributions from the transducers, leads, board etc. The magnitude of these capacitances will obviously limit the performance of this circuit.

In cases where the signal source is remote, and the capacitance of a long input cable cannot be tolerated, you can always position the charge amplifier remotely at the source itself.

4

Current-to-voltage and voltage-to-current converters

Most electronic systems are made up from voltage input and voltage output stages. There are occasions, however, when you need a current input or current output stage to simplify or enhance the performance of the system. Transconductance amplifiers are voltage input and current output stages, whereas transresistance amplifiers are current input and voltage output stages. There are many applications where the input to the system is in the form of a current, as with the leakage current of phototransistors measuring light levels or when monitoring the current output from a power supply generator and so on. Applications which might need a current output include the final stage to a moving-coil meter or the power stage to drive certain electromechanical devices such as torque motors, stepper motors or solenoid relays. For transmitting signals over long distances, as in process control, current is sometimes used (generally 4 mA to 20 mA) since this approach gives good noise immunity with the received signal being relatively unaffected by cable and contact resistances and induced noise voltages.

4.1 A simple current-to-voltage converter

The method shown in Fig. 4.1 is simple and widely used and is particularly useful for measuring the current at the input to a stage or system. The basic approach depends upon passing the current to be measured through a precision resistor and then measuring the voltage drop across that resistor, i.e. a simple application of Ohm's Law.

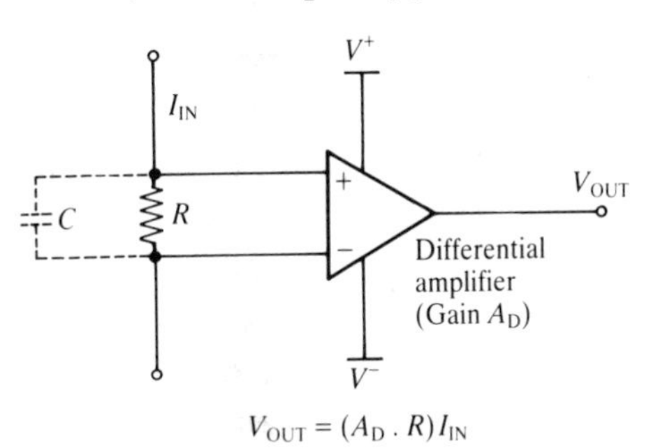

Fig. 4.1. Simple current-to-voltage converter.

The technique can be applied to current signals of virtually any size but it is more appropriate for larger currents above, say, 1 μA. The measurement of very small currents is always potentially difficult because it demands very high resistor values (which generate a considerable amount of noise) and amplifiers with very low input noise and input bias currents. If you apply this technique to converting very small currents, the amplifier must have the additional requirement of a very high input impedance. Also, the total parasitic capacitance C across R can reduce the bandwidth with the 3 dB frequency given by $1/2\pi RC$ Hz.

The amplifier can be either a differential or single-ended input type depending upon the application. Single-ended input amplifiers can only be used where one end of R is grounded. For large currents, a differential amplifier may be better since there will be little error due to voltage drops caused by current flowing in the ground. A further advantage with using a differential amplifier is that R can be placed anywhere in the current input loop and not only to ground.

For large currents, you will need to use a four-terminal resistor for good accuracy. These resistors have two terminals for the current and two other terminals for the voltage with the voltage terminals arranged inside the current terminals. This type of resistor, widely used in high accuracy resistance measurements, has the advantage that contact and wiring resistances have no effect on the measurement of the current since only the voltage drop across the required resistance is actually measured. The arrangement for using a four-terminal resistor is shown in Fig. 4.2 where C_1 and C_2 are the current terminals and P_1 and P_2 are the voltage terminals. Remember that the contact and wiring resistances can be high (several hundreds of milliohms is common) without having any marked effect on the accuracy of the current measurement. With low value resistors, it is the residual inductance, L, of the component that usually limits the bandwidth, this time the 3 dB frequency is given by $R/2\pi L$ Hz.

4.2 A current-to-voltage converter using a single op amp

The gain of the op amp ensures that the inverting input is a virtual earth with the current flowing through R_F being equal to I_{IN}. Hence the output voltage is given by $V_{OUT} = -R_F I_{IN}$. This circuit (Fig. 4.3) is useful for small currents, e.g. from tens of milliamps down to a fraction of a picoamp. The upper current limit is restricted by the output current available from the op amp. The circuit has the disadvantage that it cannot readily be used in a current loop since the input current must effectively be returned to ground.

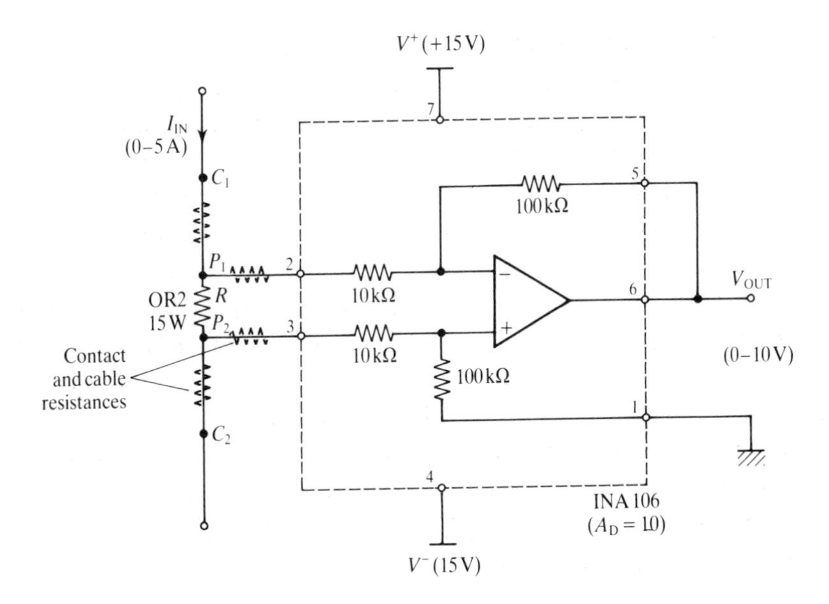

Fig. 4.2. Using a 4-terminal resistor.

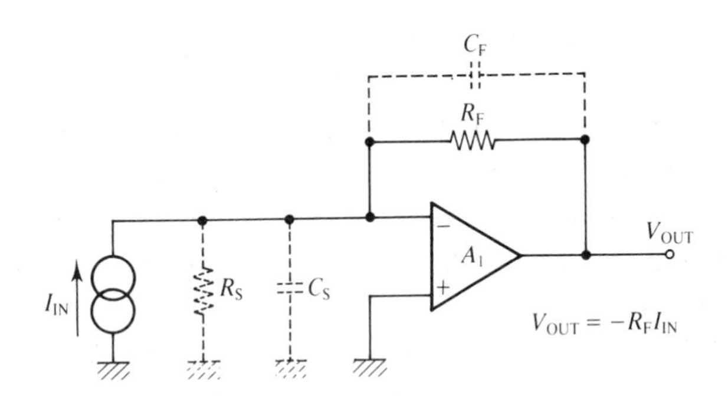

Fig. 4.3. A virtual earth current-to-voltage converter.

Transresistance

$$V_{OUT}/I_{IN} = \frac{-R_F}{1 + \dfrac{R_S + R_F}{A_V R_S}}$$

$$\simeq -R_F$$

where A_V is the op amp gain and R_S is the total effective resistance between the input and ground, which includes the source resistance and the op amp differential input resistance.

Input resistance

$$R_{IN} = \frac{R_F R_S}{R_F + (A_V + 1)R_S}$$

$$\simeq \frac{R_F}{(1 + A_V)} \quad \text{when } A_V R_S \gg R_F$$

$$\text{Output offset voltage } V_{OO} = V_{IO} + I_B R_F$$

where

V_{IO} is the input offset voltage of the op amp

and

I_B is the input bias current of the op amp.

The factors which limit the lowest detectable current are input offset voltage, input bias currents and their drifts. The following points may help to minimize circuit errors.

(1) Offset errors:

With small input currents, less than 1 μA, a FET input op amp should be used because of its low input bias currents.

Try to make $R_S \gg R_F$ otherwise the input offset voltage will be amplified.

The error due to input bias currents can be further reduced by adding a resistor equal to R_F between the non-inverting input and ground. Output offset will then be $V_{IO} + R_F I_{OS}$, where I_{OS} is the input offset bias current of the op amp. You may also need to add a decoupling capacitor (10 nF–100 nF) in parallel with this resistor to limit the noise introduced by this resistor and so prevent A_1 from oscillating.

Take great care with very low input currents to ensure that unwanted leakage currents do not cause significant errors. Use a guard ring as shown in Fig. 4.4 to ensure that leakage currents flow into the guard ring and not into the input. The guard ring should be on both sides of the circuit board. The board should be very carefully cleaned and coated otherwise surface leakage will cause problems. Finally, if very low leakage currents are required (in the pA region) PTFE stand-offs can be used for the input signal connections.

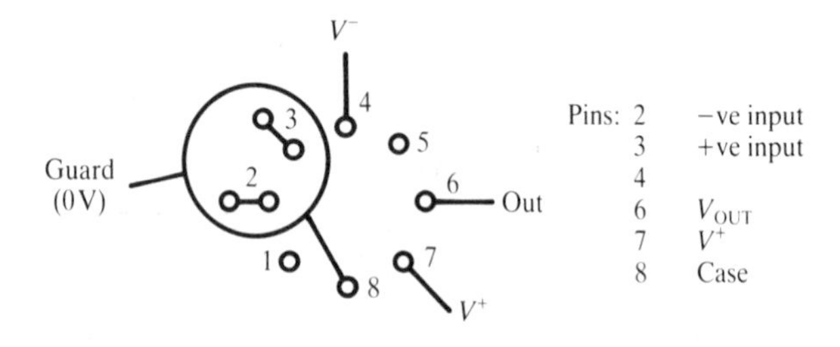

Fig. 4.4. Using a guard ring to minimize leakage currents.

To avoid problems from the bias current drifting with temperature, the amount of heat generated by the op amp itself can be limited. This means keeping the supply voltages as low as possible for the application. Also, the output of the op amp should not be driving a large load (i.e. the load impedance should be greater than 10 kΩ).

For very small input currents, offset nulling is better carried out either in the later stages of the system or by using the approach shown in Fig. 4.7, which avoids using the extremely sensitive input to the amplifier.

(2) Gain errors:

The op amp and the feedback resistor should be chosen so that $A_V R_S \gg R_F$ otherwise large errors in accuracy and linearity may result. Resistors should also be very carefully chosen to give good accuracy and keep drift low. Use high stability resistors, the best type being metal film or metal glaze. For very high value resistors (>1 GΩ), the best construction is a glass body sprayed with a silicon varnish to reduce the effects of humidity. Some even have a metal guard band wrapped around their middle.

To avoid using large value resistors, which are difficult to obtain with a highly stable performance, and which can be expensive, the T-network shown in Fig. 4.5 can be used. This network increases gain without increasing resistor values but this is at the expense of the open loop gain. Note that care must be taken in circuit layout so that R_1 is not reduced due to leakage resistances, i.e. keep points *A* and *B* well separated. Notice that this T-network does have one serious disadvantage in that it amplifies the offset

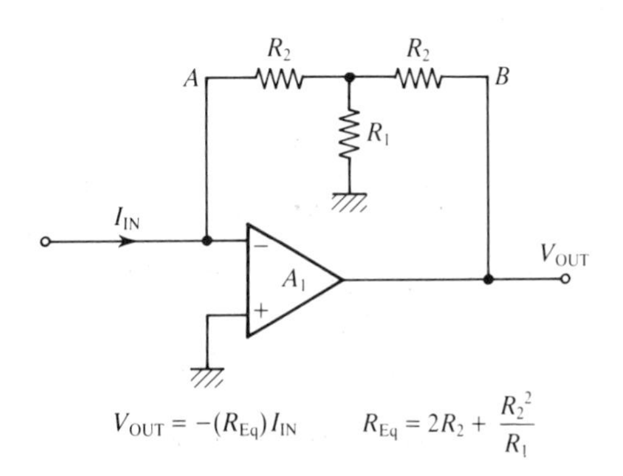

Fig. 4.5. Using a T-network.

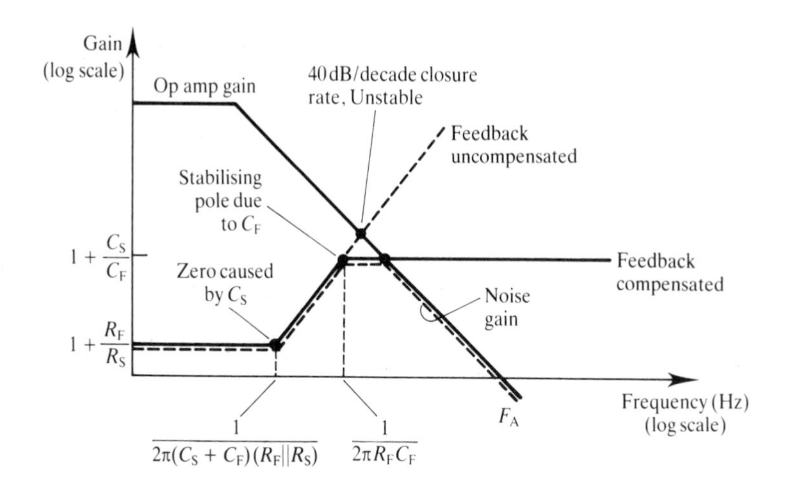

Fig. 4.6. Stability of a virtual earth current-to-voltage converter.

voltages of A_1 by $(R_2 + R_1)/R_1$; this may prejudice the use of the network in certain applications.

(3) Frequency response

Frequency instability can be caused by a finite source capacitance, C_S, particularly with long input leads. This capacitor causes a phase lag in the feedback around the op amp at higher frequencies. The problem can be avoided by placing a small value capacitor across the feedback resistor R_F and is described graphically in Fig. 4.6.

(4) Noise

The noise generated by this circuit comes from three main sources: noise generated by resistor R_F; input noise voltage of A_1 and the input noise current of A_1.

For high gain amplifiers, where $R_F > 1\ \text{M}\Omega$, the noise generated by R_F is a dominant source of noise.

The input noise voltage of the op amp is multiplied by the noise gain, which is shown in Fig. 4.6. In many cases, the noise gain is a maximum at higher frequencies which could result in a large amount of high frequency noise.

The input noise current of A_1 appears at the output multiplied by R_F.

(5) Interference

High gain I to V converters are high sensitivity, high impedance circuits. Consequently, you must shield them in a metal case or

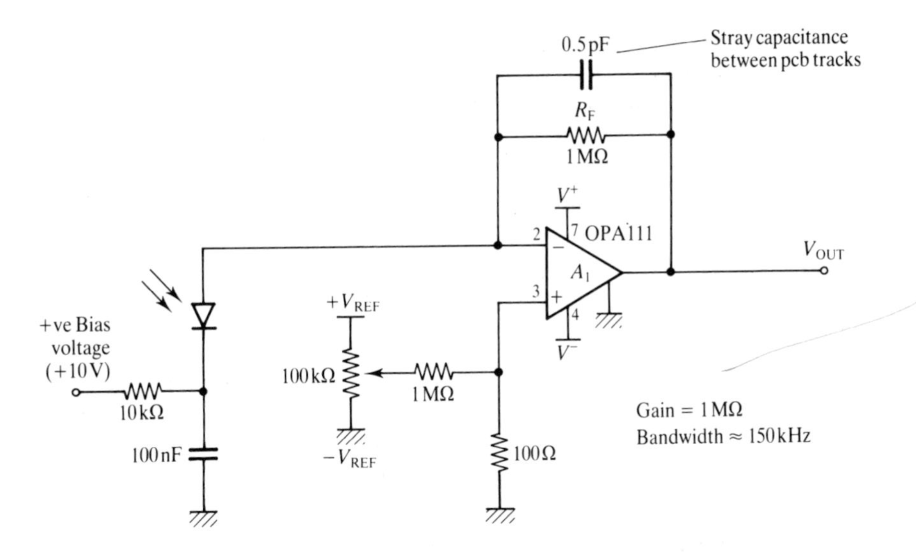

Fig. 4.7. Photodiode amplifier.

an aluminium diecast box to avoid interference. Effective power supply decoupling is also essential. As a final point, these circuits can be very sensitive to mechanical vibration.

A photodiode amplifier circuit is shown in Fig. 4.7. This circuit also incorporates an offset nulling scheme.

4.3 Voltage-to-current conversion using a single op amp

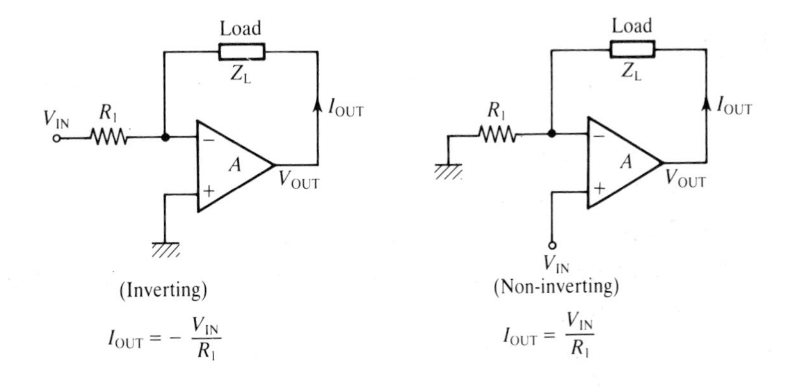

Fig. 4.8. Voltage-to-current conversion using a single op amp.

Advantages	Disadvantages
Bipolar output current Monitoring of load voltage is possible using the op amp output voltage	The load or the source must be floating

We have here (Fig. 4.8) a very simple type of voltage-to-current converter using just the one op amp. Feedback forces the input voltage and the voltage drop across R_1 to be equal. This current also flows through the load so that $I_{OUT} = V_{IN}/R_1$. The current through the load is independent of Z_L provided that the op amp operates within its limitations (for example it does not saturate).

Transconductance: Inverting $\Rightarrow -\dfrac{1}{R_1}$

Non-inverting $\Rightarrow \dfrac{1}{R_1}$

Input resistance: Inverting $= R_1$

Non-inverting $= R_{INCM}$

(i.e. R_{INCM} is the common mode input resistance of A)

Output resistance: Inverting and Non-inverting $= R_1(1 + A_V)$

Output offset current: Inverting and Non-inverting $= I_B + V_{IO}R_1$

where

V_{IO} is the input offset voltage of the op amp

and

I_B is the input bias current of the op amp.

The maximum output current is limited by both the op amp supply voltage and the impedance of the load.

For the inverting circuit: $I_{OUTMAX} = V_{SAT}/Z_L$
For the non-inverting circuit: $I_{OUTMAX} = V_{SAT}/(R_1 + Z_L)$

where

V_{SAT} is the saturation voltage of the op amp output.

Alternatively, the maximum output current could be limited by the op amp itself. To increase output current in this case, you can add a current booster to the output of the op amp as shown in Fig. 4.9.

The non-inverting circuit in Fig. 4.8 has a high input impedance since the input is fed straight into the op amp. The inverting circuit has a low input resistance equal to resistor R_1 which may be quite small. Also, with the inverting circuit, all the output current must be supplied by the source. However, to give a large transconductance without having a very small R_1, a resistor divider can be used in the feedback loop as shown in Fig. 4.9. This circuit also employs a simple current booster, which is configured using discrete components. Using this technique has the

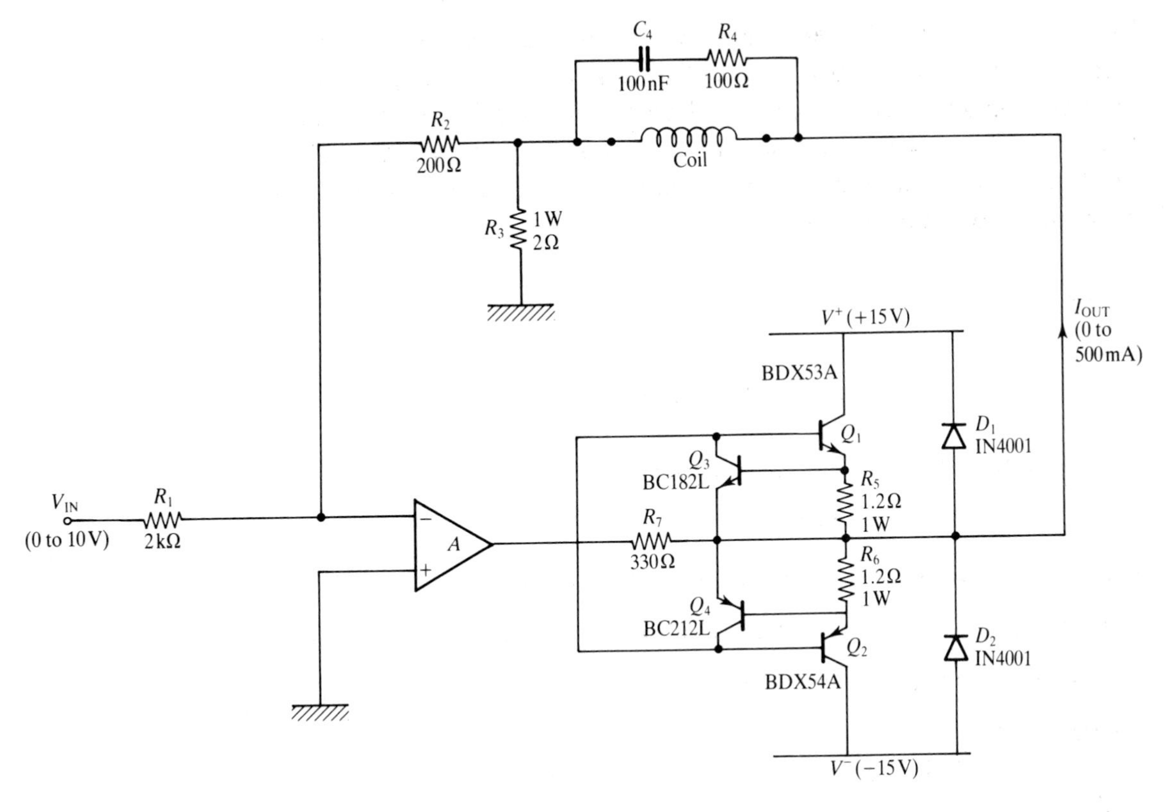

Fig. 4.9. Using a resistor divider network and a current booster.

disadvantage that it reduces the overall loop gain which has the effect of reducing linearity and gain accuracy as well as output impedance.

$$\text{The output resistance} \simeq \frac{R_3}{R_2 + R_3} \cdot A_V \cdot R_1$$

i.e.

$$\text{reduced by a factor of} \quad \frac{R_3}{R_2 + R_3}$$

When driving inductive loads (e.g. coils), take care that you do not exceed the safe operating specifications of the op amp due to the large back emfs which can be induced. Diodes should be added, in this case, to protect the op amp and other circuitry. Also, with inductive loads, frequency stability problems may arise. Inductance in the feedback adds an extra pole to the feedback network which may cause frequency instability and oscillations. To avoid this problem, you can add a compensating capacitor and resistor, which is also shown in Fig. 4.9.

Another op amp can be added to convert the basic circuit into a differential input transconductance amplifier as shown in Fig. 4.10.

For floating sources, the circuits shown in Fig. 4.11 can be used and circuits (*B*) and (*C*) also have the advantage that they can drive a grounded load. Feedback forces a voltage drop across R_1 to be equal to V_{IN}. The current flowing through R_1 must also flow through the load, hence the transconductance operation.

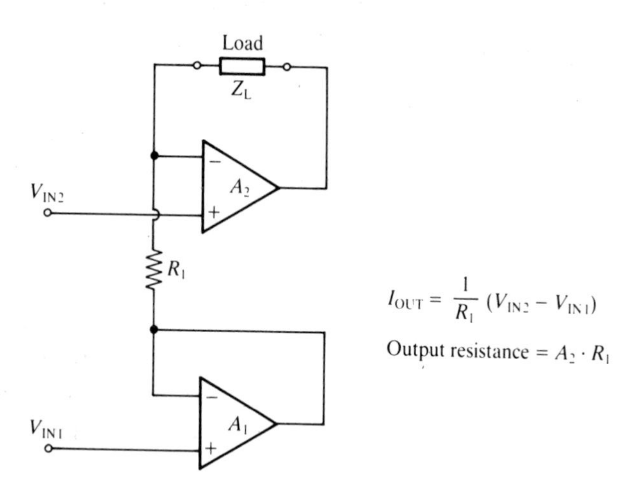

Fig. 4.10. A differential input transconductance amplifier.

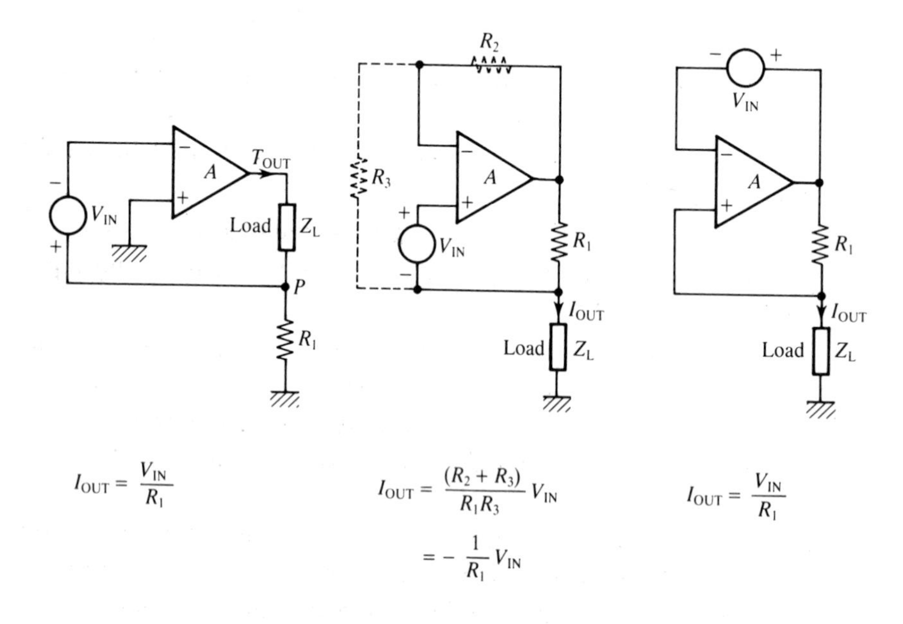

Fig. 4.11. A simple transconductance amplifier for floating sources.

The output resistance for circuit $(A) = A_V R_1$

and for circuits (B) and $(C) = R_1(A_V // \text{CMRR})$

The input offset voltage for circuits (A), (B) and $(C) = V_{IO} + I_B R_1$

where

A_V is the op amp gain

CMRR is the common mode rejection ratio of A

V_{IO} is the input offset voltage of the op amp

and I_B is the input bias current.

The output voltage for circuits (A), (B) and (C), $V_{OUT} = \left(1 + \frac{Z_L}{R_1}\right) \cdot V_{IN}$

Bear the following points in mind:

As before, the maximum current available is limited by either the output current drive of the op amp or by the saturation voltage of the op amp output. If the output current is limited by the op amp saturating, then the maximum current is given by

$$\frac{V_{Sat.}}{(R_1 + Z_1)}$$

where

$V_{Sat.}$ is the saturation point of the op amp.

With these circuits, the output voltage of the op amp can be used as an indication of load voltage where $V_{OUT} = V_L + V_{IN} \simeq V_L$ if $Z_L \gg R_1$. For direct measurement of load voltage, a buffer would be needed.

If circuit (A) has floating power supplies, you can connect point P to the system ground to give a circuit with a non-floating input and a grounded load.

The input impedance of circuit (A) is extremely high since the input impedance of the op amp is multiplied by the loop gain. In practice, stray capacitances and leakage resistances will limit the input impedance to less than 10^6 MΩ in parallel with a few picofarads.

Leakage resistance between the floating source terminals and ground will not affect circuit (C). However, circuits (A) and (B) will be affected as these leakage impedances will draw output current away from the sensing resistor R_1.

4.4 A unipolar transconductance amplifier

Advantages	Disadvantages
Can supply large current and output voltages	Unipolar input The load must normally be floating

You can only use the circuit shown in Fig. 4.12 for positive inputs and so the output can only sink current. The op amp ensures that the voltage drop across R_1 is equal to V_{IN} due to its feedback. The current flowing through R_1 must also flow through the load except for the small amount lost to the base. R_2 is added to protect the transistor against excessive base currents and diode D_1 may also be added to protect the base against large reverse bias voltages.

Transconductance: $= \alpha/R_1$ where α is the collector–emitter current ratio of the transistor.

Output impedance = the collector–base impedance of the transistor $r_c//C_{bc}$

Input offset voltage = $V_{IO} + I_B R_1$ where V_{IO} is the input offset voltage of the op amp.

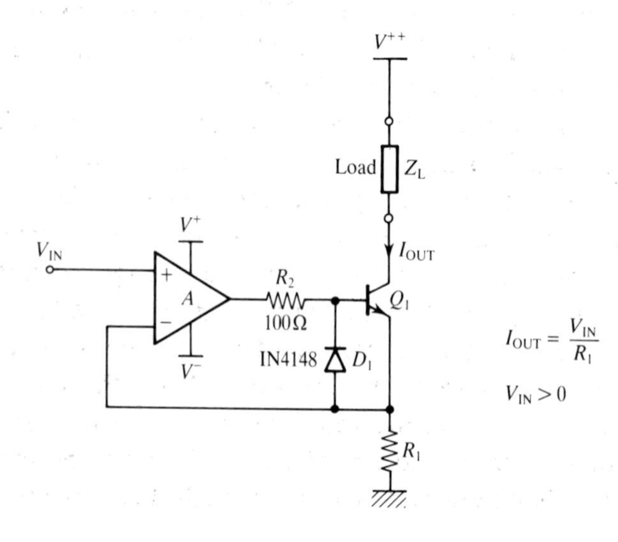

Fig. 4.12. Unipolar transconductance amplifier.

The transconductance of the amplifier is determined by resistor R_1. The value of R_1 is chosen according to the required transconductance. The transistor must be chosen so that it can withstand the maximum current, maximum voltage and maximum power required from the circuit. Note that with bipolar transistors, the manufacturers' power ratings do not apply for large collector–base voltages due to hot spots occurring in the base region (see the later note below on MOSFETs). For some transistors, their maximum operating power can be reduced to 10% at high collector–base voltages. The transistor should also have a high current gain β to avoid loading the op amp and also avoid significant gain errors or non-linearity.

The maximum available current is dependent upon the load and the supply voltage. Consequently, for large load impedances or high currents, the load can be supplied from a separate supply rail V^{++} as shown in Fig. 4.12 and the level of V^{++} can be chosen so that the maximum load current $I_{OUT(MAX)}(\simeq V^{++}/(R_1 + Z_L))$ can be delivered.

With bipolar transistors, a small gain error is caused by the α of the transistor. This gain error can be corrected by changing the value of R_1. Non-linearity will also be introduced but it can be much reduced by using a transistor with a large β or by using a Darlington configuration. MOSFETs are ideal as a replacement for the bipolar transistor since they have high power handling capabilities, introduce negligible non-linearity (because their gate current is extremely small) and do not exhibit a second breakdown characteristic common with bipolars, i.e. the secondary breakdown due to the hot spots in the base region due to large current flows. MOFSETs, by contrast, have a channel which has a positive temperature coefficient of resistance and so current flows evenly across the channel. The power rating of bipolars are much reduced at high voltages whereas MOSFETs are not similarly affected. Fig. 4.13 shows a circuit which uses a MOSFET. Note, the zener diode is added to protect the gate oxide layer.

With the transistor connected as shown earlier in Fig. 4.12, with the collector connected to the load, there is a limitation in the speed of response of the op amp to changes in input voltage, whilst the response to changes in the load voltage is fast since this is only limited by the one discrete transistor. The response to input changes is limited by the slew rate of the op amp. Connecting the emitter to the load as shown in Fig. 4.14 allows a very fast response time to changes in input due to the added loop gain from the transistor. Now, the output voltage changes of the op amp are reduced. You must take extra care with this configuration since the added gain of the transistor may cause instability. You will get the best results by using an externally frequency compensated

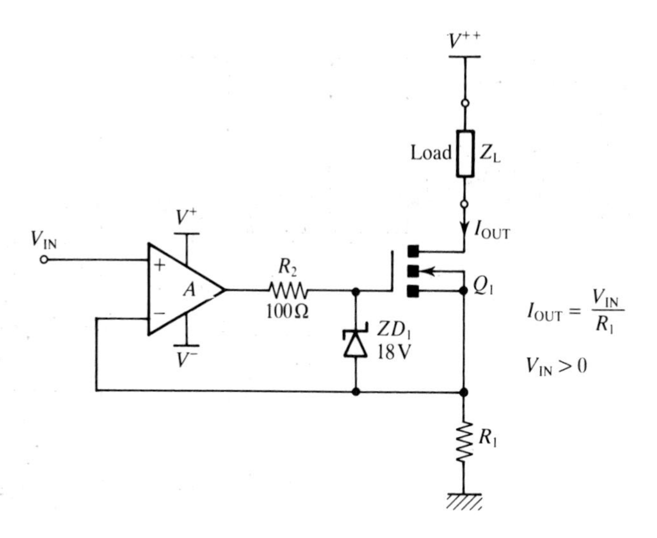

Fig. 4.13. Using MOSFETs for transconductance amplification.

op amp since its high frequency response can be changed. In Fig. 4.14 the new response to changes in the load voltage will be limited by the slew rate of the op amp. The diode D_2 is required to prevent the circuit latching with the collector–base junction forward biased. Feedback is returned to the non-inverting input of the op amp to allow the overall feedback to be negative.

There is a degree of flexibility in the configuration of the output stage as shown in Fig. 4.15. This circuit shows a high output current source (0–12 A), V–I converter. There are several points to note about this approach.

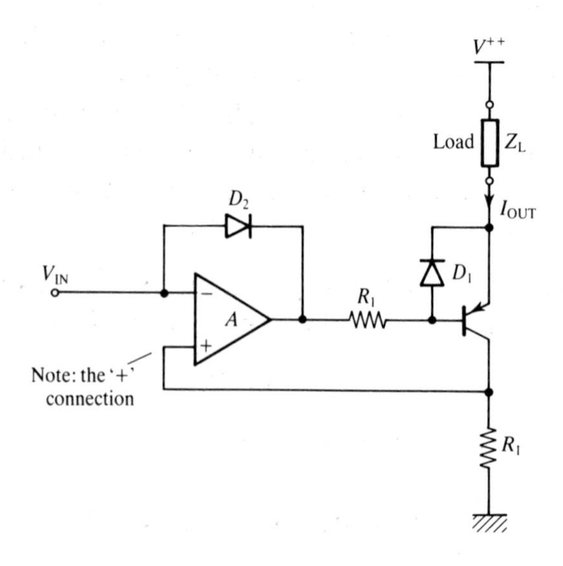

Fig. 4.14. Improving the response time.

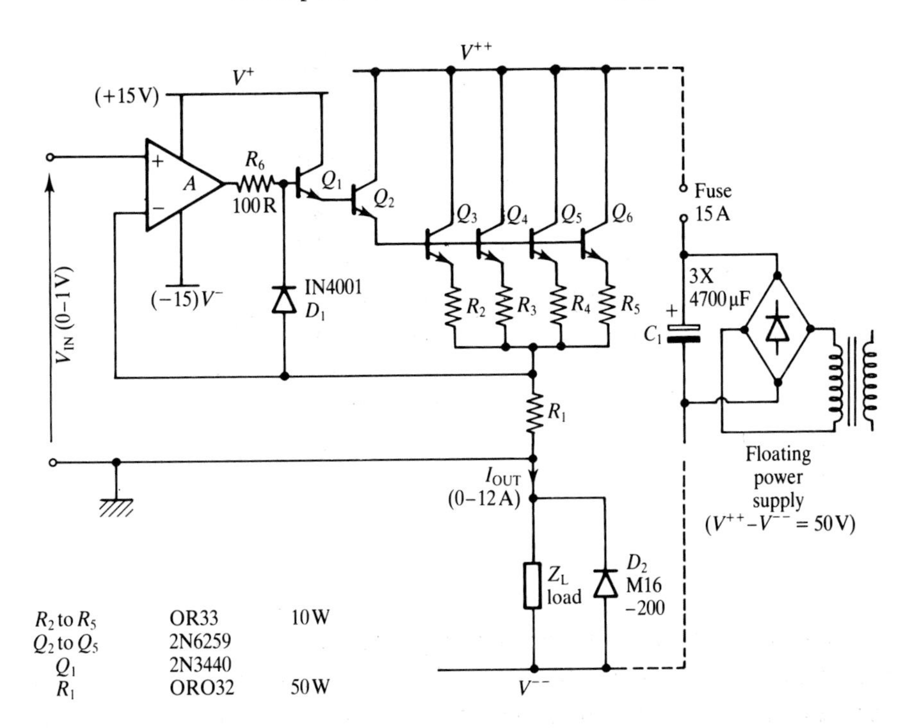

Fig. 4.15. High output current V–I converter.

(i) The current for the load is supplied with a separate floating supply (V^{++} and V^{--}). This floating supply is tied to the amplifier ground only at one point, one end of the sensing resistor R_1. This approach prevents the load current from flowing in the rest of the system and helps to eliminate interference which could otherwise arise due to load current flowing in system ground lines.

(ii) A four-terminal connection approach is used for the current sensing resistor R_1 which eliminates the effect of R_1's contact and cable resistances as described earlier in Fig. 4.2.

(iii) Several power transistors are placed in parallel to share the load current and thermal dissipation. Bipolar transistors can be paralleled quite effectively in the linear mode but only if suitable ballast resistors (R_2 to R_5) are added.

Where several loads need supplying from the same current source, matched transistors can be used as shown in Fig. 4.16.

$$I_{OUT} = V_{IN}/R_1$$

$$I_1 = I_2 = I_3 = I_4 \cdots$$

if

$$R_1 = R_2 = R_3 = R_4 \cdots$$

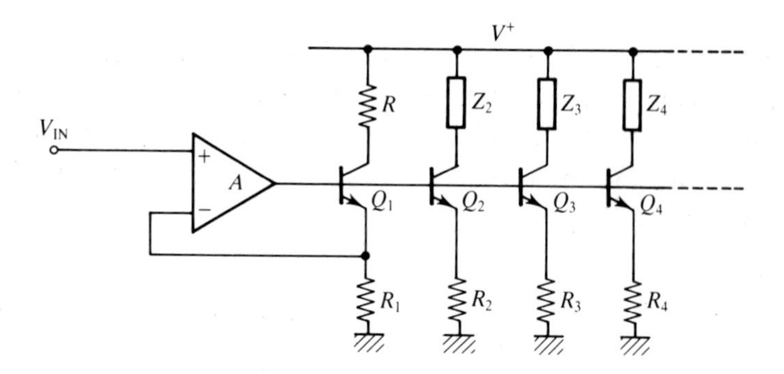

Fig. 4.16. Supplying several loads using matched transistors.

One final point: do not forget that with inductive loads you will need to protect the transistor from back emfs by using a clamping diode (e.g. D_2 in Fig. 4.15).

4.5 A differential input voltage-to-current converter

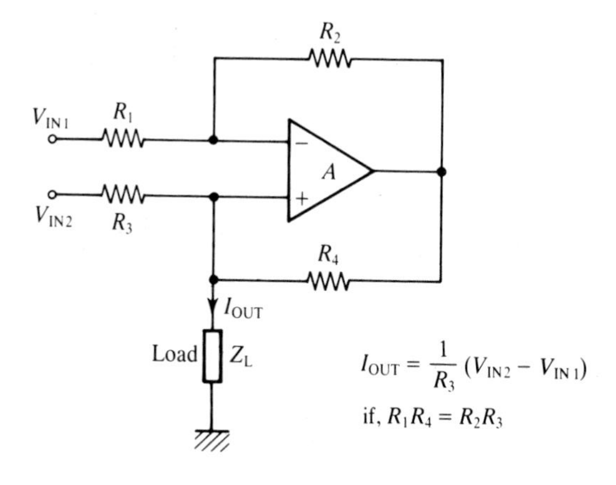

Fig. 4.17. Differential input voltage-to-current converter.

$$I_{OUT} = -V_{IN}/R_3 \quad \text{if} \quad R_1/R_2 = R_3/R_4$$

Advantages	Disadvantages
Differential input	Requires good resistor matching
Bipolar output	Limited load voltage and current
Can supply grounded loads	

The circuit in Fig. 4.17 consists of a simple modification of the single op amp instrumentation amplifier described in Chapter 1. This circuit is sometimes called a Howland voltage to current converter. You must make sure that resistor ratios are closely matched for a high CMRR and a high output impedance.

$$\text{Transconductance } I_{\text{OUT}} = \frac{1}{R_4 + Z_L\left(\dfrac{R_4}{R_3} - \dfrac{R_2}{R_1}\right)} \cdot \left(\frac{R_2}{R_1} V_{\text{IN1}} - \frac{R_4}{R_3} V_{\text{IN2}}\right)$$

$$= \frac{1}{R_3}(V_{\text{IN2}} - V_{\text{IN1}})$$

when

$$R_1 R_4 = R_2 R_3$$

Output resistance

$$\text{due to resistor mismatch} \Rightarrow R_{\text{OUT1}} = \frac{R_4 R_3 R_1}{R_4 R_1 - R_2 R_3}$$

due to finite CMRR and gain A_V of op amp $\Rightarrow R_{\text{OUT2}}$

$$= (R_4 // R_3)\left[\left(\frac{R_3}{R_3 + R_4} \cdot A_V\right) // \text{CMRR}\right]$$

Total: $R_{\text{OUT}} = R_{\text{OUT1}} // R_{\text{OUT2}}$

Common mode transconductance

$$\text{due to imperfect resistor matching} \Rightarrow \frac{-R_2 R_3 + R_1 R_4}{R_1 R_3 R_4}$$

$$\text{due to finite gain of op amp} \Rightarrow \frac{1}{R_3 A_V} + \frac{R_4}{R_3^{\,2} A_V}$$

$$\text{Total} \Rightarrow \frac{R_2}{R_1 R_4} + \frac{1}{R_3} + \frac{1}{R_3 A_V} + \frac{R_4}{R_3^{\,2} A_V}$$

$$\text{Common mode rejection ratio} = \frac{1}{R_3 \times \text{C.M. Transconductance}}$$

$$\text{Output offset current} = I_{\text{OS}} + \left(\frac{1}{R_4} + \frac{1}{R_3}\right) V_{\text{IO}}$$

where

V_{IO} is the input offset voltage of the op amp

and

I_{OS} is the input offset bias current

This circuit, although simple, has two major problems. Firstly, the resistor values have to be carefully matched so that a high output resistance and high CMRR can be achieved. For large currents, the resistor values would have to be small and it is difficult to obtain carefully matched low-value resistors. Secondly, to avoid output errors due to the finite output resistance of the op amp, Z_L must be much smaller than R_4. With $Z_L \ll R_4$ the output voltage swing from the op amp is greatly divided down before reaching the load and so the load voltage is limited to a fraction of the available op amp output. These two limitations mean that this circuit is very rarely used. With a few modifications, outlined in the design notes below, the performance of the circuit can be greatly increased.

An extra op amp can be added as shown in Fig. 4.18 which allows a significant increase in the load voltage swing capability. This circuit is similar to the single op amp differential amplifier (consider the non-inverting input and the output of A_2 as a 'ground' and it is identical) and careful resistor matching is still required but now R_1 and R_4 can be larger in value. If $R_5 \ll R_4$, so that R_4 draws very little current, then the buffer A_2 is not needed. For single-ended inputs, one of the inputs can be grounded. If maximum current is limited by the amplifier, then a current booster can be used. The output of A_2 can also be used as a guard driver to maintain a high output impedance.

To increase input impedance and improve CMRR, two extra op amps can be used as shown in Fig. 4.19. You might notice that this circuit is the same as the three op amp instrumentation amplifier as described in Chapter 1.

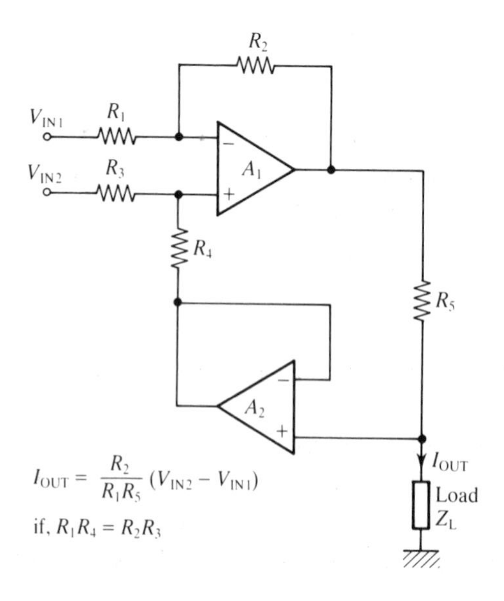

Fig. 4.18. Increasing load voltage swing capability.

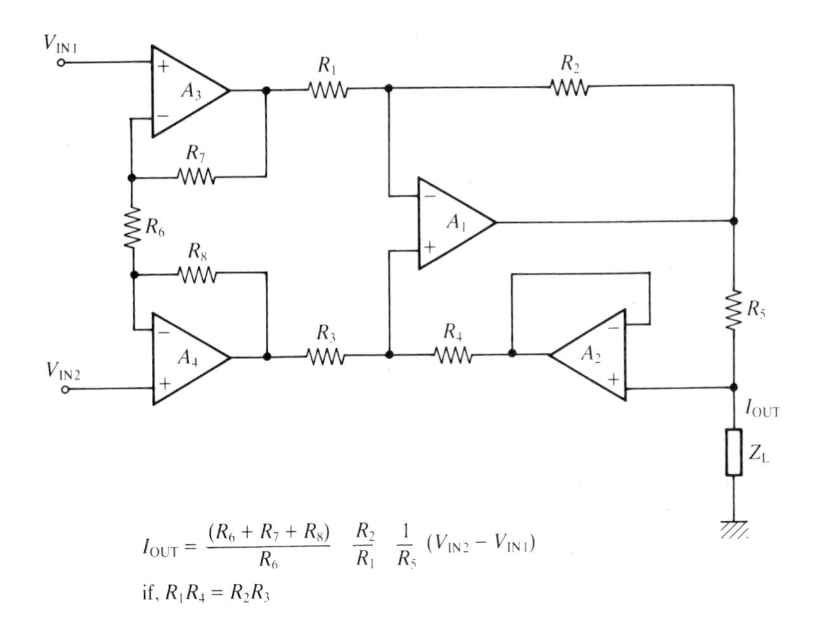

Fig. 4.19. Improving CMRR and input impedance.

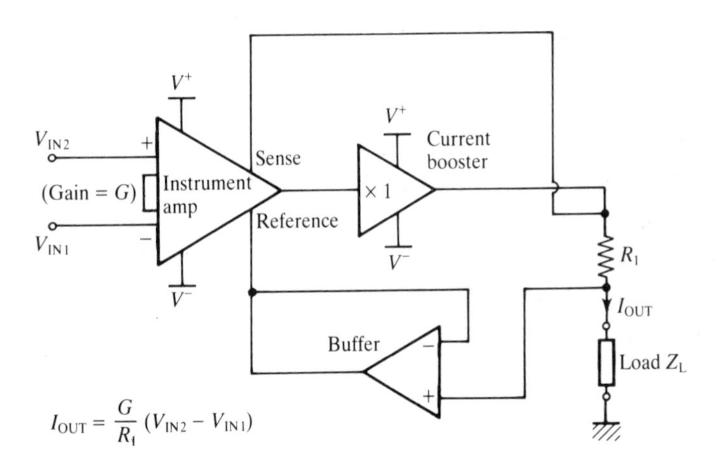

Fig. 4.20. A V to I converter using a commercial instrumentation amplifier.

Commercial instrumentation amplifier ics can be used as an excellent voltage to current converter using the same principle as that shown in Fig. 4.19. The reference terminal of the instrumentation amp should be buffered as shown in Fig. 4.20.

4.6 Operational transconductance amplifiers

Several devices are commercially available, e.g. the CA 3060, LM 13600 and the LM 13700. These devices are particularly useful since they have an extra input amplifier bias input which allows the transconductance of the main amplifier to be controlled. They can be used in many

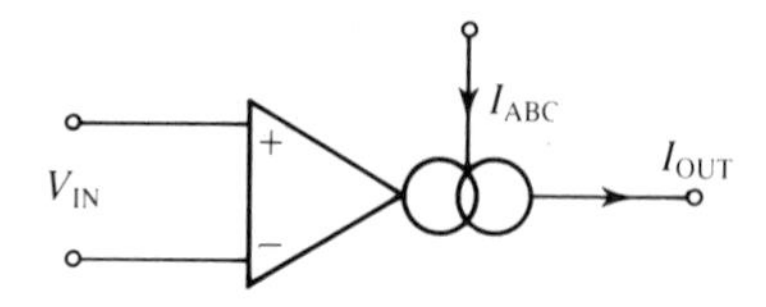

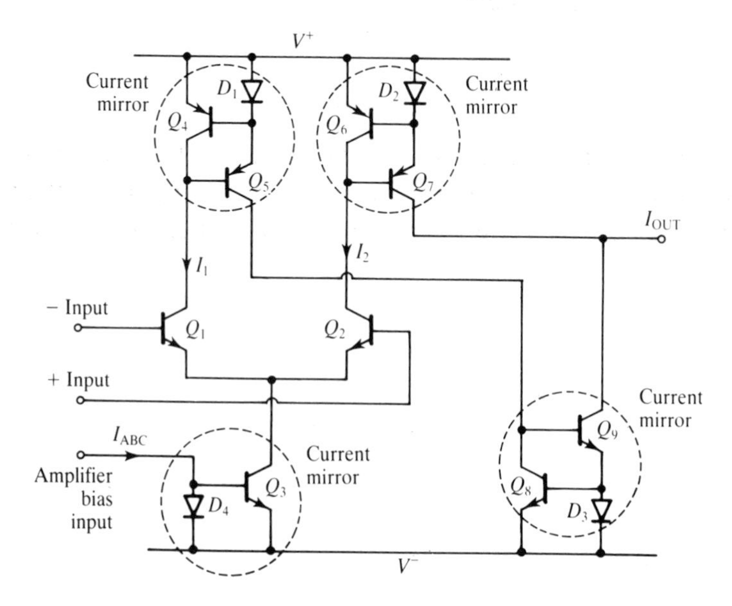

Fig. 4.21. An operational transconductance amplifier.

applications such as multipliers, voltage controlled amplifiers, voltage controlled resistors, voltage controlled filters, waveform generators and so on. The transconductance stage consists of the differential pair $Q1$ and $Q2$ (Fig. 4.21). Their operating current and hence transconductance is controlled by the current mirror of $D4$ and $Q3$. The rest of the circuitry consists of current mirrors. $D1$, $Q4$ and $Q5$ mirror the collector current of $Q1$; $D2$, $Q6$ and $Q7$ mirror the collector current of $Q2$. The current mirror $D3$, $Q8$ and $Q9$ ensure that the output current I_0 is equal to $I_2 - I_1$.

The ratio of collector currents is given in the relationship

$$V_{IN} = \frac{kT}{q_e} \log_e \left(\frac{I_2}{I_1} \right)$$

this is derived from the transistor equation where

$k =$ Boltzmann's constant

$T =$ the absolute temperature in K

and

$q_e =$ the electron charge.

Also

$$I_{ABC} = I_1 + I_2$$

and

$$I_{OUT} = I_1 - I_2$$

so that for small input signals

$$I_{OUT} = \left(\frac{I_{ABC} \cdot q_e}{2kT}\right) \cdot V_{IN}$$

The transconductance is not linear and can only be approximated for small inputs. Linearizing diodes can be used at the input to linearize the gain of the amplifier. These diodes are sometimes supplied on the ic.

5
Controlled amplifiers

This chapter deals with amplifiers whose gain can be controlled either digitally or by using an analog voltage. Digitally controlled amplifiers are common in microprocessor controlled systems where auto ranging is necessary. Voltage controlled amplifiers (VCAs) are also a very common functional block within a system, they are also a constituent part of other functional blocks such as certain oscillator circuits and automatic gain control (AGC) amplifiers.

5.1 Some approaches to voltage controlled amplification

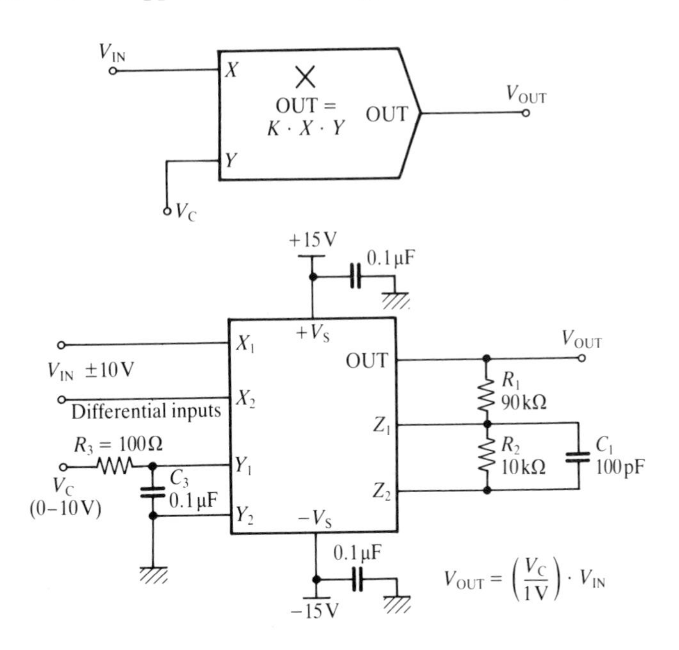

Fig. 5.1. VCA using a multiplier.

Advantages	Disadvantages
Gain is proportional to control voltages Inverting and non-inverting configurations are possible	Cost and complexity

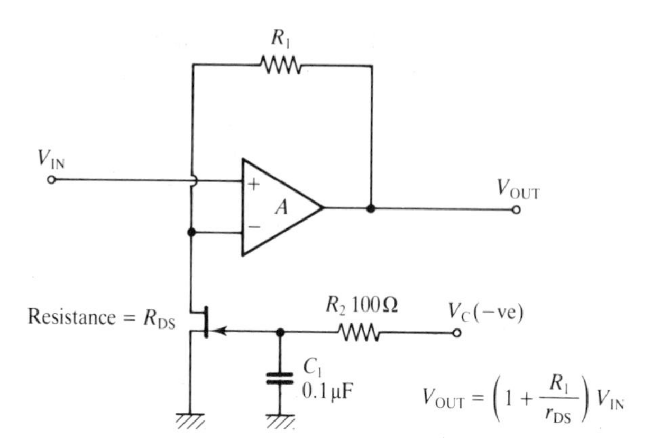

Fig. 5.2. Voltage amplification using a FET.

The circuit of Fig. 5.1 uses a multiplier which is a direct VCA (see Chapter 9 on multipliers). A positive or negative control voltage gives you an inverting or non-inverting VCA. An alternative approach to a VCA is shown in Fig. 5.2. Control of the amplifier's gain in this case is achieved by using a voltage controlled resistor (VCR) as part of the amplifier's gain network. The VCR function is achieved simply and inexpensively by using a FET with a control voltage applied to the gate of the FET. Changing V_C changes the width of the channel of the FET thereby varying the resistance r_{DS}, and hence the gain of the amplifier. When using N-channel JFETs, a negative control voltage needs to be used. The gain control using FETs is not linear. The VCR can be used in either the non-inverting or inverting configuration.

To avoid excessive distortion, you must operate the JFET safely within the non-saturated range of the source–drain characteristic. The distortion is caused due to the source–drain voltages increasing with large signal size and so changing the FETs' resistance a little and thereby modulating the gain slightly. Two points: choosing a FET with a high gate–source cut-off voltage, often called the pinch-off voltage V_P or $V_{GS(OFF)}$, will give a large dynamic range for gain control; also gain accuracy and stability will be limited mainly by the accuracy and stability of the FET used in the circuit.

You can reduce distortion by incorporating the FET into a T-network in the amplifier's feedback as shown in Fig. 5.3. This technique reduces signal swings across the FET. Distortion is further reduced by using resistors R_4 and R_5 to counteract the non-linearities of the FET. In this case, make $R_4 = R_5$ with $R_5 \gg (r_{DS}//R_2)$.

You can use devices such as transconductance amplifiers in the VCA role since they have a separate input for gain control. In the circuit shown in Fig. 5.4, the amplifier has a bias input such that increasing

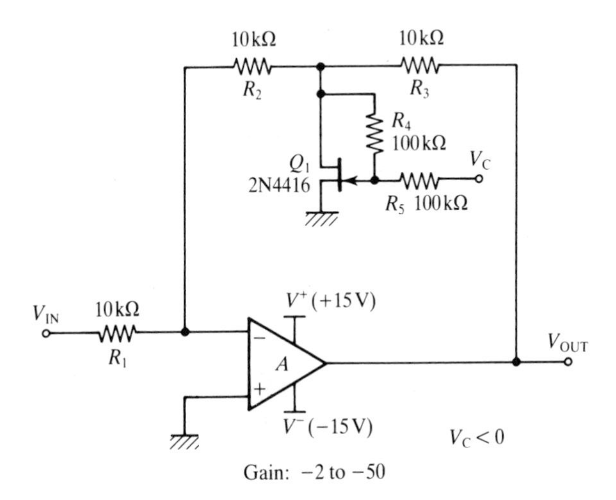

Fig. 5.3. Reducing distortion from a FET controlled amplifier.

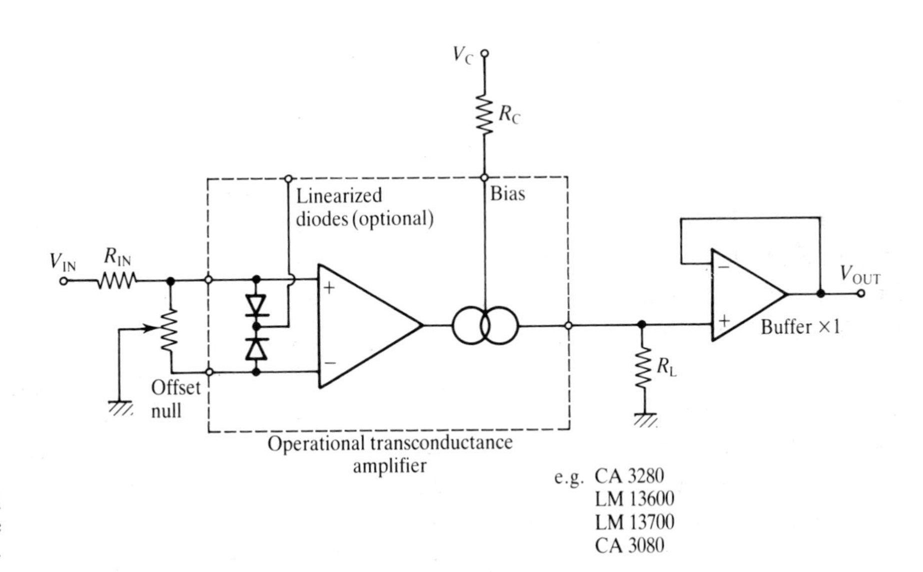

Fig. 5.4. Using operational transconductance amplifiers for VCA.

the bias current increases the amplifier's transconductance. The output current is fed through a resistor to give a voltage which is then buffered. One important plus for this approach is that gain increases linearly with control voltage. Some devices also contain linearizing diodes at their inputs which may be used to reduce distortion. A list of a few possible devices is also given in Fig. 5.4.

5.2 Commercial voltage controlled amplifiers

Several commercial VCAs are available. One of the most popular is the MC3340 which has a frequency response up to 1 MHz and gain control

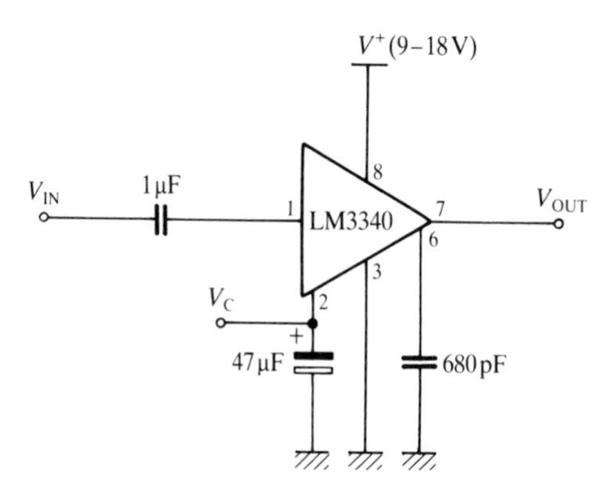

Fig. 5.5. The MC3340 VCA in an audio application.

range from +13 dB to −90 dB using a control voltage of 3.5 V–6 V dc. Alternatively, the device can be controlled using a resistor. Fig. 5.5 shows a typical audio application of this device.

5.3 Automatic gain control (AGC)

An AGC amplifier (Fig. 5.6) automatically adjusts its gain in order to maintain a predetermined average AC signal output. Fluctuations in signal output due to short term variations in signal input, whether above or below the predetermined level, are thereby avoided. The amplifier achieves this response to variations in input signal by converting a portion of the output signal to a DC level by some form of AC to DC conversion. Usually the conversion is based upon a rectifier but peak detectors or other means could be used to obtain a DC level which varies in sympathy with the input signal. An integrator varies the output level by varying the gain of the VCA until the output of the AC to DC converter equals the level required by the designer.

You can vary the response time of the AGC amplifier by changing the time constant of the integrator. The response time should be large enough not to cause an unwanted response to random fluctuations in the signal and hence distortion. In other words, the response time should

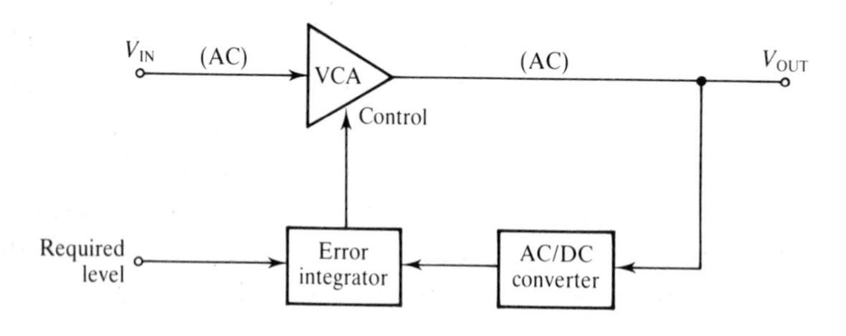

Fig. 5.6. Block diagram of an automatic gain controller.

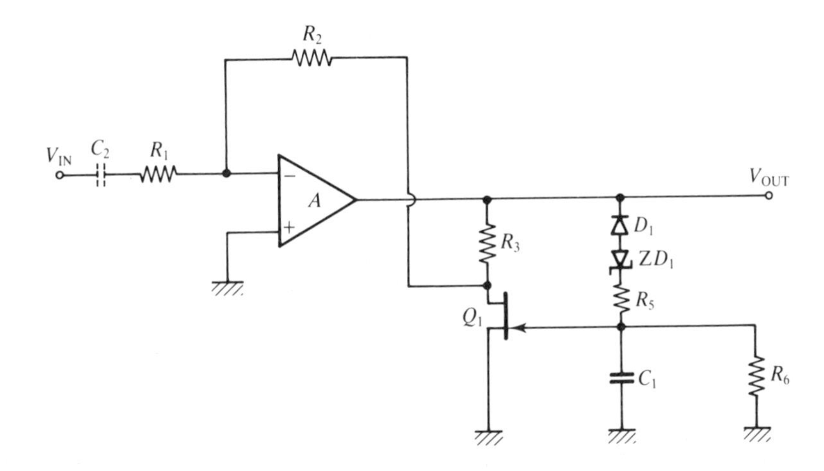

Fig. 5.7. Simple AGC circuit.

be small enough so that the lowest frequency content of the input signal is not adversely affected. Alternatively the response time should not be too long, since this will cause excessive delays.

A simple example of an AGC circuit is shown in Fig. 5.7. The Zener diode ZD_1 sets the output level. Rectification of the output signal is achieved using diode D_1. Capacitor C_1 with resistors R_5 and R_6 act as an error integrator. The remainder of the circuit uses the FET based VCR circuit described previously.

5.4 Digitally controlled amplifiers

You can build digitally controlled amplifiers very simply by using analog switches, resistors and an op amp as shown in Fig. 5.8. This circuit shows a digitally controlled amplifier with gains of -1, -4, -16 and -64 by closing either SW_1, SW_2, SW_3 or SW_4. There are several points worth mentioning about this type of circuit. First, the 'on' resistance of the analog switches (r_{on}) can cause a gain error. To overcome this potential problem the resistor values are chosen so that r_{on} is taken into account. Second, additional values of gain can be obtained if several switches are ON at the same time. In some cases, it is important to minimize the number of resistors and switches by selecting the resistor value accordingly. Third, it is sometimes necessary to ensure that A_1 always has sufficient feedback to prevent saturation (where the op amp is then effectively operating in open loop mode as a comparator), particularly during changes in gain. Otherwise, a large transient may be introduced into your system. To avoid this problem, either design the switches to have a make-before-break action or alternatively have a feedback resistor permanently connected around A_1's feedback loop.

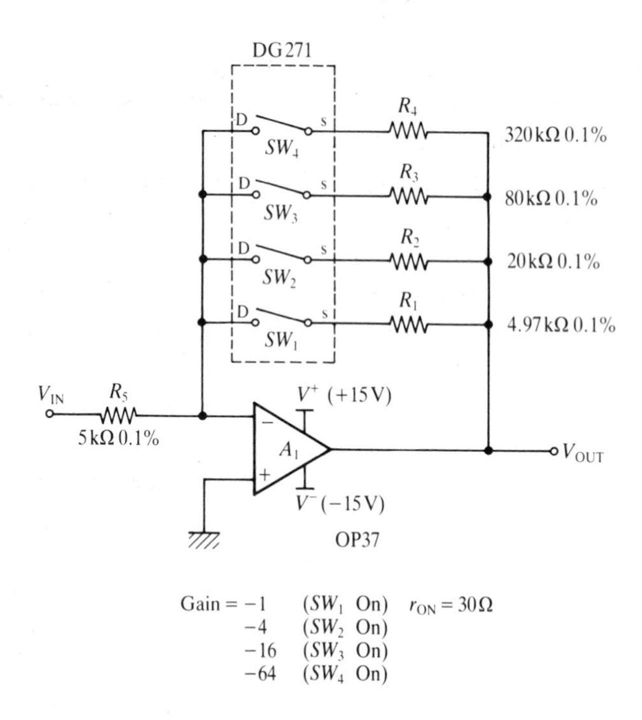

Fig. 5.8. Digitally controlled amplifier using analog switches.

There are a number of possibilities for inserting the analog switches as shown in Fig. 5.9. There is no one ideal position for the switches although C and D are the most commonly used. Consideration of the following points may help determine your choice.

(i) The ON resistance of the analog switch will affect the gain in positions A, B, C and D. With high gain amplifiers, A and B will

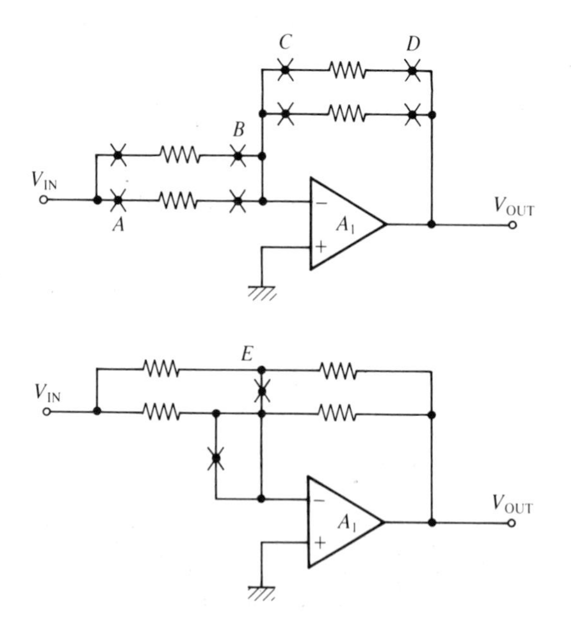

Fig. 5.9. Various analog switch positions.

be much worse than *C* and *D*. The r_{on} of the switches will have no effect in *E* as the switches are in series with the high input impedance of the op amp.

(ii) The ON resistance of many analog switches is a function of their signal voltage. Consequently, with the switches in the positions *A* and *D*, r_{on} can be modulated by the signal and distortion introduced. However, with positions *B*, *C* and *E*, the switches are operating at a virtual earth and will introduce only a negligible amount of distortion.

(iii) With position *D*, the parasitic switch capacitances and leakage currents are driven from the op amp output and so will have little effect. Similarly, with position *A* the switch capacitances and leakage currents will have little effect if the source has a low impedance. Positions *C*, *D* and particularly *E* can be very sensitive to switch leakage currents and parasitic capacitance as the switches are positioned at the most sensitive point in the circuit.

(iv) With switch positions *B*, *C* and particularly *E*, large output transients can be generated during range switching mainly because of charge being injected through the analog switches into the amplifier. Position *A* may also be sensitive to charge injection depending on the source impedance. Position *D* generally is the least affected because of the low output impedance of A_1.

An alternative approach to digitally controlled amplification uses a DAC and one or two op amps. Fig. 5.10 shows a circuit for a digitally controlled attenuator.

$$V_{OUT} = (N_{IN}/N_{MAX}) \cdot V_{IN}$$

where

$$N_{IN} = \text{value of binary input code}$$

and

$$N_{MAX} = \text{value of maximum binary code} = 2^N$$

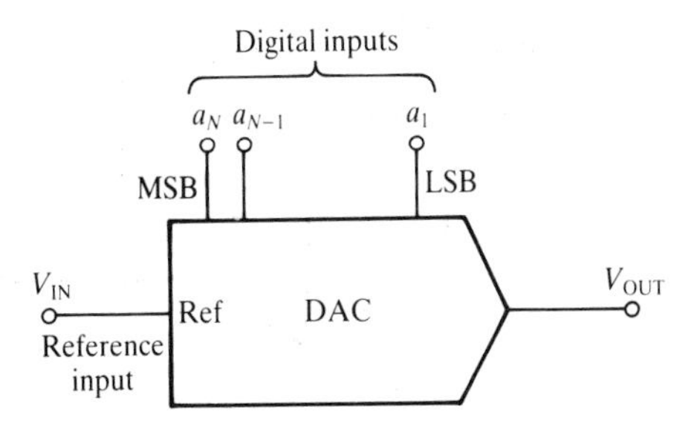

Fig. 5.10. A digital attenuator using a DAC.

For example if the binary number 11111111 is present on the input of an 8 bit DAC, the gain of the system configured as shown here equals $255/2^8 = 255/256 \simeq 1$. With a binary input of 00001111 present on the input of an 8 bit DAC the gain of the system equals $15/2^8 = 15/256 \simeq 1/16$.

The DAC must have a sufficiently fast response to changes in its reference voltage. Also, be careful since some so-called multiplying DACs will only operate with a single polarity reference input (i.e. V_{IN} must always be either +ve or −ve). Using these DACs for a DC bipolar input voltage would require a rather cumbersome offsetting arrangement, which is best avoided if at all possible. AC signals present less of a problem with these DACs as coupling capacitors can be used.

Alternatively, a DAC can be used in the feedback loop of an op amp for programmable gain as shown in Fig. 5.11 for both an inverting and a non-inverting configuration. Note that a value of $N = 0$ is not allowable with this configuration, otherwise your output will swing to the supply rails and saturate. Watch out for the gain accuracy of these circuits, particularly on the maximum gain ranges, because good gain accuracy requires a DAC with exact least significant bit weightings. One way around this problem is to use a DAC with a higher resolution than necessary (e.g. a 12 bit DAC instead of an 8 bit DAC in applications where gains between 1 and 256 are employed). In this case tie unused inputs low.

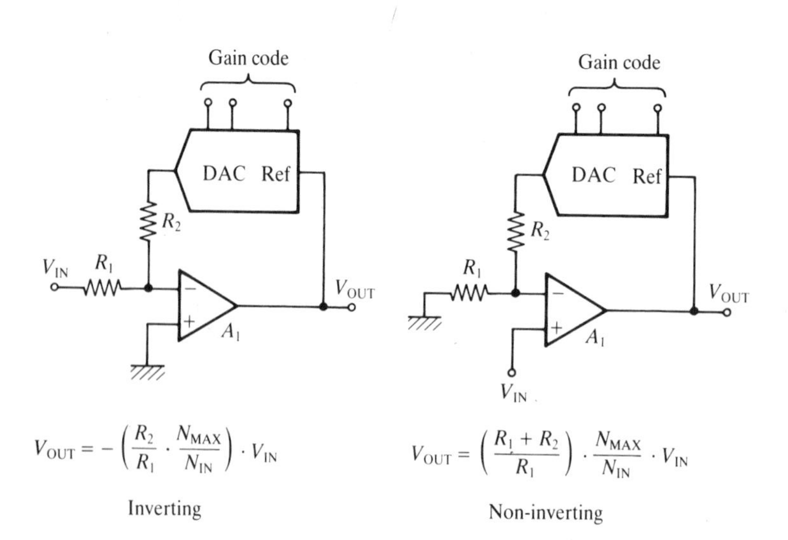

Fig. 5.11. Programmable gain using a DAC.

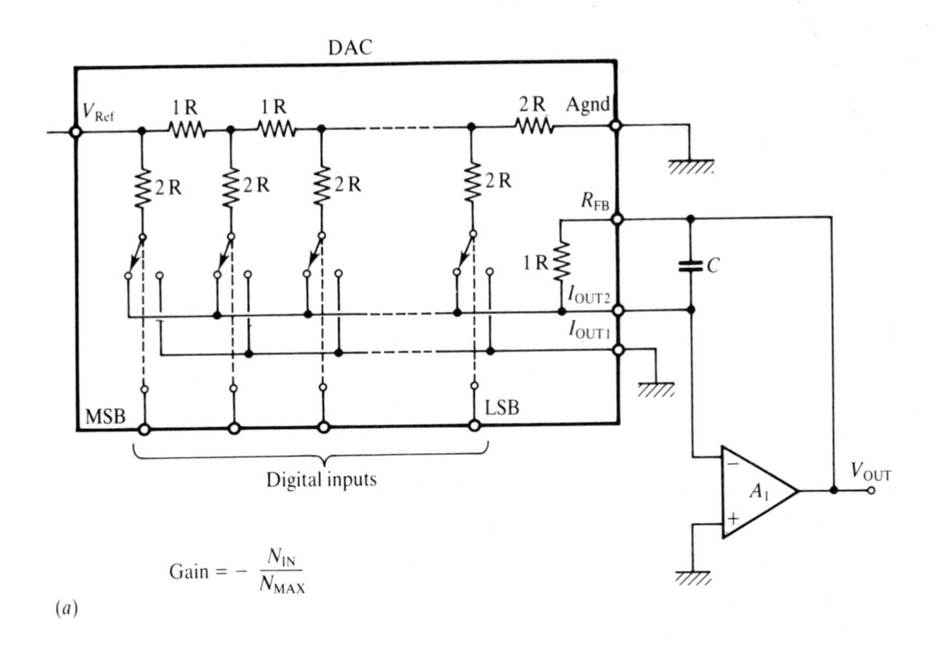

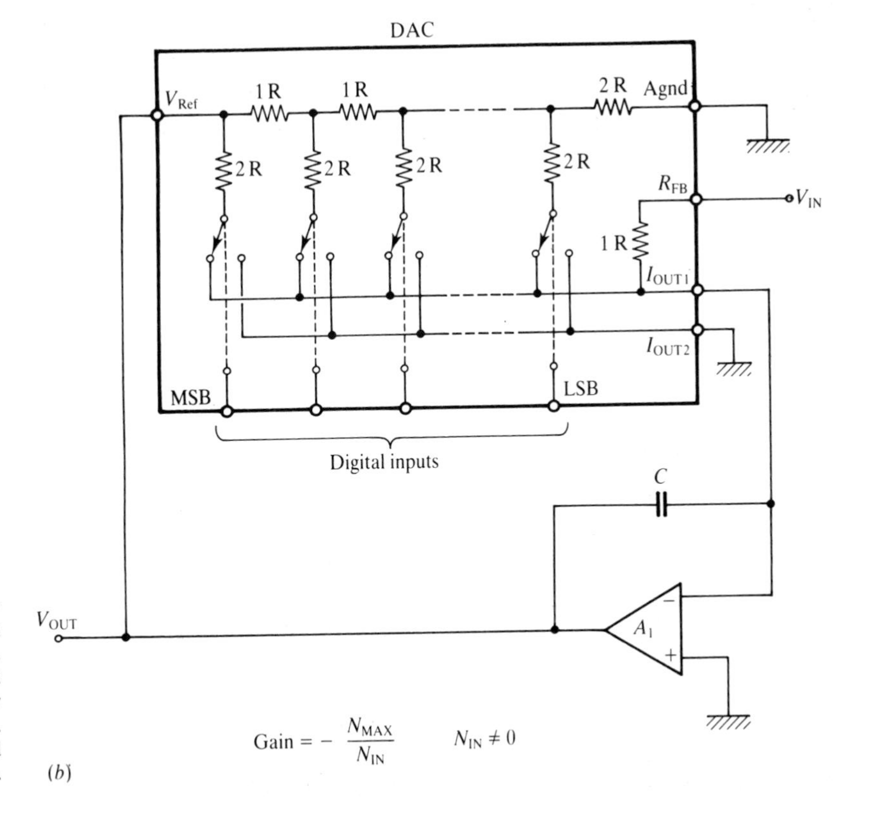

Fig. 5.12. A digitally controlled gain using an R–$2R$ ladder DAC. (*a*) Digitally controlled attenuator. (*b*) Digitally controlled amplifier. (*c*) Digitally controlled amplifier/attenuator.

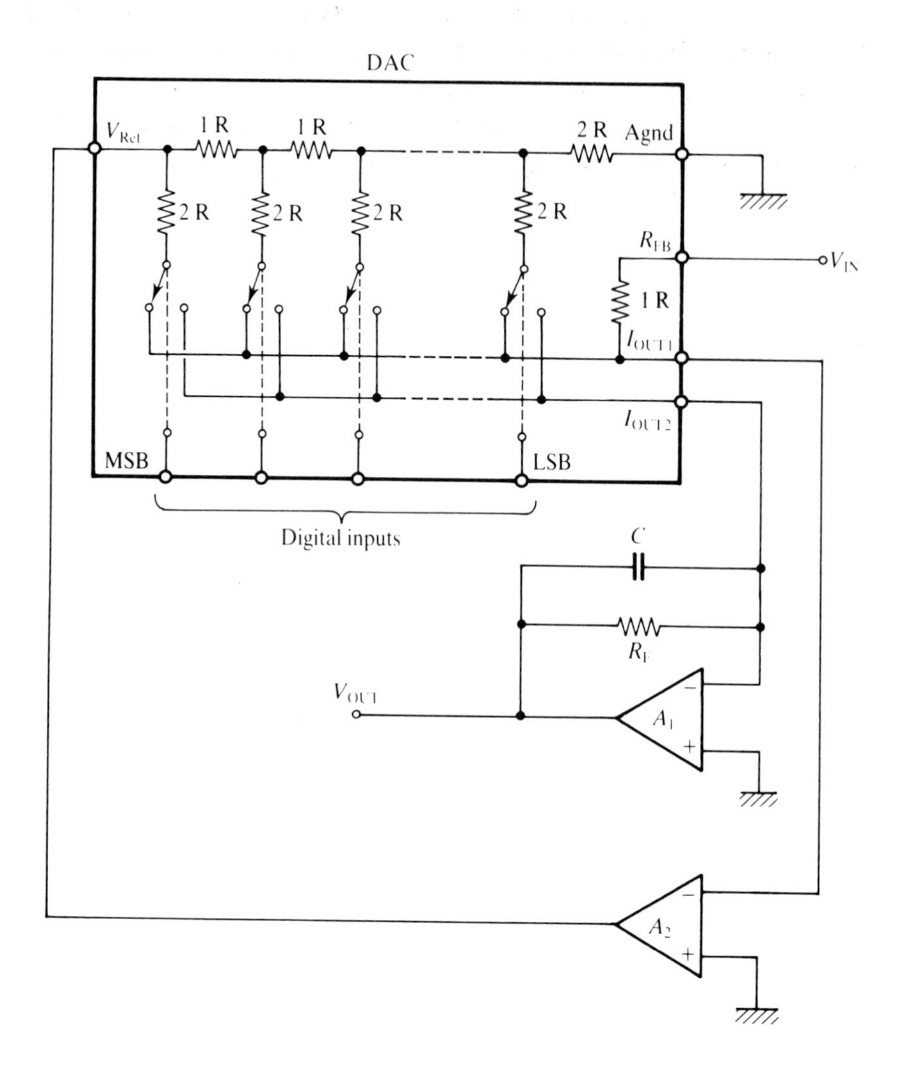

$$\text{Gain} = \frac{R_F}{R} \cdot \frac{(N_{MAX} - N_{IN} - 1)}{N_{IN}}$$

$(N_{IN} \neq 0)$

(*c*)

Fig. 5.12 (*contd.*)

The R–$2R$ ladder type of DAC is particularly suited for digital gain control applications, as shown in Fig. 5.12, since these types of DAC will usually accept positive and negative inputs at V_{Ref}. Also, all the required resistors are contained within the DAC circuitry so giving good temperature stability of gain since drifts in resistor values due to temperature changes operate almost equally on all resistors. A capacitor, C, in the tens of picofarads range might be needed to stabilize the op amp.

A last word on this: not only are DACs available for linear control of gain as described so far, but there are also devices available which allow the gain to be switched in dB steps. One device, for example, is the AD7110 which has an attenuation range from 0 to 88.5 dB in 1.5 dB steps and is intended for frequencies in the audio range.

6

Active filter design

Filters are frequency selective devices which attenuate certain bands of frequencies, whilst selectively passing other bands. Until the 1960s, filters were realized mainly using passive components, i.e. inductors, capacitors and resistors. The great problem with inductors is their size (especially at lower frequencies as they tend to be prohibitively bulky), weight, expense and purity (generally speaking capacitors have lower parasitic errors than inductors). With the development of the op amp in the 1960s, the new discipline of active filter design, based on the op amp was created. Active filters employ resistors, capacitors, along with amplifiers (the active components) and have no need of inductors. Consequently, active filters have largely replaced passive filters. Nowadays, passive filters are only used in high frequency applications (e.g. above 1 MHz) beyond the performance limits of most general purpose op amps. However, even with many high frequency applications, for example radio transmitters and receivers, traditional LCR passive filters have been replaced by crystal and SAW (surface acoustic wave) filters.

Digital filters have now replaced analog filters in many applications. Digital filters are mainly software based and so are considerably more flexible than analog filters. It is also possible to design filter functions with digital filters that would be very difficult to engineer using analog components. However, digital filters cannot replace analog filters in all situations and there will always be a wide demand for the most popular analog filter, the active RC filter, which is the subject of this chapter.

Filters can be classified according to the way in which the filter treats the whole range of frequencies as shown in idealized form in Fig. 6.1, which illustrates low pass, high pass, band pass and band reject filters (all pass filters are not shown: these filters, as their name suggests, allow all frequencies of interest to pass usually with equal gains). The purpose of the various filters is to attenuate and to apply different phase shifts to different frequency bands or to introduce a time delay between input and output.

It is impossible, with an active R–C filter, to achieve a perfect filter response of the so-called 'brick wall' variety, illustrated in Fig. 6.1, where there is a perfectly flat passband gain, infinite stop band attenuation and an infinitely sharp transition between the passband and the stopband.

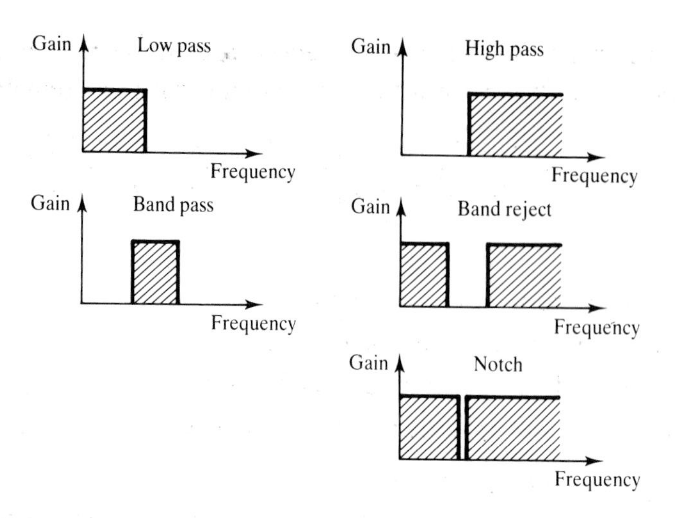

Fig. 6.1. Basic filter types.

Consequently, an active filter design is usually a compromise between the ideal response and practical limitations. This is known as the 'approximation problem'. In many cases, however, the filtering requirements are not too stringent and a basic filter circuit with a first or a second order response can be used. These are the simplest filter circuits. A catalog of such circuits is presented in Section 6.2. The filter design can be relatively easy and requires the choice of the most suitable circuit configuration and then the calculation of component values based upon your application.

Unfortunately, you may have an application in which the filtering requirements are rather demanding. This may require a more complicated filter circuit with a high order response, greater than first or second order. The design of high order filters can be quite tricky and is summarized later in this chapter.

6.1 Filter transfer functions

It is impossible in one chapter to present a comprehensive introduction to the theory behind filter responses but hopefully this section will provide a useful summary of the salient points.

As a reminder, in the following expressions:

s is the Laplace operator

where

$$s = j\omega \qquad \text{and } \omega = 2\pi f \text{ (the angular frequency).}$$

Active R–C filter networks belong to a family of linear, lumped parameter, finite networks. The response of an n-order linear lumped parameter network is given below where n-order means that the network has n-poles given by the order of the denominator polynomial:

$$T(s) = \frac{N(s)}{D(s)} = \frac{b_m s^m + b_{(m-1)} s^{m-1} + \cdots + b_1 s + b_0}{d_n s^n + d_{(n-1)} s^{n-1} + \cdots + d_1 s + d_0}$$

where

$N(s)$ is the numerator polynomial
$D(s)$ is the denominator polynomial
$d_n \cdots d_0$ and $b_m \cdots b_0$ are real coefficients
$T(s)$ is the transfer function of the network.

Note: For real networks, $n > m$.

$N(s)$ and $D(s)$ can be factorized into first and second order factors with real coefficients. Hence, the overall filter response can be made up from a cascade of first and second order filter sections. The following section summarizes the response of several first and second order transfer functions.

(i) First order response: low pass (Fig. 6.2)

This response is very simple and is of the type: $\Rightarrow \dfrac{K_{\text{LP}} \cdot \omega_0}{s + \omega_0}$

where

K_{LP} is the dc gain
ω_0 is the pole frequency, which in this case is also the frequency when the gain is 3 dB down on the passband gain.

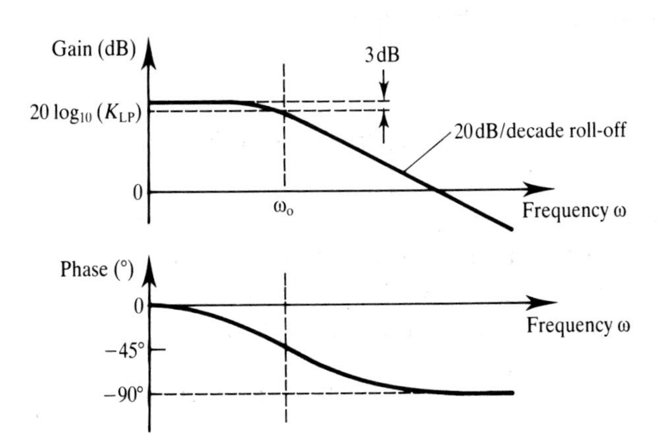

Fig. 6.2. First order lowpass frequency response.

(ii) First order response: high pass (Fig. 6.3)

This is also a simple case of the type: $\Rightarrow \dfrac{H_{\text{HP}} s}{s + \omega_0}$

where

K_{HP} is the gain at high frequencies
ω_0 is the pole frequency, which again is also the frequency when the gain is 3 dB down (= −3 dB frequency).

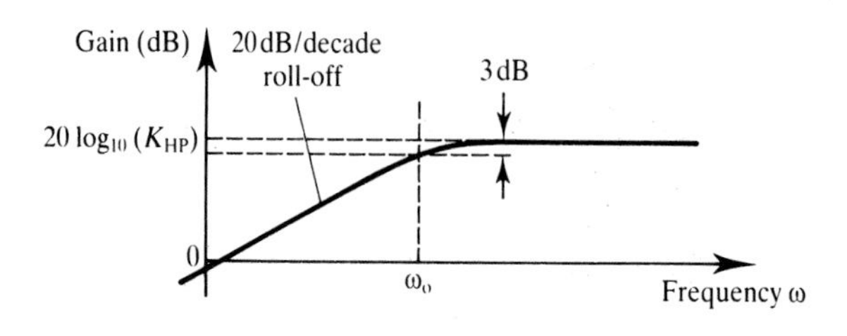

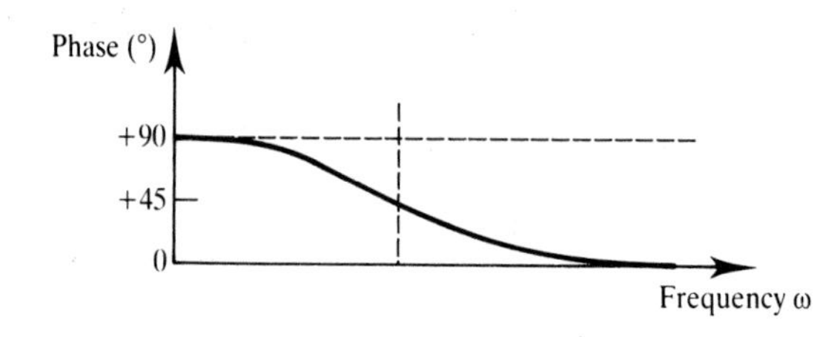

Fig. 6.3. First order highpass frequency response.

(iii) First order response: all pass (Fig. 6.4)
These are filters which have a constant gain over the entire frequency range but which introduce a phase shift or time delay.

The response of a first order all pass filter is of the type: $\Rightarrow \dfrac{K_{AP}(s-\omega_0)}{(s+\omega_0)}$

where

K_{AP} is the gain
ω_0 is the frequency at which 90° phase shift occurs.

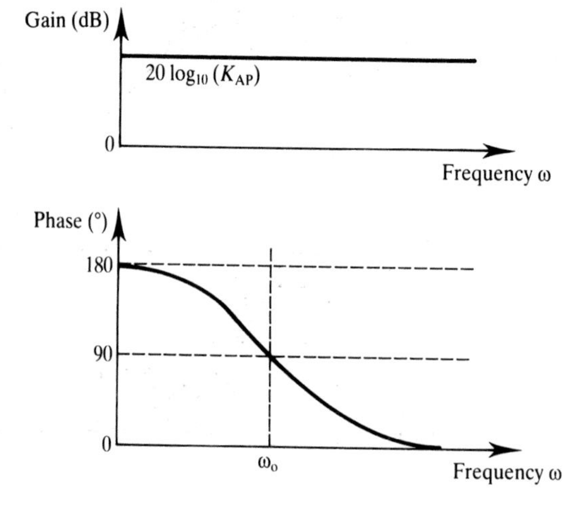

Fig. 6.4. First order all pass filter frequency response.

(iv) Second order response: low pass (Fig. 6.5)

This response is of the type: $\Rightarrow \dfrac{K_{LP}\cdot\omega_0{}^2}{s^2 + \dfrac{\omega_0}{Q_F}\cdot s + \omega_0{}^2}$

where

K_{LP} is the dc gain
ω_0 is the pole frequency
Q_F is the Q factor.

The peak gain occurs when $Q_F > \frac{1}{2}$, at

$$\omega = \omega_0\sqrt{\left(1 - \frac{1}{2Q_F{}^2}\right)} \quad (\simeq \omega_0 \text{ for high } Q_F\text{s})$$

and the peak gain equals

$$\frac{K_{LP}Q_F}{\sqrt{\left(1 - \frac{1}{4Q_F{}^2}\right)}} \quad (\simeq K_{LP}\cdot Q_F \text{ for high } Q_F\text{s})$$

with a -3 dB frequency of

$$\omega_0\left[\left(1 - \frac{1}{2Q_{F2}}\right) + \sqrt{\left(1 - \frac{1}{2Q_F{}^2}\right)^2 - 1}\right]^{\frac{1}{2}}$$

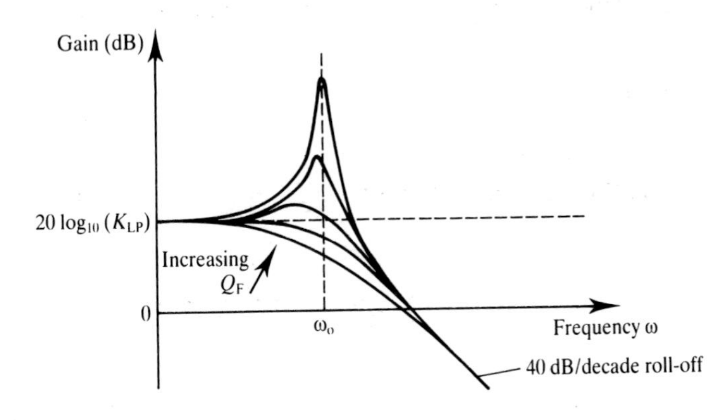

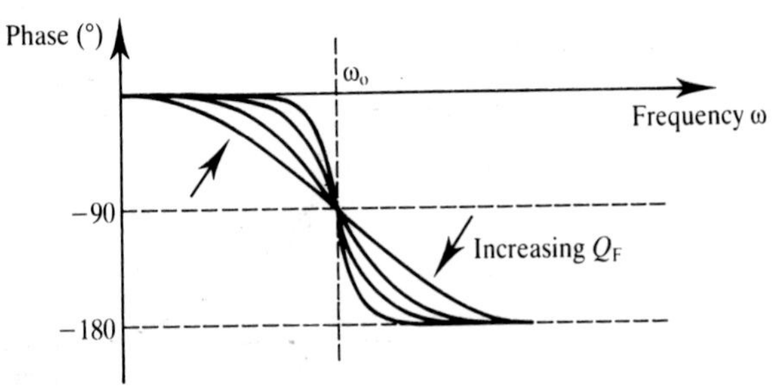

Fig. 6.5. Second order low pass frequency response.

For the low pass filter shown in Fig. 6.5, when Q_F is low (i.e. $Q_F < \frac{1}{2}$), then the poles are real and the frequency response is flat. The second order response could simply be factorized into two first order responses. As Q_F increases above $\frac{1}{2}$, a 'bump' appears in the frequency response. For high Q systems, the frequency response has a large spike.

(v) Second order response: high pass (Fig. 6.6)

This response is of the type: $\Rightarrow \dfrac{K_{HP}s^2}{s^2 + \dfrac{\omega_0}{Q_F}s + \omega_0{}^2}$

where

K_{HP} is the high pass gain
ω_0 is the pole frequency
Q_F is the Q factor.

The maximum gain for high Q systems $= K_{HP} \cdot Q_F$

Peak gain occurs when $Q_F > \frac{1}{2}$, at

$$\omega = \frac{\omega_0}{\sqrt{\left(1 - \dfrac{1}{2Q_F{}^2}\right)}} \quad (\simeq \omega_0 \text{ for high } Q_F)$$

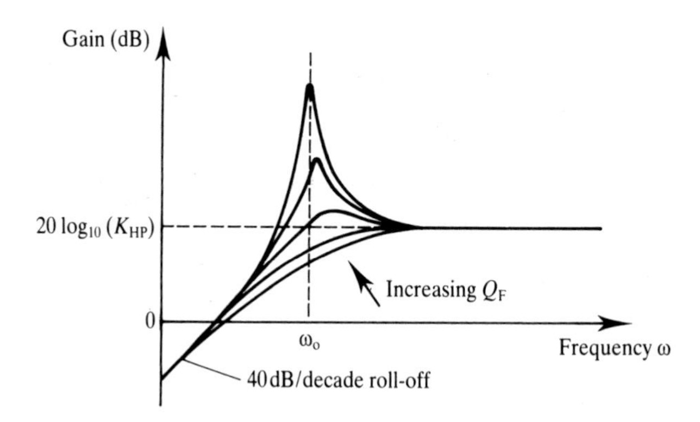

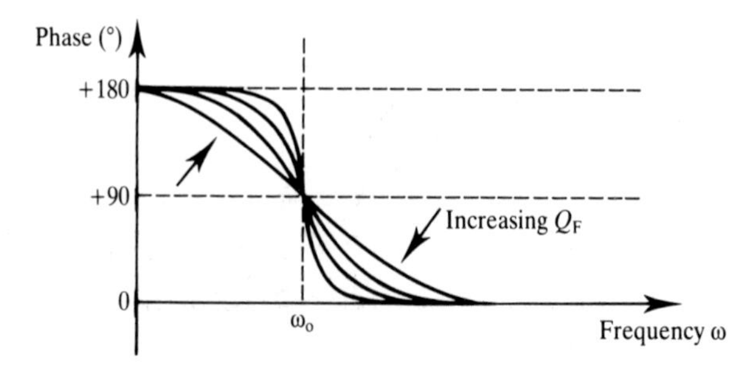

Fig. 6.6. Second order highpass frequency response.

and peak gain equals

$$\frac{K_{HP}Q_F}{\sqrt{\left(1-\dfrac{1}{4Q_F{}^2}\right)}} \quad (\simeq K_{HP}\cdot Q_F \text{ for high } Q_F)$$

The -3 dB frequency equals

$$\frac{\omega_0}{\left[\left(1-\dfrac{1}{2Q_F{}^2}\right)+\sqrt{\left(1-\dfrac{1}{2Q_F{}^2}\right)^2+1}\right]^{\frac{1}{2}}}$$

(vi) Second order bandpass response (Fig. 6.7)

This response is of the type: $\Rightarrow \dfrac{\dfrac{K_{BP}\omega_0}{Q_F}s}{s^2+\dfrac{\omega_0}{Q_F}s+\omega_0{}^2}$

which can be rearranged as

$$\frac{K_{BP}}{1+Q_F\left(\dfrac{s}{\omega_0}+\dfrac{\omega_0}{s}\right)}$$

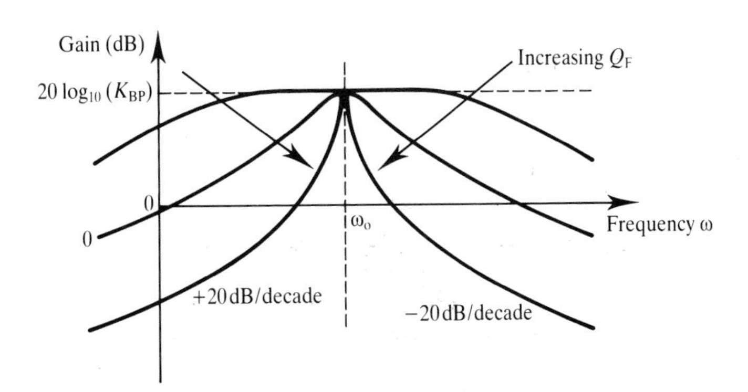

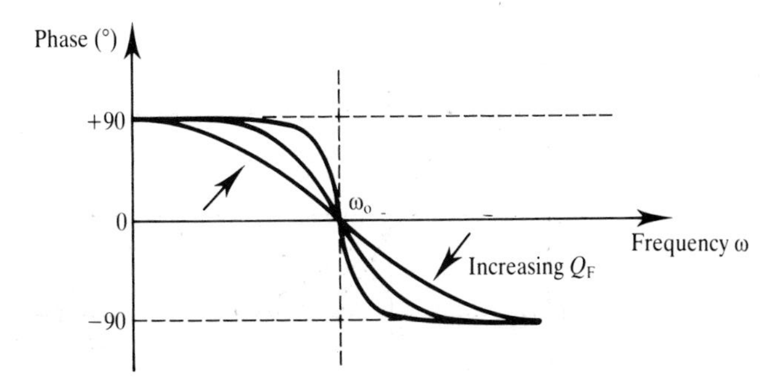

Fig. 6.7. Second order bandpass frequency response.

where

K_{BP} is the maximum gain which occurs at centre frequency ω_0
Q_F is the Q-factor

Note:
$$Q_F = \omega_0/(\omega_2 - \omega_1)$$

where ω_1 and ω_2 are the -3 dB points.

Also

$$\omega_0 = \sqrt{\omega_1 \cdot \omega_2} \qquad \omega_1 = \left[\sqrt{\left(1 + \frac{1}{4Q_F^2}\right)} - \frac{1}{2Q_F}\right]\omega_0$$

$$\omega_2 = \left[\sqrt{\left(1 + \frac{1}{4Q_F^2}\right)} + \frac{1}{2Q_F}\right]\omega_0$$

Bandwidth (3 dB) $= \omega_2 - \omega_1 = \omega_0/Q_F$.
When Q_F is low (i.e., $Q_F < \frac{1}{2}$), then the denominator polynomial of the response function can be factorised into two real poles and the gain and phase curves in Fig. 6.7 are relatively flat. For $Q_F > \frac{1}{2}$, the denominator poles are complex. As Q_F increases, the bandwidth reduces, and the gain response becomes increasingly frequency selective.

(vii) Second order response: band reject (notch) (Fig. 6.8)

This response is of the type: $\Rightarrow$
$$\frac{K_{BR}(s^2 + \omega_0^2)}{s^2 + \dfrac{\omega_0^2}{Q_F}s + \omega_0^2}$$

where

K_{BR} is the gain at dc and very high frequencies
ω_0 is the centre frequency of the notch
Q_F is the Q-factor.

The expression can be rewritten as

$$K_{BR} - \frac{\dfrac{K_{BR}\omega_0}{Q_F} \cdot s}{s^2 + \dfrac{\omega_0}{Q_F}s + \omega_0^2}$$

i.e. the constant gain K_{BR} minus a bandpass response.

The -3 dB frequencies, ω_1 and ω_2, are the same as for the second order bandpass. i.e.

$$\omega_1 = \left[\sqrt{\left(1 + \frac{1}{4Q_F^2}\right)} - \frac{1}{2Q_F}\right] \qquad \omega_2 = \left[\sqrt{\left(1 + \frac{1}{4Q_F^2}\right)} + \frac{1}{2Q_F}\right]$$

Bandwidth (3 dB) $= \omega_2 - \omega_1 = \omega_0/Q_F$.

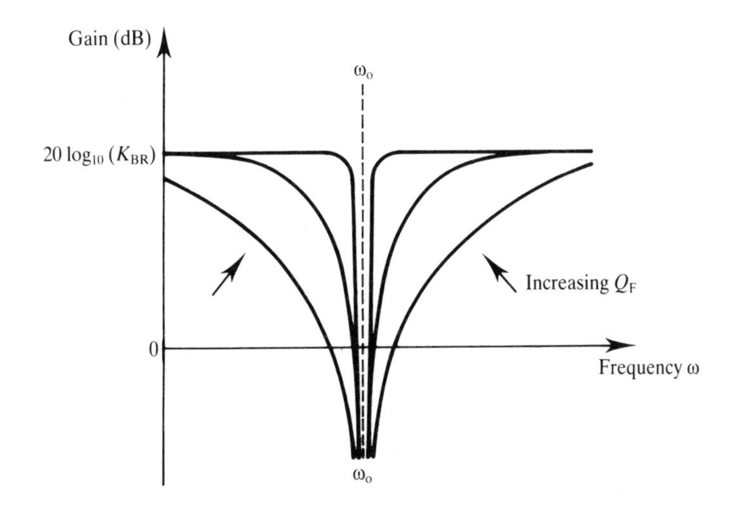

Fig. 6.8. Notch filter frequency response.

(viii) Second order response: all pass (Fig. 6.9)
The response of a second order all pass is of the type:

$$\Rightarrow \frac{K_{AP}\left(s^2 - \dfrac{\omega_0}{Q_F}s + \omega_0{}^2\right)}{\left(s^2 + \dfrac{\omega_0}{Q_F}s + \omega_0{}^2\right)}$$

This can be rearranged as:

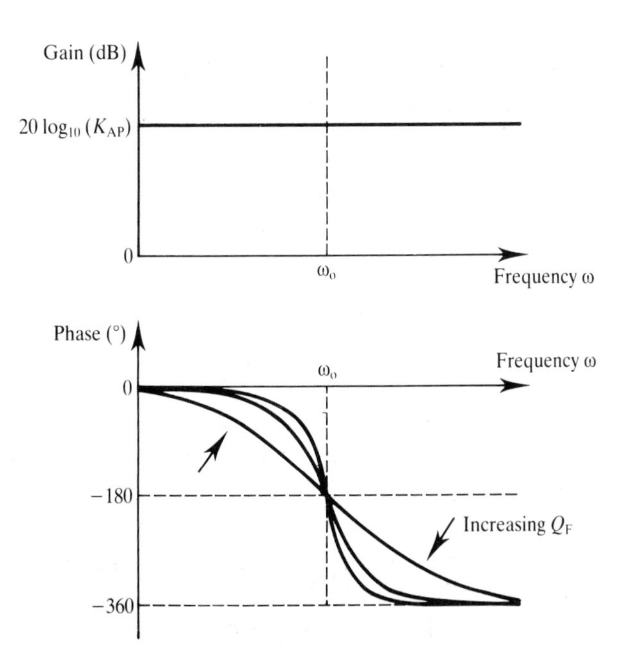

Fig. 6.9. Second order all pass frequency response.

$$K_{AP} - \frac{2K_{AP}\left(\frac{\omega_0}{Q_F}s\right)}{\left(s^2 + \frac{\omega_0}{Q_F}s + \omega_0^2\right)}$$

i.e. a constant gain minus twice the bandpass response.

6.2 Filter circuits

This section contains a collection of first and second order filter circuits. The section is structured so that all the circuits are grouped according to the various filter types: bandpass, low pass, high pass and so on. Note for this text: low Q is generally $Q_F < 2$; medium Q is typically $2 < Q_F < 20$; high Q is $Q_F > 20$.

Low pass filter circuits

(1) Single pole low pass filter (Fig. 6.10)

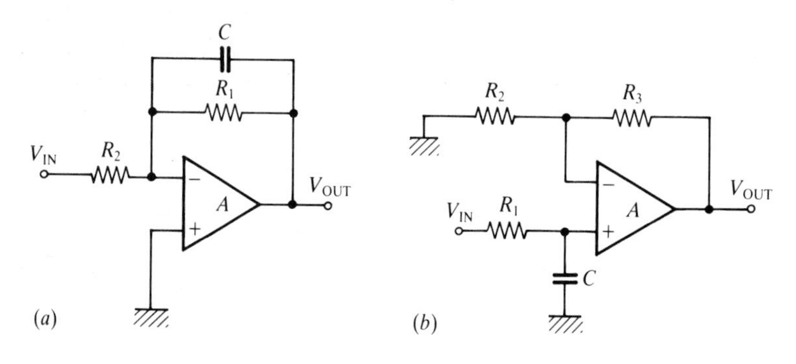

Fig. 6.10. Single pole low pass filter. (a) Inverting. (b) Non-inverting.

Transfer function: $$T(s) = \frac{K\omega_0}{s + \omega_0}$$

Pass Band Gain, K: inverting low $K = -R_1/R_2$
non-inverting $K = 1 + R_3/R_2$

Cut-off frequency ω_0: for both circuits

$$\omega_0 = 1/R_1C \qquad f_0 = \omega_0/2\pi = 1/2\pi R_1C$$

These circuits are essentially single op amp voltage amplifiers with a capacitor added to introduce the required first order frequency response. The non-inverting configuration has a high input impedance over the pass band, so avoiding loading the input.

(2) Sallen-Key low pass filter (Fig. 6.11)

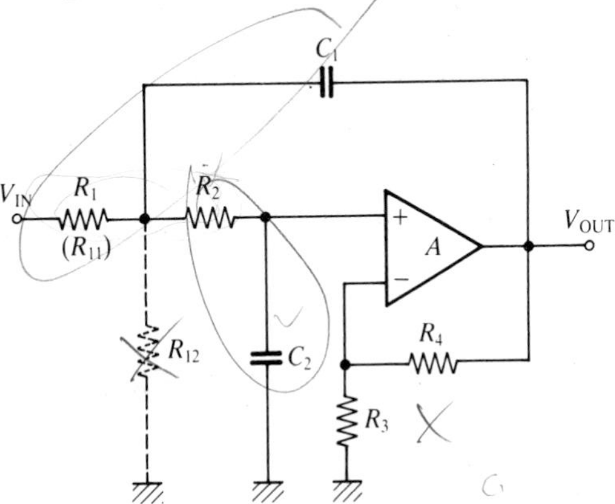

Fig. 6.11. Sallen-Key low pass filter.

General features	Advantages	Disadvantages
Two pole Low and Medium Q Non-inverting mode	High input resistance Relatively lower spread in component values	Relatively higher sensitivity to component values Limited range of realizable filters with $K > 1$ Only two parameters, ω_0 and Q_F can be tuned simply

$$\text{Transfer function} = \frac{\dfrac{1}{R_1R_2C_1C_2}\left(1+\dfrac{R_4}{R_3}\right)}{s^2+\left(\dfrac{1}{R_2C_1}+\dfrac{1}{R_1C_1}-\dfrac{R_4}{R_2R_3C_2}\right)s+\dfrac{1}{R_1R_2C_1C_2}}$$

Parameters: $K = 1 + \dfrac{R_4}{R_3}$

$$\omega_0 = 2\pi f_0 = \frac{1}{\sqrt{R_1R_2C_1C_2}}$$

$$Q_F = \frac{1}{\left(\dfrac{R_2C_2}{R_1C_1}\right)^{1/2}+\left(\dfrac{R_1C_2}{R_2C_1}\right)^{1/2}-\dfrac{R_4}{R_3}\left(\dfrac{R_1C_1}{R_2C_2}\right)^{1/2}}$$

There are several approaches to calculating the component values from the above equations, each offering a different compromise between sensitivity to component values, the spread of those values and computational complexity. The following two procedures are suggested.

Procedure 1
Let

$$R_1 = R_2 = R \qquad \text{and} \qquad C_1 = C_2 = C$$

then

$$\omega_0 = 1/RC$$

and

$$Q_F = 1/(3 - K).$$

Choose a value for either C or R. Once this value is chosen, the other value can be chosen using $\omega_0 = 1/RC$. Q_F is determined from the dc gain.

This procedure is very simple, with no spread in component values, but ω_0 and Q_F may be overly sensitive to component tolerances and there is no control over dc gain. If the dc gain is specified, then we must make $R_1 \neq R_2$ or $C_1 \neq C_2$. It is preferable to make $C_1 = C_2$ and $R_1 \neq R_2$ since resistors generally have a greater number of preferred values. Also, capacitors have higher temperature coefficients than resistors, so making the capacitors equal means that they can be of the same type and temperature variations in Q_F can be minimized.

Procedure 2
With

$$K = 1 + R_4/R_3, \qquad C_1 = C_2 = C$$

$$\omega_0 = \frac{1}{C\sqrt{R_1 R_2}}$$

$$R_2 = \frac{1}{2Q_F\omega_0 C}\left(1 + \sqrt{1 + 4Q_F{}^2(K - 2)}\right)$$ Note: real resistor values if $K > 2$ (approximately)

$$R_1 = \frac{1}{\omega_0{}^2 C^2 R_2}$$

If $1 < K < 2$, then resistor values should be made equal and values calculated for the unequal capacitors.

With

$$K = 1 + R_4/R_3 \qquad R_1 = R_2 = R \qquad \omega_0 = \frac{1}{R\sqrt{C_1 C_2}}$$

$$C_2 = \frac{1}{4Q_F R\omega_0}\left[1 + \sqrt{1 + 8(K - 1)Q_F{}^2}\right]$$

$$C_1 = \frac{1}{C_2\omega_0{}^2 R^2}$$

The dc gain can be reduced by replacing R_1 with two resistors, R_{11} and R_{12} as shown dotted in Fig. 6.11, where $R_1 = R_{11}//R_{12}$ and the gain is now given by

$$\text{gain} = \frac{R_{12}}{(R_{11} + R_{12})} \cdot \frac{(R_3 + R_4)}{R_3}$$

The use of an attenuator at the input means that the gain of the non-inverting amplifier $(1 + R_4/R_3)$ must be higher, which would give a slightly poorer high frequency and linearity performance. In addition, the output offset voltage, offset drift and noise will be higher.

Tuning procedure:
It is not possible to tune all the parameters to exact values independently. Usually, ω_0 and Q_F are tuned with R_1 and R_2. The gain can be adjusted later in the system.

Sensitivity to component values
For higher Q_F values (i.e. $Q_F > 10$) and high gain values, this circuit may produce large sensitivities to changes in component values. The sensitivities of Q_F to changes in R and C may be much greater than 1.

For high accuracy filters, there may be a problem caused by the error introduced from the finite bandwidth of the op amp. This error can be reduced by up to an order of magnitude by splitting R_2 into two resistors R_{21} and R_{22} to create a simple lead–lag compensating network as shown in Fig. 6.12.

$R_2 = R_{21} + R_{22}$ with $R_{22} = 1/2\pi f_A C_2$ where f_A is the gain-bandwidth product of the op amp.

For a simple two-pole, low pass Butterworth filter the circuit in Fig. 6.11 can be modified with $C_1 = 2C_2 = 2C$ and $R_1 = R_2 = R$. R_4 is equal to zero and R_3 is not required. This gives a two-pole low pass

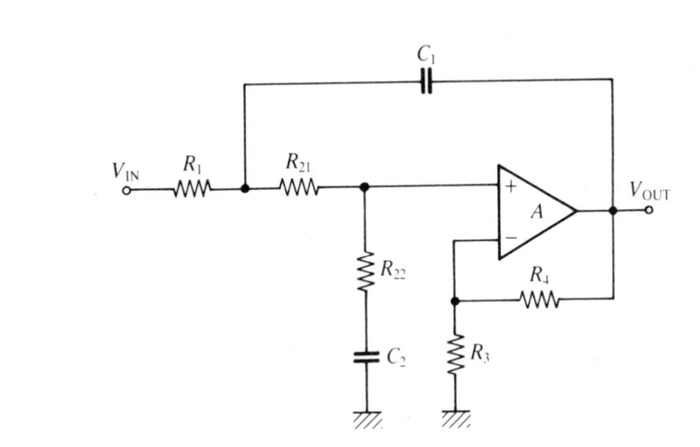

Fig. 6.12. Reducing error from the finite op amp frequency response.

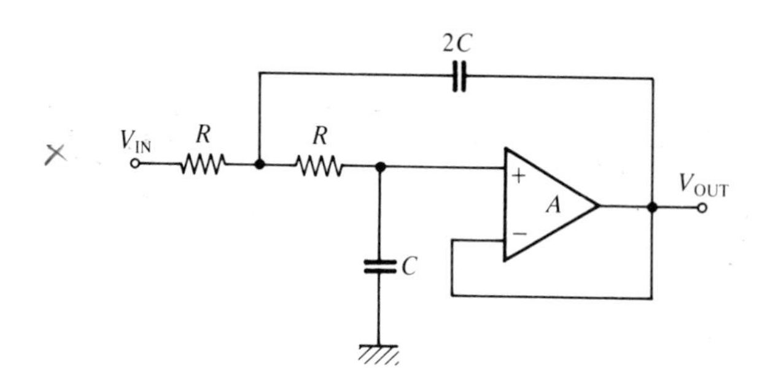

Fig. 6.13. Second order Sallen-key low pass Butterworth filter.

Butterworth with 3 dB frequency at $1/(\sqrt{2} \cdot RC)$ rad/s. The modified circuit is shown in Fig. 6.13.

(3) Multiple-loop feedback low pass filter (Fig. 6.14)

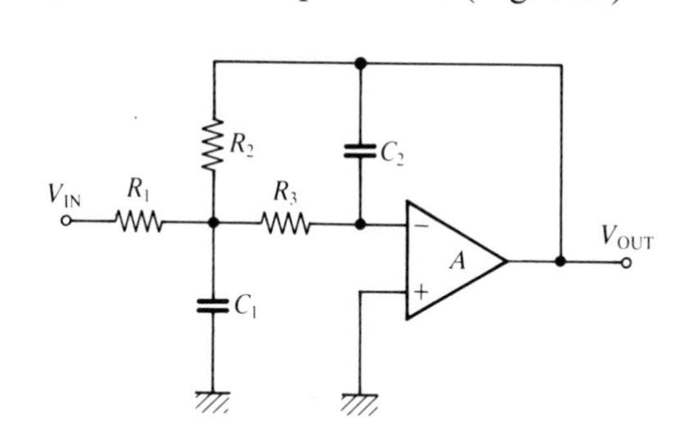

Fig. 6.14. Multiple feedback low pass filter.

General features	Advantages	Disadvantages
Two-pole Low and medium Q values Inverting mode	Can realize low pass of low pass specifications with $\|K\| < 1$ Relatively low sensitivity to component changes, with sensitivities nearly always less than one	Relatively low input resistance Only two parameters can be tuned easily: ω_0 and Q_F Large spread in component values, especially for higher Qs and gains

$$\text{Transfer function } T(s) = \frac{-\dfrac{1}{R_1R_3C_1C_2}}{s^2 + s\left(\dfrac{1}{R_1} + \dfrac{1}{R_2} + \dfrac{1}{R_3}\right)\dfrac{1}{C_1} + \dfrac{1}{R_2R_3C_1C_2}}$$

Parameters: $K = -R_2/R_1$

$$Q_F = \frac{\left(\frac{C_1}{C_2}\right)^{1/2}}{\left(\frac{R_3}{R_2}\right)^{1/2} + \left(\frac{R_2}{R_3}\right)^{1/2} + \left(\frac{R_2 R_3}{R_1^2}\right)^{1/2}}$$

$$\omega_0 = \frac{1}{\sqrt{R_2 R_3 C_1 C_2}}$$

Component values:

Choose C_1 and C_2

(note $C_1 > [4(|K| + 1)Q_F^2]C_2$ for real values)

then

$$R_2 = \frac{1}{2Q_F \omega_0 C_2}\left[1 \pm \sqrt{1 - \frac{4(|K| + 1)Q_F^2 C_2}{C_1}}\right]$$

and

$$R_1 = R_2/|K|$$

and

$$R_3 = 1/(\omega_0^2 C_1 C_2 R_2)$$

Tuning procedure:

If all three values of K, ω_0 and Q_F are critical, then tuning will be extremely difficult due to all three parameters being interdependent on the three resistors, R_1, R_2 and R_3. However, if K is chosen as a non-critical parameter, then ω_0 can be tuned with R_2 or R_3 and Q_F can be tuned with R_1.

This circuit gives relatively low sensitivities (of Q_F, ω_0 and K) to variations in component values, generally less than one. This is at the expense of a relatively large spread in component values with increasing Q_F and gain K. The range of component values increases with both gain and Q_F. Practical circuits are limited to stages with a gain–Q_F product of less than 100, i.e. $K \cdot Q_F < 100$. As the Q_F increases, the frequency response of the op amp becomes more important. The finite frequency response of the op amp causes an error in the values of ω_0 and Q_F.

(4) Zero offset low pass filter (Fig. 6.15)

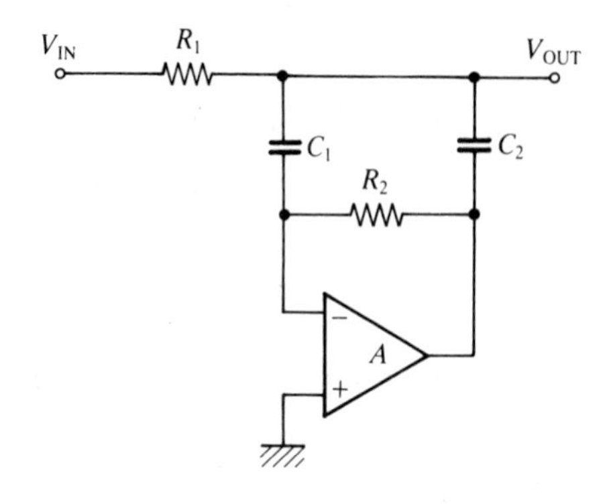

Fig. 6.15. Zero offset low pass filter.

General features	Advantages	Disadvantages
Two-pole Low and medium Q Non-inverting mode	Introduces zero offset Relatively low sensitivity to component value changes	High output resistance Fixed gain of unity Relatively poor frequency response Difficult to tune

$$\text{Transfer function } T(s) = \frac{\dfrac{1}{R_1R_2C_1C_2}}{s^2 + \dfrac{1}{R_2}\left(\dfrac{1}{C_1} + \dfrac{1}{C_2}\right)s + \dfrac{1}{R_1R_2C_1C_2}}$$

$$\text{Gain } K = 1$$

$$\text{Pole frequency } \omega_0 = \frac{1}{\sqrt{R_1R_2C_1C_2}}$$

$$Q_F = \frac{\left(\dfrac{R_2}{R_1}\right)^{1/2}}{\left(\dfrac{C_1}{C_2}\right)^{1/2} + \left(\dfrac{C_2}{C_1}\right)^{1/2}}$$

$$\text{Output resistance } R_0 = R_1$$

Component values:

Choose C_1 and C_2

then

$$R_2 = \frac{Q_F}{\omega_0}\left(\frac{1}{C_1} + \frac{1}{C_2}\right)$$

and

$$R_1 = 1/(\omega_0{}^2 C_1 C_2 R_2)$$

Choose C_1 and C_2 so that R_1 (the output resistance), is as small as possible.

Tuning procedure:
The gain is fixed at unity and cannot be altered. ω_0 and Q_F both depend upon the two resistors R_1 and R_2 and so tuning is an iterative procedure of adjusting R_1 and R_2.

This circuit has the advantage that the op amp is entirely ac-coupled to the signal path and so cannot add any offset to the signal. This is at the expense of a high output resistance equal to the value of R_1. The frequency range of the filter is limited by the op amp which introduces an extra pole and also a complex high frequency zero pair close to the imaginary axis.

(5) Current generalized immittance low pass filter (Fig. 6.16)

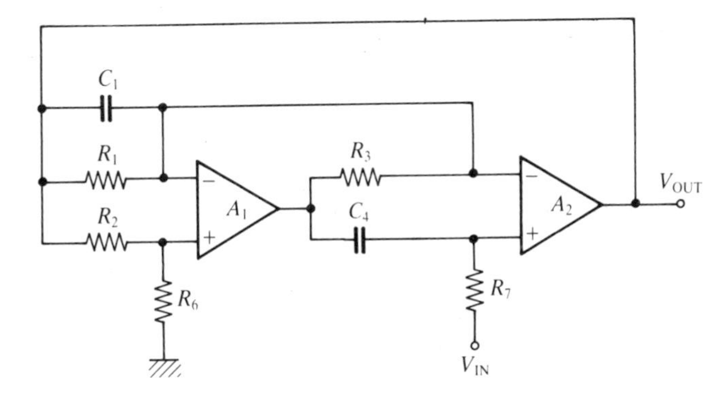

Fig. 6.16. Current generalized immittance low pass filter.

General features	Advantages	Disadvantages
Two-pole Non-inverting mode	Low to high Q values High input resistance Easily tuned High Qs realized without large spread in component values Low sensitivity (always less than 1) to changes in R and C values	Two op amps needed

$$\text{Transfer function } T(s) = \frac{\dfrac{1}{R_3 R_7 C_1 C_4}\left(1 + \dfrac{R_6}{R_2}\right)}{s^2 + s\dfrac{1}{R_1 C_1} + \dfrac{R_6}{R_2 R_3 R_7 C_1 C_4}}$$

Parameters:

$$\text{Gain } K = 1 + R_2/R_6$$

$$\text{Pole frequency } \omega_0 = \sqrt{\frac{R_6}{R_2 R_3 R_7 C_1 C_4}}$$

$$Q_F = R_1 C_1 \omega_0$$

Component values:
The equations are uncomplicated and there are five resistors and two capacitors involved and calculating the component values is a relatively easy matter.

Tuning procedure:
(i) Tune the gain with R_2
(ii) Tune the frequency ω_0 with R_7
(iii) Tune Q_F with R_1

This circuit is very good for high Q_F applications since it has a low sensitivity to component values and can be easily tuned.

High pass filter circuits

(1) Single pole high pass filter (Fig. 6.17)

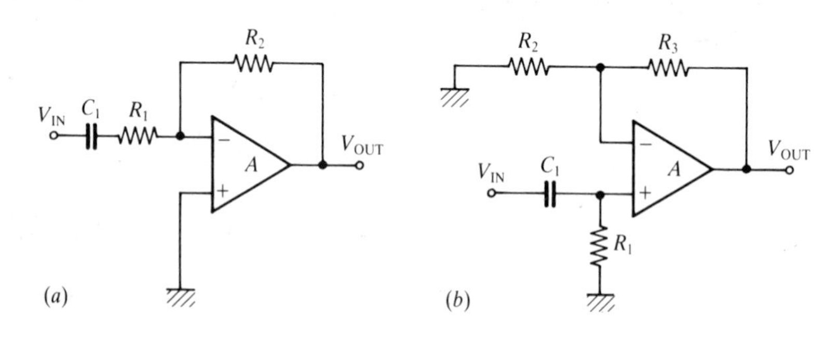

Fig. 6.17. Single pole high pass filters. (*a*) Inverting. (*b*) Non-inverting.

These circuits are just simple inverting and non-inverting voltage amplifiers with an added capacitor to create the low frequency attenuation.

$$\text{Transfer function } T(s) = \frac{Ks}{s + \omega_0}$$

Parameters: $\text{Gain} = 1 + R_3/R_2$ (non-inverting)

$= -R_2/R_1$ (inverting)

$$\text{Pole frequency } \omega_0 = 1/(R_1C_1)$$

(2) Sallen-Key high pass filter (Fig. 6.18)

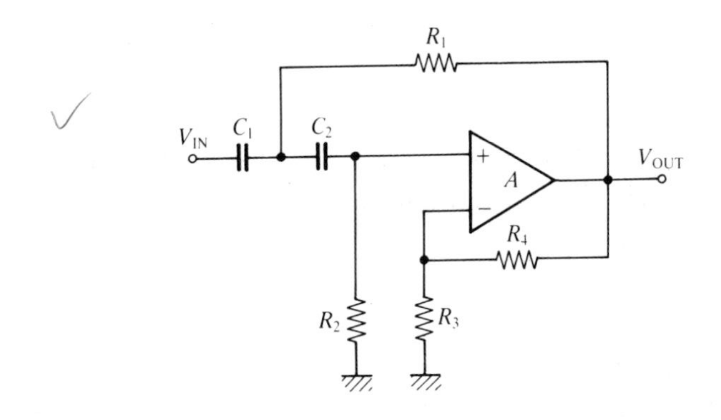

Fig. 6.18. Sallen-Key high pass filter.

General features	Advantages	Disadvantages
Two-pole Non-inverting mode Low and medium Q values	Relatively low spread in component values	Relatively higher sensitivity of Q_F to changes in component values Cannot realize complete range of values of K, ω_0 and Q_F

$$\text{Transfer function } T(s) = \frac{\left(1 + \frac{R_4}{R_3}\right)s^2}{s^2 + \left(\frac{1}{R_2C_1} + \frac{1}{R_2C_2} - \frac{R_4}{R_1R_3C_1}\right)s + \frac{1}{R_1R_2C_1C_2}}$$

Parameters: $K = 1 + R_4/R_3$

$$\omega_0 = \frac{1}{\sqrt{R_1R_2C_1C_2}}$$

$$Q_F = \frac{1}{\left(\frac{R_1C_2}{R_2C_1}\right)^{1/2} + \left(\frac{R_1C_1}{R_2C_2}\right)^{1/2} - \frac{R_4}{R_3}\left(\frac{R_2C_2}{R_1C_1}\right)^{1/2}}$$

Component values:

If an accurate gain is not important, then choose $C_1 = C_2 = C$ and let $R_1 = R_2 = R$

then

$$K = 3 - 1/Q_F$$

and

$$R = 1/\omega_0 C$$

Tuning procedure:
It is difficult to tune all three parameters, K, ω_0 and Q_F, since they are interdependent on resistor values. Parameters ω_0 and Q_F can be adjusted as follows.

(i) To tune ω_0 use R_1 or R_2
(ii) To tune Q use R_3 or R_4

The gain is better adjusted elsewhere in the system if possible. Alternatively, the gain can be adjusted by adding a capacitor dividing network at the input as shown in Fig. 6.19, in which:

$$K = \frac{C_{11}}{C_{11} + C_{12}}\left(1 + \frac{R_4}{R_3}\right)$$

$C_1 = C_{11} + C_{12}$ and can be used in the previous equations.

If K is specified, then choose $C_1 = C_2 = C$

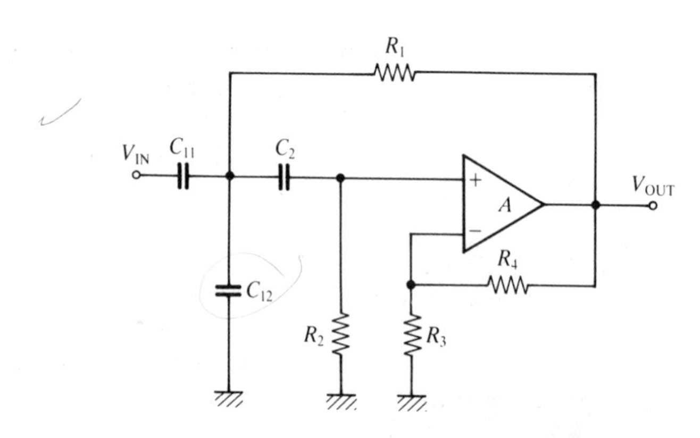

Fig. 6.19. Adjusting the gain of the Sallen-Key high pass filter.

then

$$K = 1 + R_4/R_3$$

$$R_1 = \frac{1}{4Q_F\omega_0 C}[1 + \sqrt{1 + 8(K-1)Q_F{}^2}]$$

$$R_2 = \frac{1}{\omega_0^2 C^2 R_1}$$

For filters requiring high Q_F and gain values (e.g. greater than 10), this circuit may produce large sensitivities to component values.

To make a two-pole high pass Butterworth filter, as shown in Fig. 6.20: $C_1 = C_2 = C$; $2R_1 = R_2 = 2R$; $R_4 = 0$; and R_3 omitted.

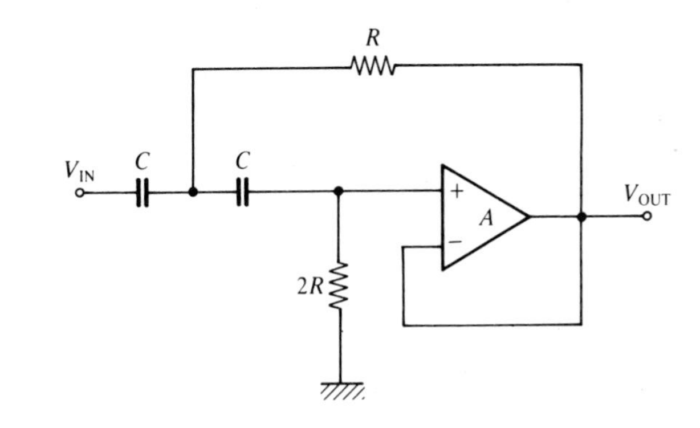

Fig. 6.20. Two-pole Butterworth high pass filter.

(3) Multiple loop feedback high pass filters (Fig. 6.21)

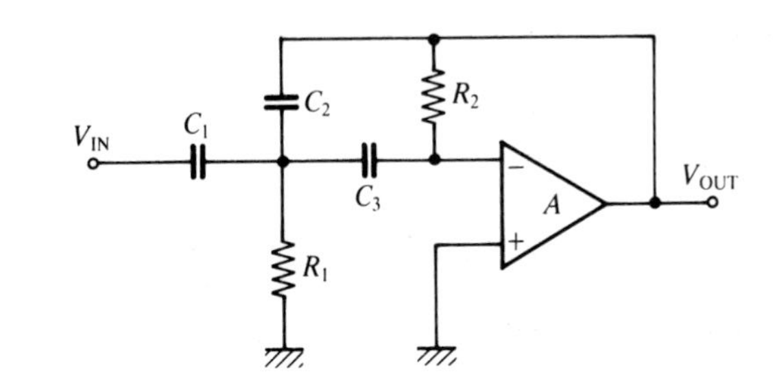

Fig. 6.21. Multiple loop feedback high pass filter.

General features	Advantages	Disadvantages
Two-pole Low and medium Q values	Can realize values for K less than unity	High spread in component values Tuning is difficult
Inverting mode	Relatively low sensitivity to component value changes	The gain is a ratio of capacitor values and is not as stable as resistor ratios Requires 3 capacitors

$$\text{Transfer function } T(s) = \frac{-\left(\dfrac{C_1}{C_2}\right)s^2}{s^2 + \left(\dfrac{C_1}{C_2C_3} + \dfrac{1}{C_2} + \dfrac{1}{C_3}\right)\dfrac{1}{R_2}s + \dfrac{1}{R_1R_2C_2C_3}}$$

$$\text{Gain } |K| = C_1/C_2$$

$$\text{Pole frequency } \omega_0 = \frac{1}{\sqrt{R_1R_2C_2C_3}}$$

$$Q_F = \frac{\left(\dfrac{R_2}{R_1}\right)^{1/2}}{\dfrac{C_1}{\sqrt{C_2C_3}} + \left(\dfrac{C_3}{C_2}\right)^{1/2} + \left(\dfrac{C_2}{C_3}\right)^{1/2}}$$

Component values:

Choose $C_1 = C_2 = C$ for a convenient value of C
then

$$C_2 = C/|K|$$

and

$$R_1 = \frac{|K|}{R_2{\omega_0}^2C^2}, \qquad R_2 = \frac{Q_F(2|K| + 1)}{\omega_0 C}$$

Tuning procedure:
Tuning is difficult since there are only two resistors and both Q and ω_0 depend simultaneously on both resistor values. Tuning Q_F and ω_0 is consequently an iterative procedure. The gain is a ratio of capacitor values and so cannot be tuned by varying either resistor. With this particular circuit a trimming capacitor might be considered to make the tuning easier.

This circuit has the advantage that the three parameters K, Q_F and ω_0, are less sensitive to changes in R and C values but this advantage is at the expense of a wide spread in component values for moderate gains and Q values. As Q_F and K increase, component values acquire a larger spread. This spread limits the KQ_F product to less than 100. The finite frequency response of the op amp causes considerable errors in very high Q_F circuits and at higher frequencies where the gain of the op amp is small.

Another disadvantage of this circuit is that the gain is a ratio of capacitors which are, generally, not as temperature stable as resistors. Capacitors are usually more expensive and bigger than resistors and are better avoided if possible: with three capacitors, this configuration uses more than other circuits.

(4) Current generalized immittance high pass filter (Fig. 6.22)

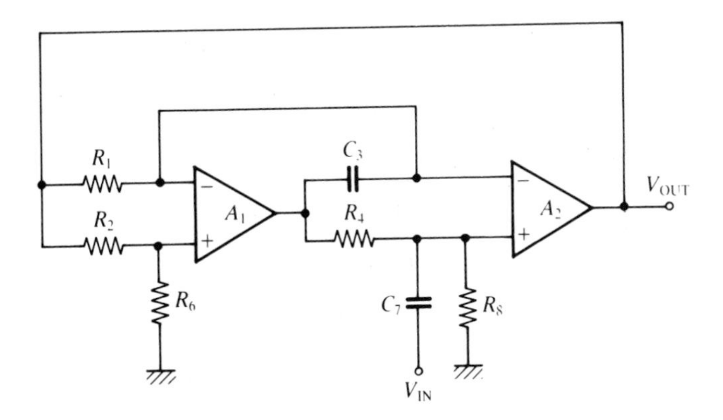

Fig. 6.22. Current generalized immittance high pass filter.

General features	Advantages	Disadvantages
Two-pole Non-inverting mode	Can realize low to high Q values Low sensitivity to component changes with the sensitivity of Q, K and ω_0 being less than unity Easily tuned with independent tuning possible for Q, K and ω_0 High Q values realized without large spread in component values	Two op amps are required

$$\text{Transfer function } T(s) = \frac{s^2\left(1 + \dfrac{R_2}{R_6}\right)}{s^2 + \dfrac{1}{R_8C_7}s + \dfrac{R_2}{R_1R_4R_6C_3C_7}}$$

Parameters: $K = 1 + R_2/R_6$

$$\omega_0 = \sqrt{\frac{R_2}{R_1R_4R_6C_3C_7}}$$

$$Q_F = R_8C_7\omega_0$$

Component values:
With five resistors and two capacitors to work with, calculating component values is relatively easy.

Tuning procedure:
(i) Tune the gain with R_2
(ii) Tune ω_0 with R_4
(iii) Tune Q_F with R_8

Each of the above parameters can be tuned independently of the others.
This filter is suitable for high Q applications since it has low sensitivity to changes in component values, is easily tuned and does not have a large spread in component values. This high performance is achieved at the expense of an extra op amp and more resistors. The extra op amp may cause problems in very sensitive applications due to extra noise and drift.

Bandpass filter circuits

(1) Real pole bandpass filter (Fig. 6.23)

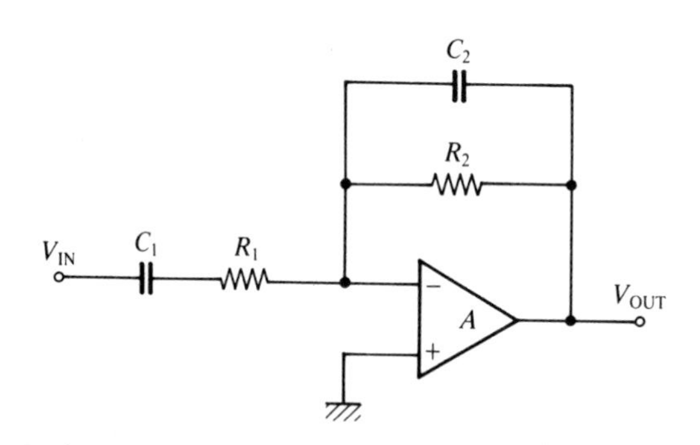

Fig. 6.23. Real pole bandpass filter.

$$\text{Transfer function } T(s) = \frac{-\dfrac{1}{R_1C_2}s}{\left(s+\dfrac{1}{R_1C_1}\right)\left(s+\dfrac{1}{R_2C_2}\right)}$$

Pole frequencies $= 1/R_1C_1$ and $1/R_2C_2$ Rad/s (ω_1 and ω_2)

The circuit of Fig. 6.23 is simply an inverting amplifier with two capacitors added to provide high and low frequency attenuation as shown in Fig. 6.24 which is best used as a wide bandpass filter with the high and low frequency breakpoints separated by at least a decade in frequency. In this case, the two pole frequencies will be almost equal to the 3 dB frequencies. Also, it is preferable to make $1/(R_1C_1)$ the low frequency 3 dB point and $1/(R_2C_2)$ the high frequency 3 dB point. Consequently, the passband gain will be given by $-R_2/R_1$, i.e. a ratio

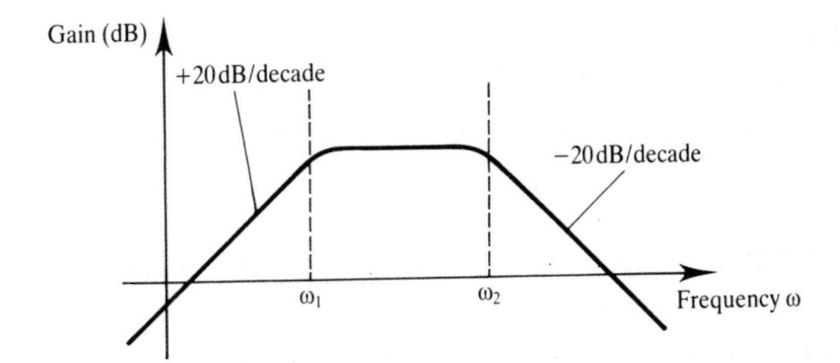

Fig. 6.24. Real pole bandpass filter frequency response.

of resistor values. Note, however, if $1/R_2C_2$ was the low 3 dB frequency and $1/(R_1C_1)$ was the high 3 dB frequency, then the passband gain will be given by $-C_1/C_2$, i.e. a ratio of capacitor values and you may remember that capacitors are often not as temperature stable as resistors.

(2) Multiple loop feedback bandpass filters (Fig. 6.25)

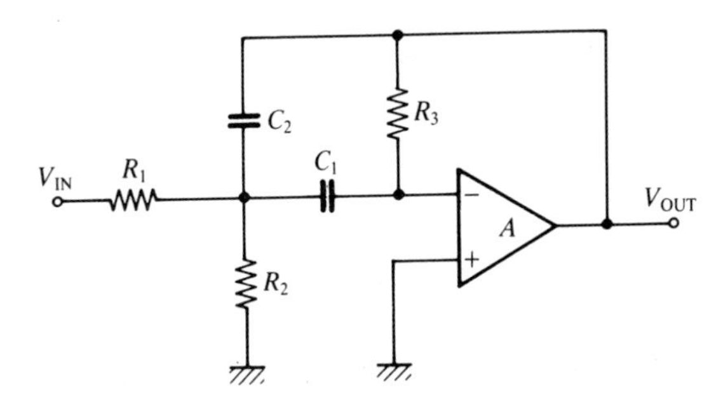

Fig. 6.25. Multiple loop feedback bandpass filter.

General features	Advantages	Disadvantages
Inverting mode	A simple modification using positive feedback will allow Q values up to 20 to be realized Low sensitivity to component values High spread in component values may be reduced by using positive feedback	$\|K\| < 2Q^2$ to realize the filter in this circuit

$$\text{Transfer function } T(s) = \frac{-\dfrac{1}{R_1C_2}s}{s^2 + \left(\dfrac{1}{C_1} + \dfrac{1}{C_2}\right)\cdot\dfrac{1}{R_3}s + \dfrac{1}{R_3C_1C_2}\left(\dfrac{1}{R_1} + \dfrac{1}{R_2}\right)}$$

Parameters: $$|K| = \frac{R_3}{R_1} \cdot \frac{C_1}{(C_1 + C_2)}$$

$$\omega_0 = \sqrt{\frac{1}{R_3 C_1 C_2}\left(\frac{1}{R_1} + \frac{1}{R_2}\right)}$$

$$Q_F = \frac{\left(\frac{R_3}{R_2}\left(1 + \frac{R_2}{R_1}\right)\right)^{1/2}}{\left(\frac{C_1}{C_2}\right)^{1/2} + \left(\frac{C_2}{C_1}\right)^{1/2}}$$

$$\Delta\omega = 3 \text{ dB bandwidth} = \frac{1}{R_3}\left(\frac{1}{C_1} + \frac{1}{C_2}\right)$$

Component selection:

Choose $C_1 = C_2 = C$

$$R_1 = \frac{Q_F}{|K| C \omega_0}$$

$$R_2 = \frac{Q_F}{(2Q_F^2 - |K|)\omega_0 C}$$

$$R_3 = \frac{2Q_F}{C\omega_0}$$

For realizable filters $|K| < 2Q_F^2$

Tuning procedure:

Tuning this circuit is difficult since ω_0 and Q_F are interdependent on the same resistor values. Since this circuit is only used for low Q_F values, however, the tuning of such circuits is not very critical. For higher Q_F values, where tuning is critical, use positive feedback as shown in Fig. 6.26. Note, the center frequency can be varied independently of $\Delta\omega$ using either R_1 or R_2 (but this will also change K).

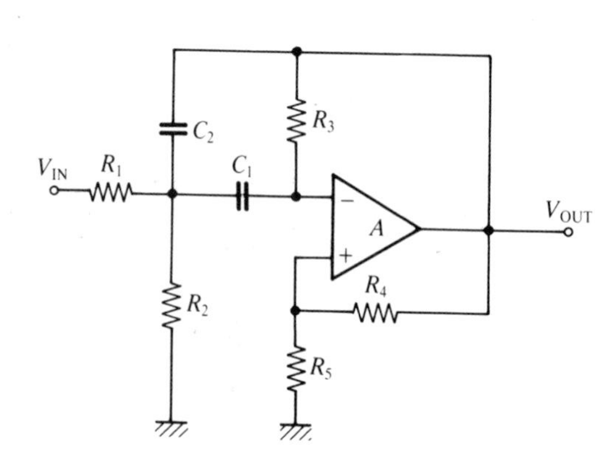

Fig. 6.26. Using positive feedback with LF bandpass filters.

The transfer function of the circuit is now:

$$T(s) = \frac{-(1+A)\dfrac{1}{R_1C_2}s}{s^2 + \left[\dfrac{1}{R_3}\left(\dfrac{1}{C_1}+\dfrac{1}{C_2}\right) - \dfrac{A}{C_2}\left(\dfrac{1}{R_1}+\dfrac{1}{R_2}\right)\right]s + \dfrac{1}{R_3C_1C_2}\cdot\left(\dfrac{1}{R_1}+\dfrac{1}{R_2}\right)}$$

where

$$A = \frac{R_5}{R_4}$$

To calculate component values;

Select $C_1 = C_2 = C$

The actual values of the components may then be found by an iterative process due to the complexity of some of the equations.

Choose a value for A: $0 < A < 1$.

$$R_1 = \frac{(1+A)Q_F}{|K|\omega_0 C}$$

$$R_3 = \frac{1}{2AQ_F\omega_0 C}(\sqrt{1+8AQ_F{}^2} - 1)$$

$$1/R_2 = R_3C^2\omega_0{}^2 - \frac{1}{R_1}$$

If the component values are too widely spread, then increase A. If a negative value is obtained for R_2, try another value for A.

(3) Current generalized immittance bandpass filters (Fig. 6.27)

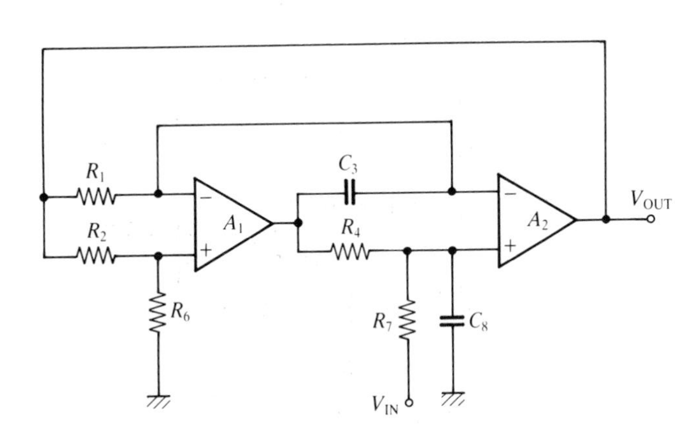

Fig. 6.27. Current generalized immittance bandpass filter.

General features	Advantages	Disadvantages
Low to high Q values Non-inverting mode	Low sensitivity in Q, K and ω_0 (always less than one) to changes in component values Easily tuned High Q values without a large spread in component values	Two op amps required

$$\text{Transfer function } T(s) = \frac{s\left(1 + \dfrac{R_2}{R_6}\right)\dfrac{1}{R_7C_8}}{s^2 + s\dfrac{1}{R_7C_8} + \dfrac{R_2}{R_1R_4R_6C_3C_8}}$$

Parameters:

$$K = 1 + R_2/R_6$$

$$\omega_0 = \sqrt{\frac{R_2}{R_1R_4R_6C_3C_8}}$$

$$Q_F = R_7C_8\omega_0$$

$$\Delta\omega = 3\text{ dB bandwidth} = 1/R_7C_8.$$

Component selection:
Component values are easy to calculate since there are 5 resistors and 2 capacitors involved. Also, the equations are relatively simple. The parameters Q_F, K and ω_0 can be easily tuned in the following order:

(i) K with R_2
(ii) ω_0 with R_4
(iii) Q_F with R_7

This circuit is very good in high Q_F applications because of its low sensitivity to changes in component values, its low spread of component values and the ease with which it is tuned. It does require two op amps, however, which may cause problems in small signal or high accuracy applications due to the extra noise and drift from the extra op amp. It may be useful to note that the 3 dB bandwidth $\Delta\omega$ can be varied independently of ω_0 using R_7. Similarly, ω_0 can be varied independently of $\Delta\omega$ and K with either R_1 or R_4.

Band reject filters

(1) Multiple loop feedback band reject filter (Fig. 6.28)

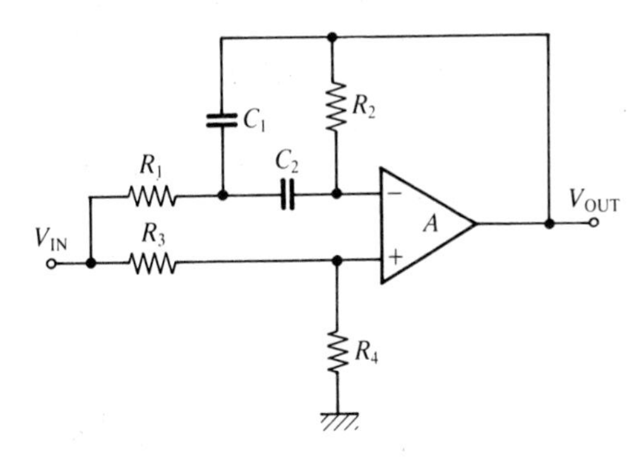

Fig. 6.28. Multiple loop feedback band reject filter.

General features	Advantages	Disadvantages
Non-inverting mode	Uses a single op amp Only 2 capacitors needed	Requires close component value matching Low Q values Not easily tuned Attenuates pass band signals

Transfer function

$$T(s) = \frac{K\left[s^2 + s\left(\frac{1}{R_2C_1} + \frac{1}{R_2C_2} - \frac{R_3}{R_4}\cdot\frac{1}{R_1C_1}\right) + \frac{1}{R_1R_2C_1C_2}\right]}{\left[s^2 + s\left(\frac{1}{R_2C_1} + \frac{1}{R_2C_2}\right) + \frac{1}{R_1R_2C_1C_2}\right]}$$

Notch frequency: $$\omega_0 = \frac{1}{\sqrt{R_1R_2C_1C_2}}$$

$$K = \frac{R_4}{R_3 + R_4}$$

$$Q_F = \frac{\left(\frac{R_2}{R_1}\right)^{1/2}}{\left(\frac{C_2}{C_1}\right)^{1/2} + \left(\frac{C_1}{C_2}\right)^{1/2}}$$

Relationships between components:

$$\frac{1}{R_2C_1} + \frac{1}{R_2C_1} = \frac{R_3}{R_4} \cdot \frac{1}{R_1C_1} \quad \text{for zero transmission at } \omega_0$$

This notch filter is very similar to the multiple loop feedback bandpass filter described above. It can be considered as a bandpass filter around the inverting input subtracted from a constant gain mode from R_3 and R_4 around the non-inverting input. For zero transmission at the notch frequency, the components must be related as shown above. There may be a large error due to component tolerances, drift and aging. This error will increase with Q_F. Consequently, the circuit is only useful for low Q_F values. The circuit also attenuates pass band signals and cannot have a gain greater than one.

Tuning this circuit can be difficult since the component values are inter-related. A simple procedure could be as follows:

(i) tune ω_0 with R_1 or R_3
(ii) tune for zero transmission with R_3 or R_4

This procedure gives no control over Q_F and K and the values obtained for these must be accepted. Otherwise, where full control over parameters is required, a complicated iterative procedure must be used.

The configuration shown in Fig. 6.28 can also be used to realize the all pass function if the component relationship is changed to,

$$\frac{R_3}{R_4} \cdot \frac{1}{R_1C_1} = \frac{2}{R_2}\left(\frac{1}{C_1} + \frac{1}{C_2}\right)$$

(2) Twin-T band reject filter (Fig. 6.29)

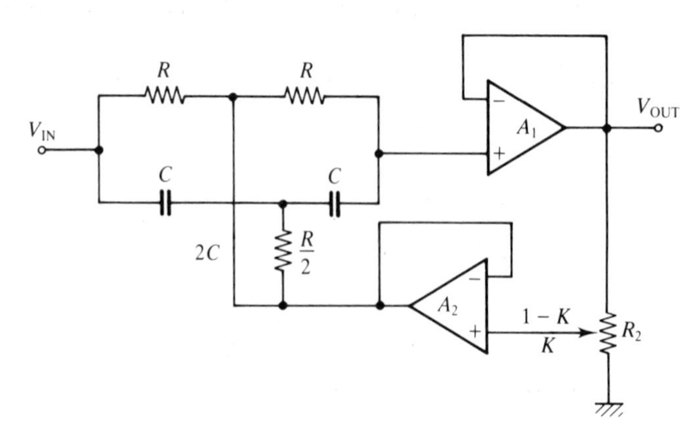

Fig. 6.29. Twin-T band reject filter.

General features	Advantages	Disadvantages
Low and medium Q values Non-inverting mode	High input resistance Adjustable Q with R_2	Fixed gain of unity Requires careful matching of resistors and capacitors Needs 2 op amps and 3 capacitors

$$\text{Transfer function } T(s) = \frac{1 + s^2R^2C^2}{1 + 4RC(1 - K)s + s^2R^2C^2}$$

$$\text{Pass band gain} = 1$$

$$\omega_0 = 1/CR$$

$$Q_F = 1/4(1 - K)$$

This circuit is quite popular despite requiring 2 op amps, 3 capacitors and closely matched resistor and capacitor values. Q_F can be adjusted using pot R_2, where $K = 1$ gives a maximum value of Q_F. The depth of the notch is also affected by the pot position. As a result, R_2 may require adjustment to strike a compromise between notch depth and Q-factor, i.e. notch width.

The depth of the notch is very sensitive to the matching of resistor and capacitor values. A slight mismatch in component values not only reduces the depth of the notch but also introduces a high frequency pole-zero pair.

(3) Current generalized immittance band reject filters (Fig. 6.30)

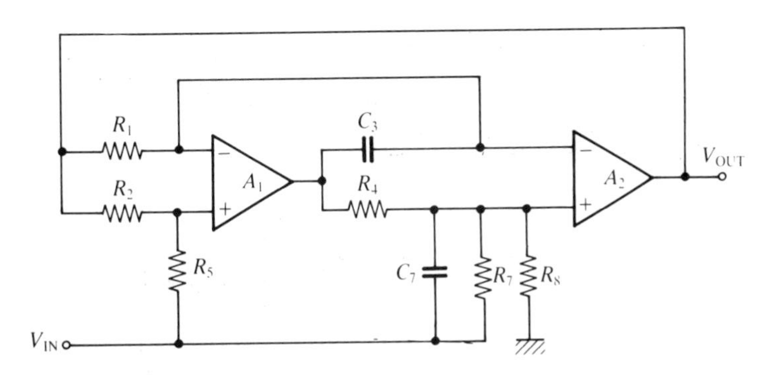

Fig. 6.30. Current generalized immittance band reject filter.

Features	Advantages	Disadvantages
Non-inverting mode	Low to high Q_F values possible Low sensitivity to component values Relatively easily tuned High Q_F values can be realized without a large spread in component values	Requires 2 op amps Fixed gain of unity Requires resistors to be matched in value

$$\text{Transfer function } T(s) = \frac{s^2 + s\dfrac{R_2}{C_7}\left(\dfrac{1}{R_2R_7} - \dfrac{1}{R_5R_8}\right) + \dfrac{R_2}{R_1R_4R_5C_3C_7}}{s^2 + s\cdot\dfrac{1}{C_7}\left(\dfrac{1}{R_7} + \dfrac{1}{R_8}\right) + \dfrac{R_2}{R_1R_4R_5C_3C_7}}$$

Notch frequency:

$$\omega_0 = \sqrt{\frac{R_2}{R_1R_4R_5C_3C_7}}$$

$$Q_F = \frac{R_7R_8C_7\omega_0}{(R_7 + R_8)}$$

$$\text{Gain} = 1$$

The required resistor relationship for minimum transmission at ω_0:

$$R_2R_7 = R_5R_8$$

This circuit gives a high Q_F but at the expense of an extra op amp. It has the advantages of good frequency response, low sensitivity to component values, is relatively easy to tune and gives a high Q_F without a large spread in component values. However, the extra op amp introduces more noise and drift which may cause problems in small signal or high accuracy applications.

Tuning procedure:

(i) tune ω_0 with R_4
(ii) tune for zero transmission at ω_0 and Q with R_7 and R_8, iteratively

(4) Notch filter with unequal high frequency and dc gains (Fig. 6.31)

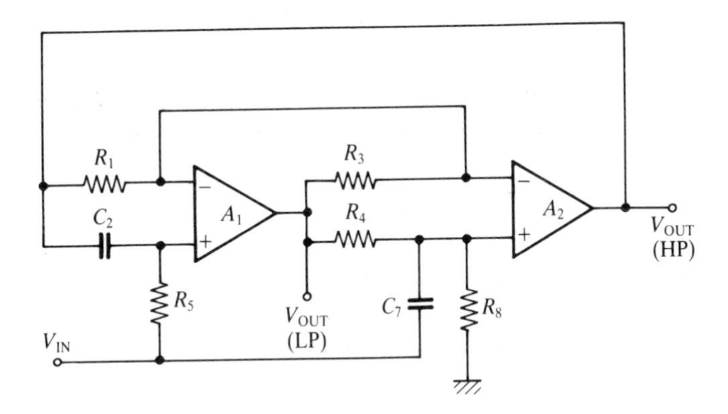

Fig. 6.31. Notch filter with unequal high frequency and DC gains.

Features	Advantages	Disadvantages
Non-inverting mode DC gain is not equal to hf gain	Low to high Q_F values possible Requires only 2 capacitors Low sensitivity to components Easily tuned High Q values are realized without a large spread in component values	Requires 2 op amps

$$\text{Transfer function } T(s) = \frac{s^2 + \omega_z^{\,2}}{s^2 + \dfrac{\omega_p}{Q_F} + \omega_p^{\,2}}$$

$$\omega_p = \sqrt{\frac{R_3}{R_1 R_4 R_5 C_2 C_7}}$$

$$Q_F = R_8 C_7 \omega_p$$

$$\omega_{zlp} = \omega_p \sqrt{\left(1 + \frac{R_4}{R_8}\right)} \Rightarrow \text{Low pass output notch frequency}$$

$$\omega_{zhp} = \omega_p \sqrt{\left(1 - \frac{R_1 R_4}{R_3 R_8}\right)} \Rightarrow \text{High pass output notch frequency}$$

Gain at DC:	Low pass output	$1 + R_4/R_8$
	High pass output	$1 - \dfrac{R_1 R_4}{R_3 R_8}$
Gain at HF:	Unity for both outputs.	

This circuit is particularly useful for applications which require a notch filter and also require different gains above and below the notch frequency (i.e. dc and hf). The equations are simple and the calculation of component values is a relatively straightforward matter.

Tuning is an iterative procedure as follows.

(i) tune ω_z with R_4
(ii) tune ω_p with R_5
(iii) tune Q_F with R_8

All pass filter circuits

(1) Single pole all pass filter (Fig. 6.32)

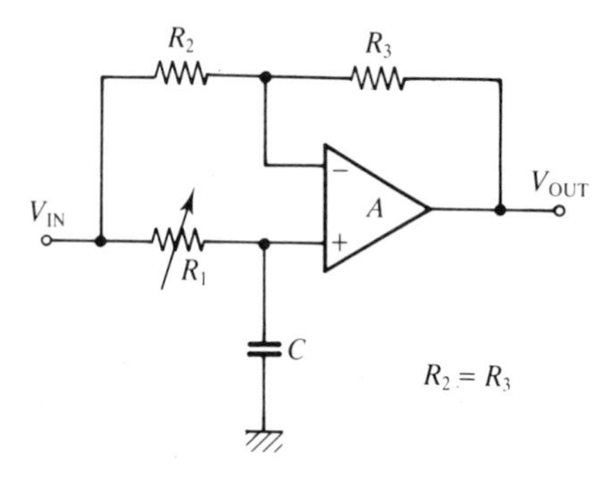

Fig. 6.32. Single pole all pass filter.

Features
Single pole
0°–180° or 180–360° operation
Unity gain
Single op amp
Single capacitor
Requires two equal resistors

In this circuit, at low frequencies, C has no effect and so the circuit acts like a buffer with unity gain $+1$. At high frequencies, C acts like a short circuit with the op amp connected as an inverting amplifier with gain R_3/R_2. Consequently for equal low and high frequency gains, the resistors R_2 and R_3 are made equal. Phase shifts swing from 0° at low frequencies to $-180°$ at high frequencies. If R_1 is made variable, the circuit will act as an adjustable phase shifter.

$$\text{Transfer function } T(s) = \frac{1 - \dfrac{R_3}{R_2} \cdot R_1 C_S}{1 + R_1 C_S}$$

Phase shift ($R_2 = R_3$): $\varphi = -2 \tan^{-1}(R \cdot C \cdot \omega)$

R_2 and R_3 must be equal to give a flat all pass transfer function with a constant gain independent of frequency. Ideally, these two resistors are better as a matched pair. At higher frequencies, a phase shift error will be introduced due to the finite bandwidth of the op amp. For good phase accuracy, choose a fast op amp with a high bandwidth.

The circuit can give phase shifts from 0° to 180° as the frequency increases from dc or as R_1 increases from 0. To obtain frequencies from 180° to 0°, interchange C and R_1. With this change, phase shift = 180° − 2 $\tan^{-1}(R \cdot C \cdot \omega)$, if $R_2 = R_3$.

(2) Current generalized immittance all pass filters (Fig. 6.33)

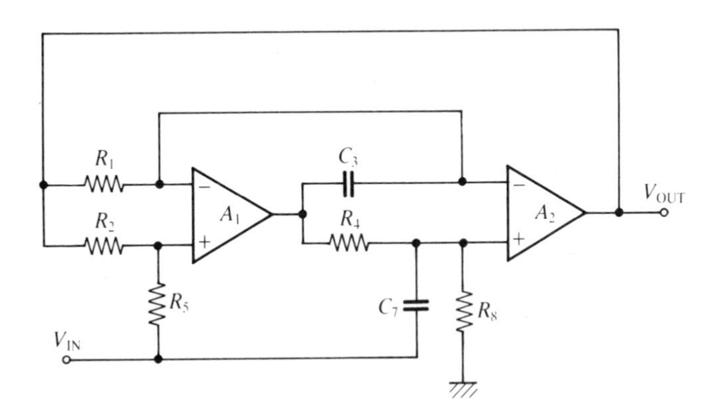

Fig. 6.33. Current generalized immittance all pass filter.

Features	Advantages	Disadvantages
Two-pole Phase shifts 0°–360°	Low to high Q values possible Requires only two equal resistors $R_2 = R_5$ for flat gain response Easily tuned Low sensitivity to component changes High Q values realized without large spread in component values	Requires two op amps Fixed gain of unity

$$\text{Transfer function } T(s) = \frac{s^2 - s\cdot\dfrac{R_2}{R_5R_8C_7} + \dfrac{R_2}{R_4R_1R_5C_3C_7}}{s^2 + s\cdot\dfrac{1}{R_8C_7} + \dfrac{R_2}{R_1R_4R_5C_3C_7}}$$

Parameters: $$\omega_0 = \sqrt{\frac{R_2}{R_1R_4R_5C_3C_7}}$$

$$Q_F = \omega_0 C_7 R_8$$

$$K = 1$$

Required relationship for resistor values:

$$R_2 = R_5$$

This circuit realizes high Q_F values and low sensitivities to component changes at the expense of an extra op amp.

Tuning procedure:
(i) tune ω_0 with R_4 or R_5
(ii) tune Q_F with R_8

State variable filters

These circuits (Fig. 6.34) have several outputs offering simultaneous high pass, low pass and bandpass outputs. The circuits are called state variable configurations after the technique used in analog computers for

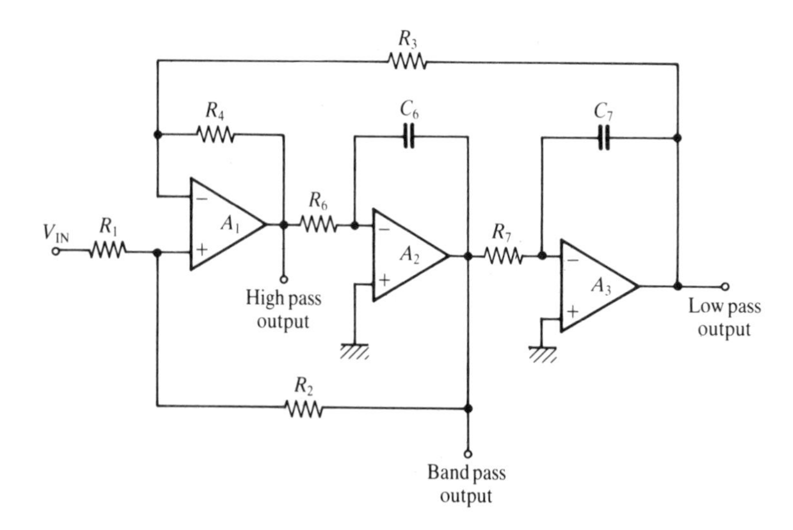

Fig. 6.34. General purpose state variable filter.

solving differential equations. The circuits are built up from integrators and summers.

Features	Advantages	Disadvantages
Two-pole	Low to high Q_F values (>50) are possible Low sensitivity to changes in components Easily tuned Simultaneous low pass, high pass and bandpass outputs Relatively insensitive to op amp frequency response	Requires three op amps

Transfer functions: high pass, low pass and bandpass:

$$T_{LP}(s) = \frac{K_{LP}\omega_0^2}{s^2 + s\dfrac{\omega_0}{Q_F} + \omega_0^2}$$

$$T_{BP}(s) = \frac{-K_{BP}\left(\dfrac{\omega_0}{Q_F}\right)s}{s^2 + s\dfrac{\omega_0}{Q_F} + \omega_0^2}$$

$$T_{HP}(s) = \frac{K_{HP}s^2}{s^2 + s\dfrac{\omega_0}{Q_F} + \omega_0^2}$$

where

$$\omega_0 = \sqrt{\frac{R_4}{R_3 T_1 T_2}}$$

$$Q_F = \frac{\left(1 + \dfrac{R_2}{R_1}\right)}{\left(1 + \dfrac{R_4}{R_3}\right)} \cdot \sqrt{\frac{R_4 T_1}{R_3 T_2}}$$

$$K_{HP} = \frac{1 + \dfrac{R_4}{R_3}}{1 + \dfrac{R_1}{R_2}}$$

$$K_{BP} = \frac{R_2}{R_1}$$

$$K_{LP} = \frac{1 + \dfrac{R_3}{R_4}}{1 + \dfrac{R_1}{R_2}}$$

and where

$$T_1 = R_6C_6 \qquad T_2 = R_7C_7$$

This circuit consists of a summer (A_1, R_1, R_2, R_3 and R_4) and two integrators built from, respectively, A_2, R_6, C_6 and A_2, R_7, C_7 with time constants of T_1 and T_2. As can be seen from the above equations, the component values are relatively easy to calculate. A possible tuning procedure is shown below.

(i) tune ω_0 with R_4
(ii) tune Q_F with R_1 or R_2

If K must also be tuned then the tuning is a more complicated procedure.

With high Q_F circuits, care must be taken to ensure that all the outputs of every op amp are not saturating and causing distortion. Also, the finite bandwidth of the op amps limits the high frequency response and also causes errors in the values of Q_F and ω_0. If all the op amps used are of the same type with the same gain–bandwidth product, ω_a in rad/s then

$$\frac{Q_F'}{Q_F} = \left(1 + 4Q_F\cdot\frac{\omega_0}{\omega_A}\right) \qquad \text{and} \qquad \frac{\omega_0'}{\omega_0} = 1 - \frac{3}{2}\cdot\frac{\omega_0}{\omega_A}$$

where ω_0' and Q_F' are the actual values of pole frequency and Q-factor obtained. The finite bandwidth of the op amps has the effect of increasing the value of Q. With high Q circuits (>50), there may be a possibility of the circuit being unstable and bursting into oscillation. There can be some frequency compensation provided by splitting the input resistor of one of the integrators, either R_6 or R_7, into two parts with some part of its value R_{COMP} placed in series with its feedback capacitor as shown in Fig. 6.35. Note, $R_{COMP} = 1/(C\cdot\omega_A)$.

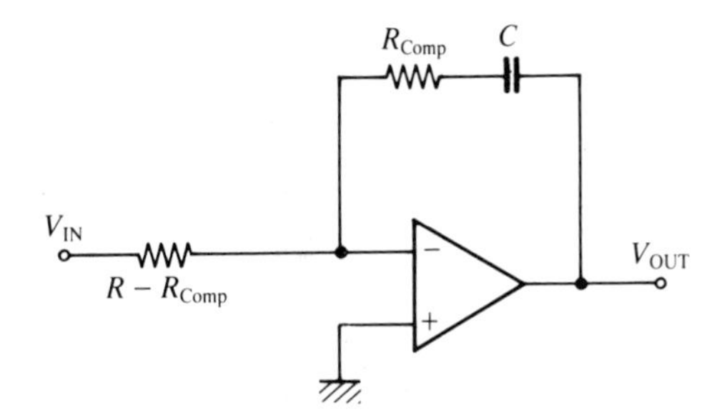

Fig. 6.35. Frequency compensation for high Q circuits.

If only low pass and bandpass outputs are required, then the circuit can be simplified by omitting the summer to leave only the two integrators as shown in Fig. 6.36. The equations for this circuit are shown below.

$$T_{LP}(s) = \frac{K_{LP}{\omega_0}^2}{s^2 + \dfrac{\omega_0}{Q_F}s + {\omega_0}^2}$$

$$T_{BP}(s) = \frac{-K_{BP}\dfrac{\omega_0}{Q_F}s}{s^2 + \dfrac{\omega_0}{Q_F}s + {\omega_0}^2}$$

$$\omega_0 = \sqrt{\frac{1}{C_1C_2R_2}\cdot\left(\frac{1}{R_1} + \frac{1}{R_3}\right)}$$

$$Q_F = R_2C_2\omega_0$$

$$K_{LP} = \frac{R_3}{R_1 + R_3}$$

$$K_{BP} = \frac{R_2C_2}{R_1C_1}$$

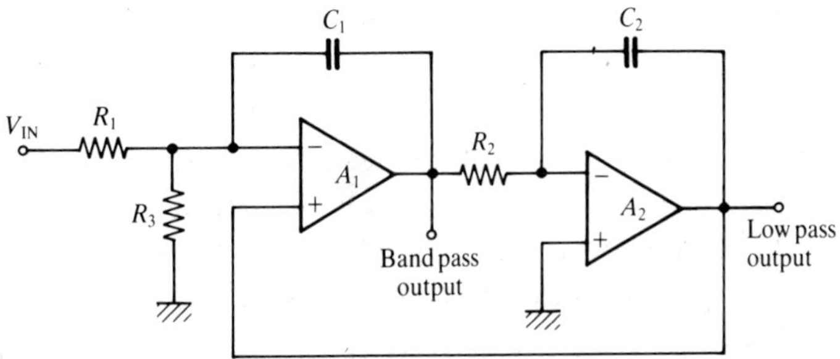

Fig. 6.36. Low pass and bandpass outputs.

Adding an extra op amp enables a notch filter to be realized. The extra op amp forms an adder to add high pass and low pass responses as shown in Fig. 6.37. A quad op amp ic could be used to realize the active elements. The equations for this circuit are listed below (note $R_3 = R_4$ and $R_8 = R_{10}$).

$$T_{BR} = \frac{-K_{BR}(s^2 + \omega_0{}^2)}{s^2 + \dfrac{\omega_0}{Q_F}s + \omega_0{}^2}$$

where

$$T_1 = R_6C_6 \quad \text{and} \quad T_2 = R_7C_7$$

$$\omega_0 = \frac{1}{\sqrt{T_1T_2}} \qquad Q_F = \frac{1}{2}\left(1 + \frac{R_2}{R_1}\right)\sqrt{\frac{T_1}{T_2}}$$

$$K_{BR} = \frac{R_9}{\left(1 + \dfrac{R_1}{R_2}\right)\cdot R_8}$$

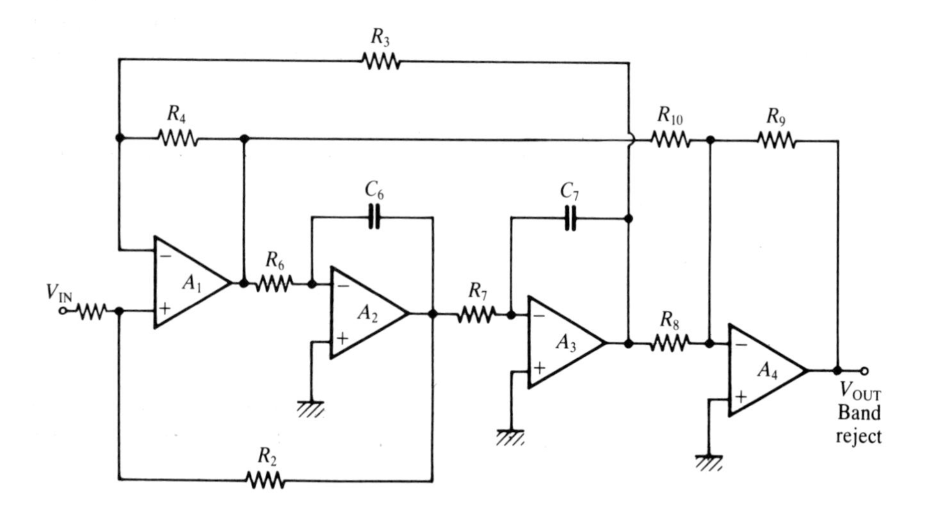

Fig. 6.37. Realizing a notch filter using a state variable filter.

Summary

The filter circuits described previously have been grouped according to function, (i.e. low pass, high pass, bandpass etc.) but they could also have been grouped according to the type of circuit configuration used

Table 6.1. *Summary of filter characteristics*

Type	Sallen-Key	Multi-loop feedback	Current generalized immittance	State variable
Functions	H, L, BP, A, N	H, L, BP, A, N	H, L, BP, A, N	Simultaneous L, H + BP possible
Number of op amps	1	1	2	3
Q values	Low–medium (<20)	Low–medium (<20)	Low–high (up to 100)	Low–high (up to 100)
Sensitivity to component values	High	Low–medium	Low	Low
Spread in component values	Moderate	High	Low–medium	Low
Tuning	Easy to tune $\omega_0 + Q_F$	Easy to tune ω_0 and Q_F independently but not K	Good: can tune all three parameters (K, Q and ω_0) without iteration	Good; can tune three parameters (K, Q_F and ω_0) without iteration

for each filter. Table 6.1 summarizes the general properties for the various two-pole configurations.

6.3 Controlled filters

Controlled filters are filters whose characteristics (cut-off frequency, Q_F value, bandwidth and so on) can be controlled electronically. The electronic control can be:

(i) digitally with a binary code applied to a D–A converter
(ii) an analog voltage via an analog multiplier or a FET as a voltage controlled resistor
(iii) frequency or pulse width modulation using a switched capacitor technique.

Controlled filters are usually based upon the state variable configuration described previously where the control is achieved by varying the integrator time constants. The state variable configuration is

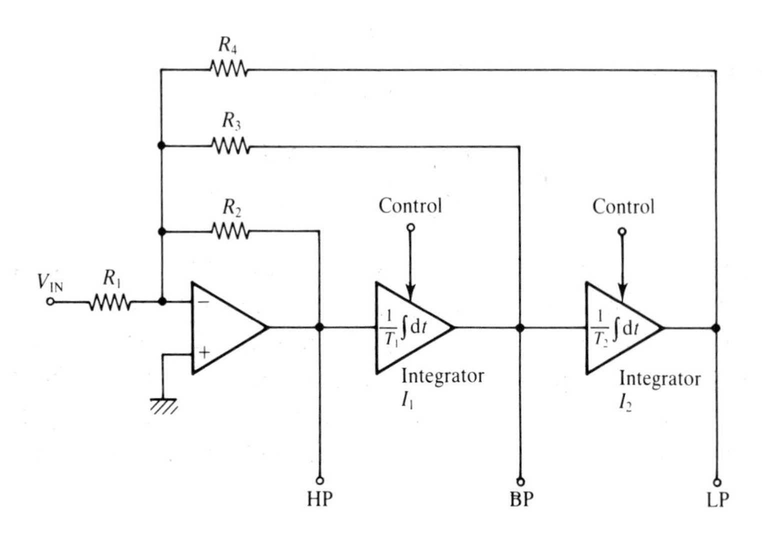

Fig. 6.38. A controlled filter.

particularly attractive because it has several outputs (e.g. low pass, band pass and high pass) and because filter parameters such as Q_F and ω_0 can be varied independently of one another. Fig. 6.38 outlines a two-pole controlled filter based upon this technique. This circuit is somewhat simplified and a fuller schematic requires the integrator details which are described more fully in the following pages.

The equations for this circuit are as follows.

Low pass
$$T_{LP}(s) = \frac{-\dfrac{R_2}{R_1 T_1 T_2}}{s^2 + s\left(\dfrac{R_2}{T_1 R_3}\right) + \dfrac{R_2}{R_4 T_1 T_2}}$$

Bandpass
$$T_{BP}(s) = \frac{-\dfrac{R_2}{R_1 T_1} s}{s^2 + s\left(\dfrac{R_2}{T_1 R_3}\right) + \dfrac{R_2}{R_4 T_1 T_2}}$$

High pass
$$T_{HP}(s) = \frac{-\dfrac{R_2}{R_1} s^2}{s^2 + s\left(\dfrac{R_2}{T_1 R_3}\right) + \dfrac{R_2}{R_4 T_1 T_2}}$$

Gains:

Low pass $K_{LP} = -R_4/R_1$

Bandpass $K_{BP} = -R_3/R_1$

High pass $K_{HP} = -R_2/R_1$

Parameters: $$\omega_0 = \sqrt{\frac{R_2}{R_4 T_1 T_2}}$$

$$Q_F = \frac{T_1 R_3}{R_2}\omega_0 = \sqrt{\frac{R_3{}^2}{R_2 R_4}\cdot\frac{T_1}{T_2}}$$

From these equations you can see that varying the integrator time constants T_1 and T_2 allows the Q_F, ω_0 and bandwidth to be controlled without varying the filter gains, K_{lp}, K_{bp} and K_{hp}. If T_1 and T_2 are varied together, in direct proportion to each other, then ω_0 can be varied without upsetting Q_F. Varying T_2 alone, allows the centre frequency of a bandpass or notch filter to be varied without changing its bandwidth, ω_0/Q_F.

Voltage controlled filters

A voltage controlled filter can be realized by configuring the variable time constant integrators in the schematic circuit as a standard integrator and an analog multiplier as shown in Fig. 6.39. Note that V_c must be

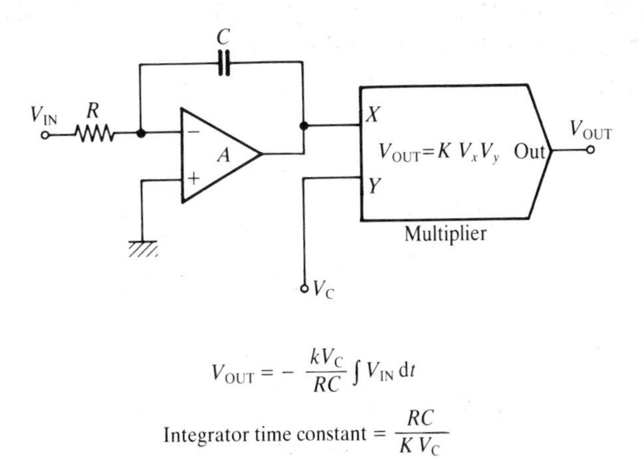

$$V_{OUT} = -\frac{kV_C}{RC}\int V_{IN}\,dt$$

$$\text{Integrator time constant} = \frac{RC}{K\,V_C}$$

Fig. 6.39. Configuring a voltage controlled filter.

negative to make a non-inverting integrator. Of course, the integrator may introduce extra errors such as dc offset and noise and the multiplier must be fast enough to ensure that it does not affect the frequency response of the filter. Also, the linearity of the multiplier will determine the linearity of the control function and the amount of distortion introduced to the filter.

The state variable circuit shown earlier can be simplified to single pole filters as shown in the following schematic circuits. A single pole low pass voltage controlled filter is shown in Fig. 6.40.

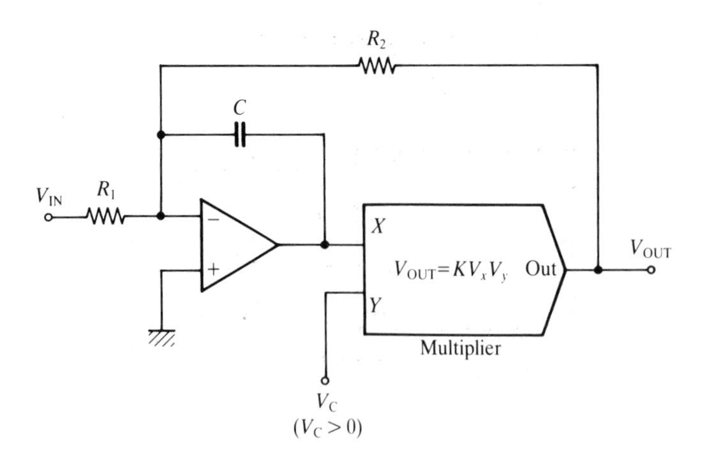

Fig. 6.40. Simplified low pass voltage controlled filter.

Transfer function: $$T(s) = \frac{-\dfrac{R_2}{R_1}}{1 + s\left(\dfrac{R_2C}{KV_c}\right)}$$

$$\text{DC gain} = \frac{R_2}{R_1}$$

$$\text{Bandwidth} = \frac{KV_c}{2\pi RC} \quad \text{Hz}$$

This circuit can be very simply configured using, for example, a single operational transconductance amplifier. Alternatively, the analog multiplier can be replaced with a multiplying DAC.

Digitally controlled filters

The integrators in a digitally controlled filter can be realized by combining an ordinary integrator and a D-to-A converter. The integrator can be positioned either in front of or after the DAC, but the DAC must be of the multiplying type. Typically, an R–$2R$ multiplying DAC is used as shown in Fig. 6.41. In this example, the output I to V stage of the DAC is configured to give the integrating action.

$$V_{OUT} = -\frac{\text{(CODE)}}{RC}\int V_{IN}\,dt$$

$$\text{Integrator time constant} = \frac{RC}{\text{(CODE)}}$$

where (CODE) is the ratio determined by the digital inputs. For

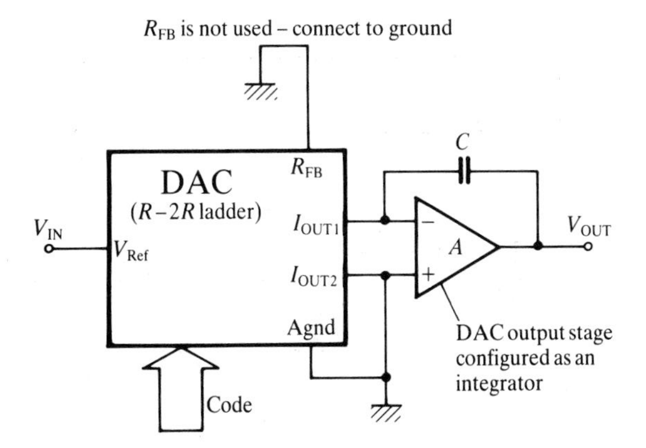

Fig. 6.41. An integrator for a digitally controlled filter.

example: for an 8-bit DAC; with 0000 0000 input then CODE = 0; 1000 0000 input then CODE = $\frac{1}{2}$; 1111 1111 input then CODE = 255/256. Linearity of control will depend upon the linearity of the DAC. Also, glitches will be introduced as the DAC switches from one level to another which may be troublesome in some high Q_F filters.

Switched capacitor techniques

The switched capacitor technique employs switches, capacitors and op amps to realize a variety of analog circuit functions such as analog/digital converters (ADCs and DACs) and instrumentation stages as well as analog filters. The basic principle behind the technique is to simulate components such as resistors by switching charge rapidly into and out of a capacitor. The technique is particularly suited to very large scale integration (VLSI) MOS technology because it allows analog circuit functions to be integrated on an ic using standard elements such as MOS transistors and small value (pF) capacitors. Consequently, traditional components such as very stable, low tolerance or high value resistors, inductors and high value capacitors which are difficult to fabricate on an ic can be avoided.

A simple switched capacitor element is shown in Fig. 6.42. The two switches are switched ON and OFF in alternate half cycles with a so called non-overlapping action (i.e. break before make), which ensures the switches are never simultaneously ON.

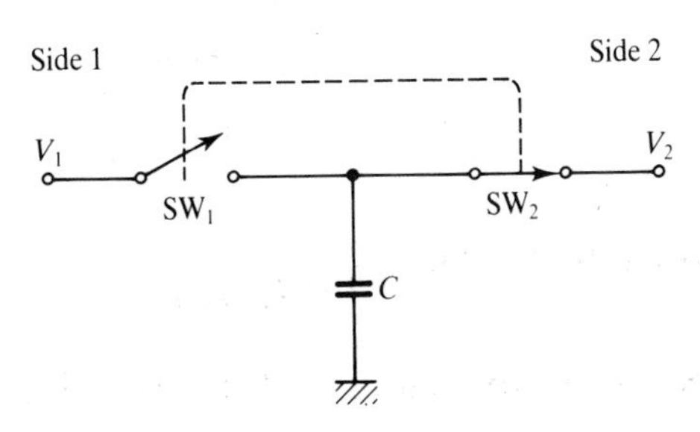

Fig. 6.42. Simple switched capacitor element.

When SW_1 is ON (SW_2 is OFF), the capacitor charges to V_1. When the capacitor is swopped to V_2 (SW_1 is OFF, SW_2 is ON) an amount of charge ΔQ is transferred from side 1 to side 2, where

$$\Delta Q = C\,\Delta V = C(V_1 - V_2)$$

If the charge is moved from side 1 to side 2 at a rate of f_{CL} times per second, then the average current flowing between V_1 and V_2 is

$$I = f_{CL}\,\Delta Q = f_{CL}C(V_1 - V_2)$$

As a result, provided that the maximum signal frequency content in V_1 or V_2 is considerably less than the clock frequency, then the switched capacitor element can be represented by a resistor, R_{EQ}, with a value;

$$R_{EQ} = (V_1 - V_2)/I = 1/f_{CL}C$$

This is effectively a digitally controlled resistor which can be used as the basis of a number of digitally controlled analog circuits although here we are only concerned with digitally controlled filters.

The main features of the switched capacitor technique are listed below.

(i) Resistors can be eliminated. In ic fabrication, resistors are generally avoided if possible, because of high tolerances, large temperature coefficients and big die area.

(ii) Analog circuit characteristics such as filter cut-off frequency can be made directly proportional to f_{CL}.

(iii) Circuit characteristics can be made dependent on a ratio of capacitors, which can be fabricated on an ic with low tolerances ($\approx 0.1\%$) and high stability.

(iv) Large value equivalent resistors can be fabricated. For example, C can be fabricated in the range of a fraction of a pF to one hundred pF (1 pF–10 pF is most popular). If $C = 5$ pF and $f_{CL} = 1$ kHz, then $R_{EQ} = 200$ MΩ.

However, the technique does have drawbacks.

(i) Digital feedthrough of the clock into the analog circuit can result in several mVs of clock frequency ripple superimposed upon the analog output. Fortunately, this ripple is usually much higher in frequency than the analog signal and can be easily reduced with a simple RC filter.

(ii) Large dc offsets can be caused because of charge injection and leakage currents across the switches. In some cases, this offset can approach 100 mV.

A simplified circuit of an integrator based on the switched capacitor technique is shown in Fig. 6.43, which can be used as part of a digitally controlled state variable filter. In this circuit, the input resistor is replaced by a switched capacitor element of Fig. 6.42. The response of this integrator is

$$V_{OUT} = -\frac{f_{CL} C_1}{C_2} \int V_{IN}\, dt$$

and the integrator time constant is $T = C_2/C_1 f_{CL}$.

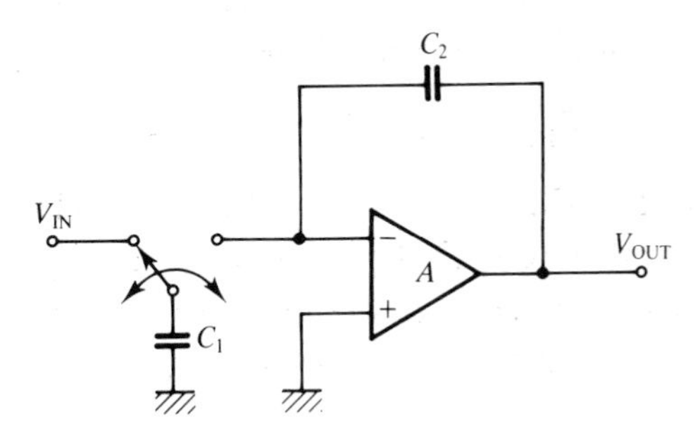

Fig. 6.43. Switched capacitor integrator.

The clock frequency is limited by the speed of the op amp and also by the time constant $r_{ON}C$ (C = switched capacitor value; r_{ON} = switch On resistance). As a result, the maximum practical value of f_{CL} is limited to around a few MHz. The minimum value of f_{CL} is limited by offsets introduced by switch charge injection and leakage currents. As a result, the minimum practical value of f_{CL} is around a few hundred Hz.

Switched capacitor circuits employ small value capacitors (i.e. less than 100 pF) and consequently are susceptible to errors from charge injection and stray capacitance. Charge injection effects can be minimized by careful switch design using matched transistors to ensure that the charge injected by one switch transistor is cancelled by another. Fig. 6.44(*a*) shows the switched capacitor integrator of Fig. 6.43, but redrawn to show all the main stray capacitances associated with the switches and capacitor. C_1' and C_6' have no effect because these simply load the input and output respectively. Similarly C_4' and C_5' are across a virtual earth and so have little effect. However, C_2' and C_3' are connected in parallel with C and so will modify the time constant of the integrator and cause an error in the performance of the filter. This error can be avoided using a dual switching configuration as shown in Fig. 6.44(*b*). Here the effects of the troublesome stray capacitances C_2' and C_3' are avoided by adding SW_3 and SW_4 and reconfiguring C_1. Now, the troublesome stray capacitances are shorted out each cycle and cannot contribute to the charge transfer. This circuit is now stray insensitive. Fig. 6.44 also shows two stray insensitive integrators in schematic form configured for inverting Fig. 6.44(*c*) and non-inverting Fig. 6.44(*d*) operation.

There are many switched capacitor filter ics on the market. One of the most popular low-cost devices is the MF10 which contains two two-pole general purpose filter building blocks. All capacitors and switches and active elements are contained within a 20 pin DIL package. It requires only a few extra resistors and a clock to realize filters with $Q_F f_0$ products of 200 kHz and signal frequencies up to 20 kHz.

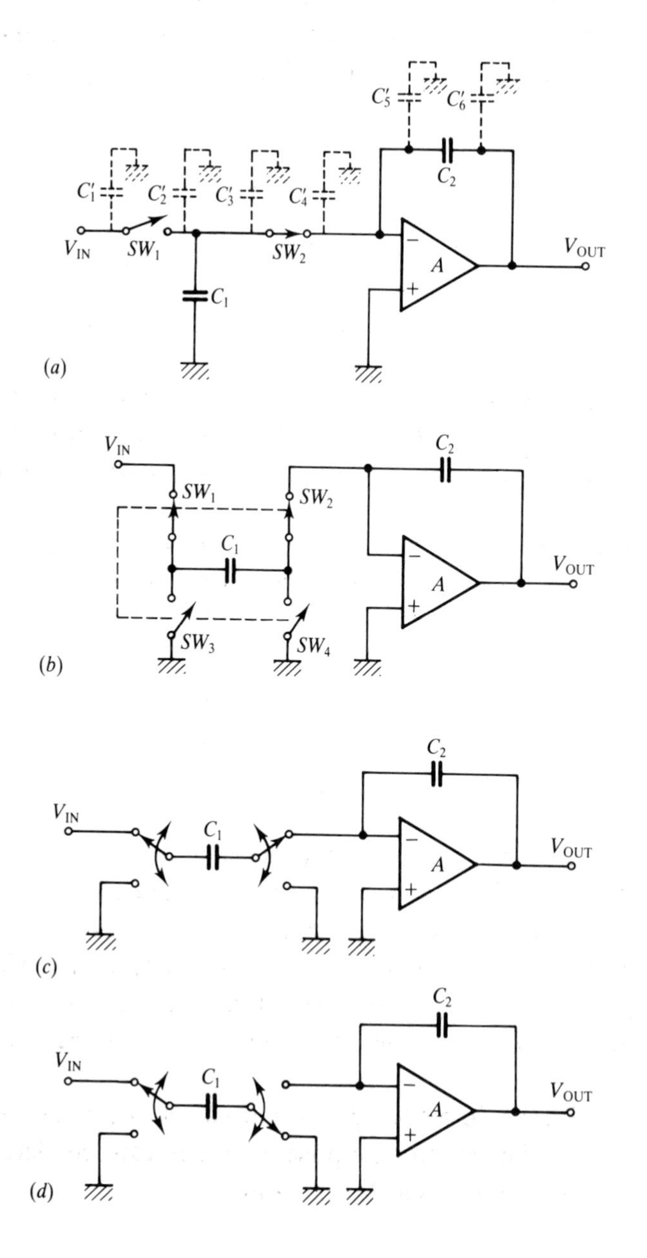

Fig. 6.44. Stray insensitive switched capacitor circuits. (*a*) Switched capacitor circuit with parasitic capacitances. (*b*) Stray insensitive switching configuration. (*c*) Inverting operation. (*d*) Non-inverting operation.

6.4 Practical aspects of filter circuit design

Sensitivity

Filter sensitivity is concerned with how much a particular filter parameter is affected by small changes in circuit components due to effects such as temperature drifts, tolerance range and ageing. In the worst case, a small change in a filter component value could cause a large change in the performance of the filter.

For the first and second order filter circuits described earlier, the main circuit parameters such as pole frequency, ω_0, Q-factor, Q_F and passband gain K can be expressed in terms of the circuit components as shown below.

$$K = f_K(R_1, R_2 \cdots R_n, C_1, C_2 \cdots C_m)$$

$$Q_F = f_Q(R_1, R_2 \cdots R_n, C_1, C_2 \cdots C_m)$$

$$\omega_0 = f_{\omega 0}(R_1, R_2 \cdots R_n, C_1, C_2 \cdots C_m)$$

where $R_1, R_2 \cdots R_n$ and $C_1, C_2 \cdots C_m$ are the circuit resistor and capacitor values. In addition, parasitic effects, such as stray capacitances and amplifier limitations, could also be incorporated into these expressions.

The sensitivity of a filter parameter, f_K for example, to a variation in a particular component x_i is defined mathematically as

$$S^f_{x_i} = \frac{\dfrac{\Delta f}{f}}{\dfrac{\Delta x_i}{x_i}} \simeq \frac{\partial f}{\partial x_i} \cdot \frac{x_i}{f}$$

where $S^f_{x_i}$ is the sensitivity of f to changes in x_i.

Sensivities of K, ω_0 and Q_F can be determined by partially differentiating their expression with respect to the particular component, multiplying by the component value and then dividing by K, ω_0 or Q_F as the case may be. A sensitivity of unity means that a 1% change in component value will cause a 1% change in the filter parameter.

The overall error in a filter parameter (Q_F in the example below) to changes in all the circuit component values can be determined by summing the individual sensitivities using

$$\frac{\Delta Q_F}{Q_F} \simeq \left(|S^{Q_F}_{R_1}| \cdot \frac{\Delta R_1}{R_2}\right) + \left(|S^{Q_F}_{R_2}| \cdot \frac{\Delta R_2}{R_2}\right) + \cdots + \left(|S^{Q_F}_{R_n}| \cdot \frac{\Delta R_n}{R_n}\right)$$

$$+ \left(|S^{Q_F}_{C_1}| \cdot \frac{\Delta C_1}{C_1}\right) + \left(|S^{Q_F}_{C_2}| \cdot \frac{\Delta C_2}{C_2}\right) + \cdots + \left(|S^{Q_F}_{C_n}| \cdot \frac{\Delta C_n}{C_n}\right)$$

where Q_F could be replaced by any other filter parameter. This expression assumes the worst case error with all the errors from each component being additive.

A knowledge of filter sensitivity allows the designer to determine the required component tolerances to achieve a particular specification. Sensitivities highlight those component errors which are critical to filter performance and can be informative on whether tuning is necessary.

To be thorough, the sensitivities of all the K, Q_F and ω_0 parameters to variations in every component value would have to be derived and calculated. Although this is a relatively straightforward task as K, Q_F and ω_0 can usually be explicitly expressed in terms of component values, it is an extremely tedious one since many component values are involved. Fortunately, with a little experience, the expressions for K, Q and ω_0 can often be visually examined to identify whichever component values might cause large sensitivities.

Tuning

Tuning a filter involves trimming or adjusting the value of a component or components to achieve the required response. It may be needed to correct errors in response caused by component tolerances, component imperfections and parasitic capacitances and resistances within the circuit. Tuning is often not needed in low accuracy applications where a large degree of error is acceptable.

Tuning is mainly achieved by adjusting a resistor rather than a capacitor. With hybrid and integrated filters, resistors can be laser trimmed as part of the manufacturing process whereas capacitor values cannot be so easily changed. With the filter circuits employing discrete resistors and capacitors, trimming resistors can be obtained to span a large range of values from hundreds of ohms to millions of ohms whereas trimming capacitors can usually only be obtained in small values, typically less than 100 pF.

Circuits are tuned to achieve the required pass band gain, K, Q_F and cut-off frequency ω_0. In some circuits, it is not possible to independently tune K, Q_F and ω_0 by using separate resistors for each parameter. One

Table 6.2. *Functional and deterministic tuning*

Functional tuning	Deterministic tuning
Functional tuning is done with the circuit in operation. It involves measuring the required parameters and adjusting component values until all the parameters are at an acceptable value	This involves the measurement of each individual component and might also include parasitic effects. This measurement takes place prior to component insertion or before the circuit is made operational. The required value of trimming resistor, derived from equations, allows the correct trimmer to be inserted.
Simple and directly applicable but labour intensive and not easily automated	Requires considerable initial work on equation derivation, with extensive computation, but easy to automate once set up.

resistor, consequently, may affect two or more parameters at the same time. Tuning must therefore be done in a certain order, so that previously-set values are not changed, and on components which minimally affect other parameters in the circuit. Sometimes an iterative tuning procedure is necessary, where repeated sequences of adjustments are carried out, due to the inability of adjusting one circuit parameter independently of others. For the majority of the filter circuits in the previous sections a possible tuning sequence was suggested.

There are two main approaches to tuning; functional and deterministic. These two approaches are compared in Table 6.2. As can be seen, functional tuning is more suited to small number applications, whereas deterministic tuning is more applicable to large scale operations. In some cases, a combination of functional and deterministic tuning is used where an approximate value for trimming resistors is found first by deterministic methods and then the value is fine-tuned if necessary by functional tuning.

With functional tuning, it is necessary to be able to measure the Q_F and ω_0 of the various two-pole stages. It is often better to use phase shift to measure these two parameters, rather than gain, since phase shift gives a more accurate measure of Q_F and ω_0. The procedure for measuring Q_F and ω_0 is shown in Table 6.3 for the different types of filters.

Table 6.3. *Measurement of filter parameters for tuning*

Filter	Transfer function	Measurement of parameters
Low pass (first order)	$\dfrac{K_{LP}\cdot\omega_0}{s+\omega_0}$	ω_0 = Frequency when phase is $-45°$ Or ω_0 = Frequency for which gain has reduced by 3 dB (i.e. $\div\sqrt{2}$)
Low pass (second order)	$\dfrac{K_{LP}\cdot\omega_0{}^2}{s^2+\dfrac{\omega_0}{Q_F}\cdot s+\omega_0{}^2}$	ω_0 = Frequency when phase is $-90°$ Q_F = Ratio of the gain at ω_0 to the gain at dc
High pass (first order)	$\dfrac{K_{HP}\cdot s}{s+\omega_0}$	ω_0 = Frequency when phase is $+45°$ Or ω_0 = Frequency at which the gain is 3 dB less than the hf value
High pass (second order)	$\dfrac{K_{HP}\cdot s^2}{s^2+\dfrac{\omega_0}{Q_F}\cdot s+\omega_0{}^2}$	ω_0 = Frequency for which the phase is $+90°$ Q_F = Ratio of the gain at ω_0 to the pass band gain
Bandpass	$\dfrac{\left(\dfrac{K_{BP}\omega_0}{Q_F}\right)\cdot s}{s^2+\dfrac{\omega_0}{Q_F}\cdot s+\omega_0{}^2}$	ω_0 = Frequency where phase shift is 0° $\Delta\omega$ = Frequency interval $\pm 45°$ phase shifts Or $\Delta\omega$ = 3 dB bandwidth $Q_F = \dfrac{\omega_0}{\Delta\omega}$
Band reject	$\dfrac{K_{BR}(s^2+\omega_0{}^2)}{s^2+\left(\dfrac{\omega_0}{Q_F}\right)s+\omega_0{}^2}$	ω_0 = Frequency for minimum gain (notch frequency) $\Delta\omega$ = Frequency interval between $\pm 45°$ phase shifts $Q_F = \dfrac{\omega_0}{\Delta\omega}$
All pass (first order)	$\dfrac{K_{AP}(s-\omega_0{}')}{(s+\omega_0)}$	If the dc and hf gains are equal then $\omega_0 = \omega_0{}'$ If $\omega_0 = \omega_0{}'$, then the ω_0 equals the frequency when the phase shift is $-90°$
All pass (second order)	$\dfrac{K_{AP}\left(s^2-\dfrac{\omega_0{}'}{Q_F{}'}s+\omega_0{}^2\right)}{\left(s^2+\dfrac{\omega_0}{Q_F}s+\omega_0{}^2\right)}$	If $\omega_0 = \omega_0{}'$, then the dc gain and hf gain will be equal. Frequency, ω_0, can then be measured as the frequency which gives 180° phase shift If $Q_F = Q_F{}'$ and $\omega_0 = \omega_0{}'$, then the filter will have a constant gain independent of frequency If $\omega_0 = \omega_0{}'$ and $Q_F = Q_F{}'$, then $\Delta\omega$ = Frequency interval between 90° and 270° phase shifts $Q_F = \dfrac{\omega_0}{\Delta\omega}$

Component selection

Op amps

(*a*) *Frequency response* The op amp chosen must be fast enough (i.e. have a high enough gain–bandwidth product), otherwise the filter performance will depend upon the dynamics of the op amp and an error will be introduced in the desired response. The higher the frequency response of the op amp, the less the error. As a general rule of thumb, choose an op amp with a GBP greater than $50 \times Q^2$ times the pass band gain for multiple feedback circuits, over the pass band of the filter. With the other circuits, choose an op amp with a GBP greater than $100 \times Q$ times the desired gain for the filter, over the entire pass band of the filter.

A major concern with the dynamic response of the op amp is that the filter may be unstable and oscillate due to the op amp being poorly compensated with an inadequate phase margin. Some op amps (usually the fast ones) need to be externally frequency compensated with a small capacitor whose value is tailored to the application. With an internally compensated op amp, a good indication of its closed loop stability is whether the op amp is stable under the worst-case conditions of unity gain and a capacitive load.

(*b*) *Input and output impedance* The output impedance of the op amp should be low enough. Generally, op amps have output impedances less than 100 Ω. However, if the op amp is supplying a large load, particularly a capacitive load, as is likely with some filter circuits, then a buffer may be required.

The input impedance should be much greater than other circuit impedances. In some circuits with large resistor values (e.g. MΩ) a FET input op amp may be necessary, in which case watch out for the input capacitance of the op amp, which can be tens of pF.

(*c*) *Input offset voltage and bias currents* These do not affect the frequency response of the filter but they do introduce a dc error voltage at the output. If the filter can be ac coupled, as with high pass and bandpass filters, then offsets may not be important provided they do not cause saturation.

Resistors

The smallest values of resistors are limited either by the maximum op amp output current (typically 30 mA) or by finite connection resistances (typically less than 0.1 Ω) or by power considerations. The maximum values of resistors are limited by the op amp input impedance, the effects of stray capacitances and by leakage resistance or by op amp bias currents. FET input op amps generally enable larger values of resistors to be used. Large value resistors (above 100 KΩ say) should be avoided at higher frequencies (above 10 kHz say) due to the effects of stray capacitances.

If discrete resistors are used, it is better to use metal film resistors and avoid carbon composite types. This is because metal film resistors have lower temperature coefficients (50 ppm typically) and have better long term stability. For every high precision application, metal foil or wirewound resistors can be used. Finally, if the ratio of resistor values is important, try to use resistors in the same package so that temperature variations in resistor values tend to cancel.

Capacitors

The lowest value of capacitors which can be used is often limited by circuit strays (which can easily be a few pF) or by leakage resistances and currents at the lower frequencies. As a rule of thumb, take care using capacitors less than 100 pF particularly at frequencies below 1 kHz (as their impedances start to rise into the MΩ range). The highest value of capacitance is usually limited by the physical size and the cost. Capacitors start to get big and expensive above 1 μF. The following capacitor types are commonly used in filters:

Ceramic ... small size and cost. Values from 10 pF to 10 000 pF. NPO ceramic capacitors are particularly recommended due to their very low temperature coefficients and excellent stability.

Metallized polycarbonate ... are very popular for larger values from 0.001 μF to 10 μF and generally perform better than other types over these ranges of capacitor values.

Mica ... values from 10 pF to 10 000 pF. Bigger than other types. Excellent stability can be obtained in tolerances down to 0.5%. Usually more expensive.

Polystyrene ... values up to 10 000 pF. Very low temperature coefficients, very high insulation resistance but they are affected by high temperatures (the specifications end at 70°C) and the polystyrene dielectric may be melted due to bad soldering techniques.

Stability of filters

In some cases, particularly with high Q circuits, the filter may be unstable and oscillate. Firstly, if oscillation is happening, short the input of the filter to ground, as often a low source impedance is a requirement for stability. If the oscillations stop when this short is applied and restarts when the short is removed, then the source impedance is probably too high and needs reducing.

If the circuit continues to oscillate, with the short applied to the input, check that a high impedance probe is being used since the probe itself could cause the filter to oscillate.

If the previous two approaches fail to prevent the oscillations then the circuit is inherently unstable. If the oscillation is at a frequency much higher than the pass band, then the instability is most probably caused by the dynamics of the op amps being used. Check that compensation and decoupling capacitors have been added and try replacing the op amp with a different type.

If the oscillation is near the pole frequency of the filter, then it is most probably caused by the discrete network of components around the op amp being improperly tuned. This is more of a problem with high Q filters. In this case, component values may need to be changed. If all else fails, use a circuit which is better suited for high Q applications.

6.5 Designing high order filters

So far this chapter has been mainly concerned with the design of simple first order and second order filter circuits for applications which do not have stringent specifications. Unfortunately, filters are often required with responses which more closely resemble that of a 'brick wall' with an infinite stop band attenuation and an infinitely sharp transition between pass band and stop band. However, an exact brick-wall response is impossible to achieve with active R–C filters and so the transfer function of the filter must approximate to this ideal case. This is called the approximation problem. It is usual to form a specification for a filter type which gives acceptable levels for error in pass band gain, minimum attenuation in the stop band and a maximum value for the frequency width of the region between pass band and stop band. These parameters are shown in the example in Fig. 6.45 for a low pass filter. In addition to a specified gain response, a particular type of phase response or step response may also be important.

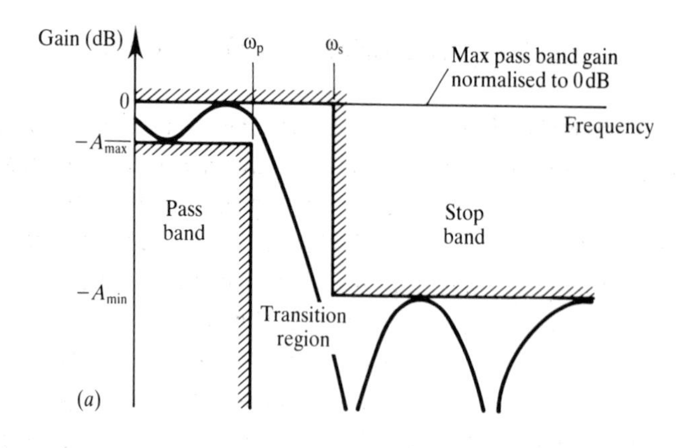

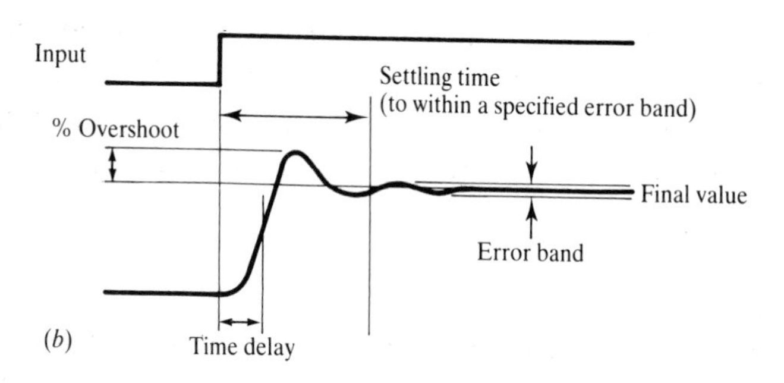

Fig. 6.45. Filter specification parameters. (*a*) Gain–frequency response. (*b*) Step response.

To achieve this approximation, a higher order filter (i.e. greater than second order) may be required. The higher the order of the filter, the more closely is it able to approximate to the ideal brick-wall response. However, as the order of the filter increases, so does its complexity, size and cost. Consequently, there is often a compromise to be struck between the desired performance and the circuit complexity.

The design of these filters generally consists of two main stages.

(i) Determination of the required mathematical transfer function.
(ii) Circuit design to implement this mathematical function.

The following sections describe each of these stages. However, it is not possible in the space available to describe these stages in full detail. The intention here is to provide a review of the procedure which will hopefully be sufficient for most applications. With more exacting applications, consult the bibliography.

Determining the required transfer function

There are a number of different groups of transfer functions with different properties and consequently different advantages and disadvantages. We will concentrate on three of the most popular transfer functions; Butterworth, Chebyshev and Bessel. These are generally sufficient for most applications. Other filter groups will be described, but you will have to consult the other literature for further information.

The determination of the most suitable transfer function can be done in a number of ways. It may be possible to choose a transfer function based solely on the description of the various filter groups given below. Alternatively, it may be possible to choose a transfer function from the graphs of gain/frequency, phase/frequency, or step response. However, for a more formal analytical approach, again consult the references.

The following filter groups are all Normalized with a pass band gain of unity (0 dB) and a cut-off frequency at 1 rad/s. For other types of filters (i.e. bandpass or high pass etc.) Transformations and Scaling are required. These processes are described in a later section.

Butterworth (or maximally flat)

A Butterworth filter is characterized by a frequency response which is maximally flat in gain over its pass band with no ripples in gain over the stop band as shown in Fig. 6.46. Since there are no ripples (i.e. maxima and minima) in the gain response, each value of gain occurs only once at a particular frequency. This property refers to the *monotonicity* of the filter response. The attenuation rate beyond the pass band is $20n$ dB/decade where n is the order of the filter.

The maximally flat gain response over the pass band is obtained at the expense of phase linearity. Non-linearity in the phase response causes distortion since different frequencies take different amounts of time to pass through the filter. Consequently, there is an overshoot and some ringing in response to a step input which increases with the order of the filter. Butterworth low pass filters are all-pole, i.e. there are no zeros.

The Butterworth filter is a good general purpose filter since it has maximally flat gain response over its pass band combined with moderate phase linearity, a reasonable step response and a reasonably steep attenuation slope. These properties make it one of the most commonly used filter groups.

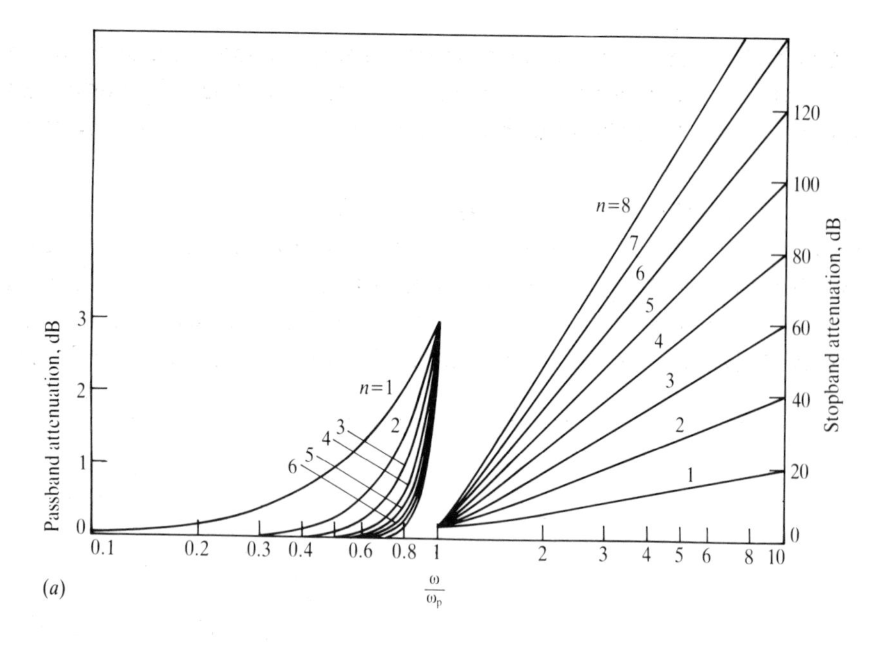

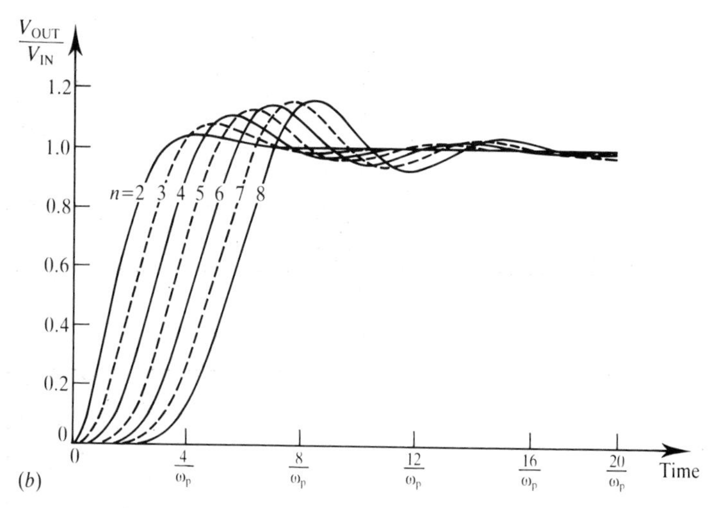

Fig. 6.46. Butterworth filter response. (*a*) Gain–frequency response. (*b*) Step response.

Chebyshev (or equi-ripple) (Fig. 6.47)

A Chebyshev filter is characterized by very steep attenuation slopes combined with a ripple in the gain of the filter over the pass band frequencies. This ripple over the pass band is introduced so that a very sharp cut-off outside the pass band can be achieved. The ripple is usually between 0.1 dB and 3 dB, the gain response, consequently, is no longer monotonic.

The increased attenuation slope is also at the expense of a further increased phase response non-linearity over the pass band. Ringing and overshoot due to a step input are, consequently, further increased. Chebyshev filters are all-pole with no zeros.

Chebyshev filters are popular where a sharp cut-off is required in an application. The phase response can be made more linear by adding an all-pass network to the filter but this will cause increased time delays.

Bessel (linear phase or Thomson) (Fig. 6.48)
Bessel filters have a phase response which most closely approximates the ideal of a perfectly linear response. With a linear phase response, all frequencies in the pass band pass through the filter with equal time delays. This characteristic, however, applies only to the low pass Bessel

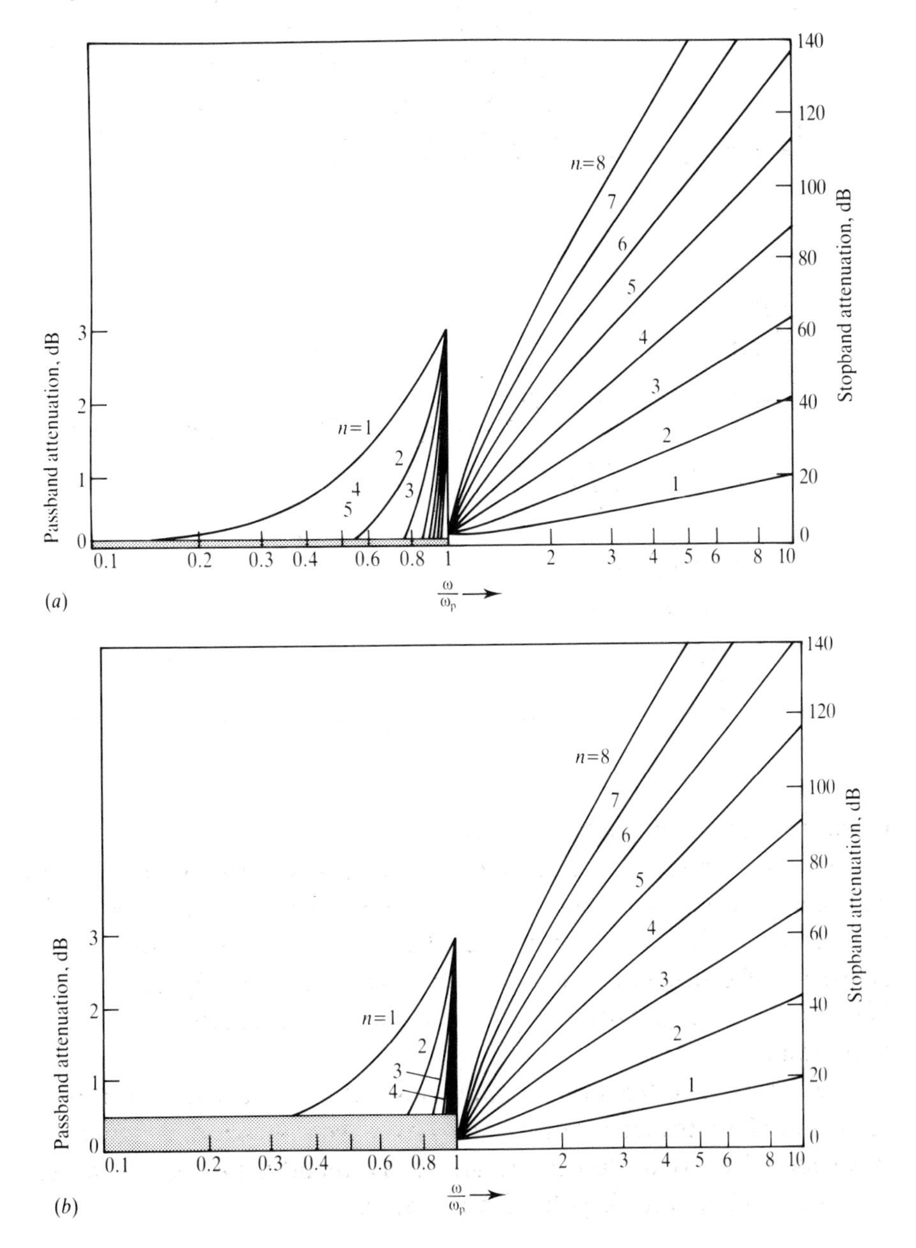

Fig. 6.47. Chebyshev filter response. (*a*) Gain–frequency response with 0.1 dB ripple. (*b*) Gain–frequency response with 0.5 dB ripple.

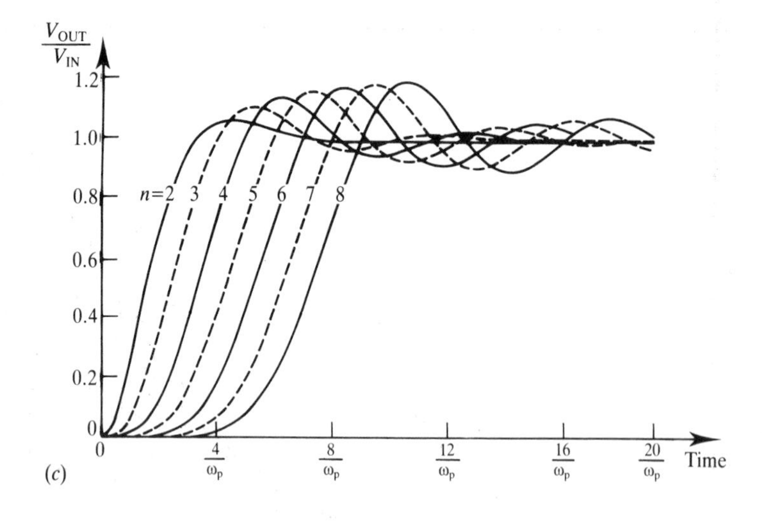

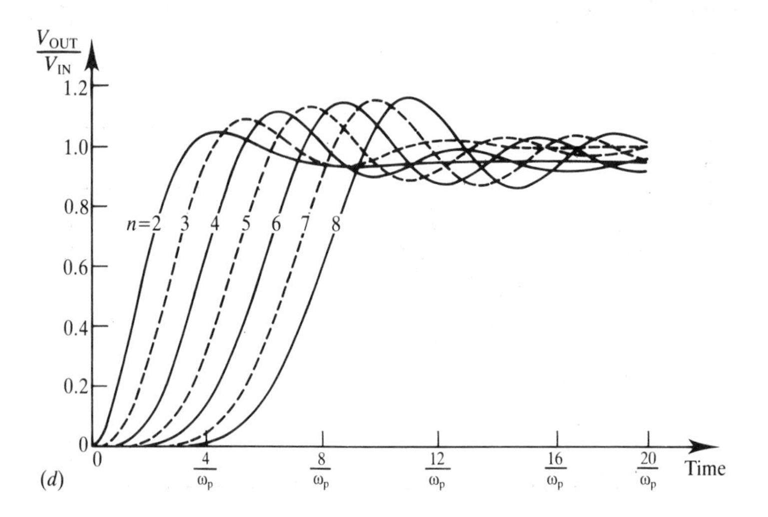

Fig. 6.47 (*contd.*). Chebyshev filter response. (*c*) Step response with 0.1 dB ripple. (*d*) Step response with 0.1 dB ripple.

filter, the other types of Bessel filters – bandpass, notch and high pass – do not have the same phase response (the linear phase property of the low pass Bessel filter is not preserved when the low pass Bessel filter response is transformed into these other filter types).

Bessel filters have a very small overshoot in their response to a step input. This makes a Bessel filter attractive for applications where pulses are being transmitted with minimal distortion. The excellent phase response of this filter group is obtained at the expense of the filter's gain response. The gain response, furthermore, is not maximally flat over the pass band and does not have a steep cut-off. The gain response is monotonic. Bessel filters are all-pole.

This filter is commonly used as a low pass filter, where it is important to preserve the shape of the signal. It can also be used to create time delays.

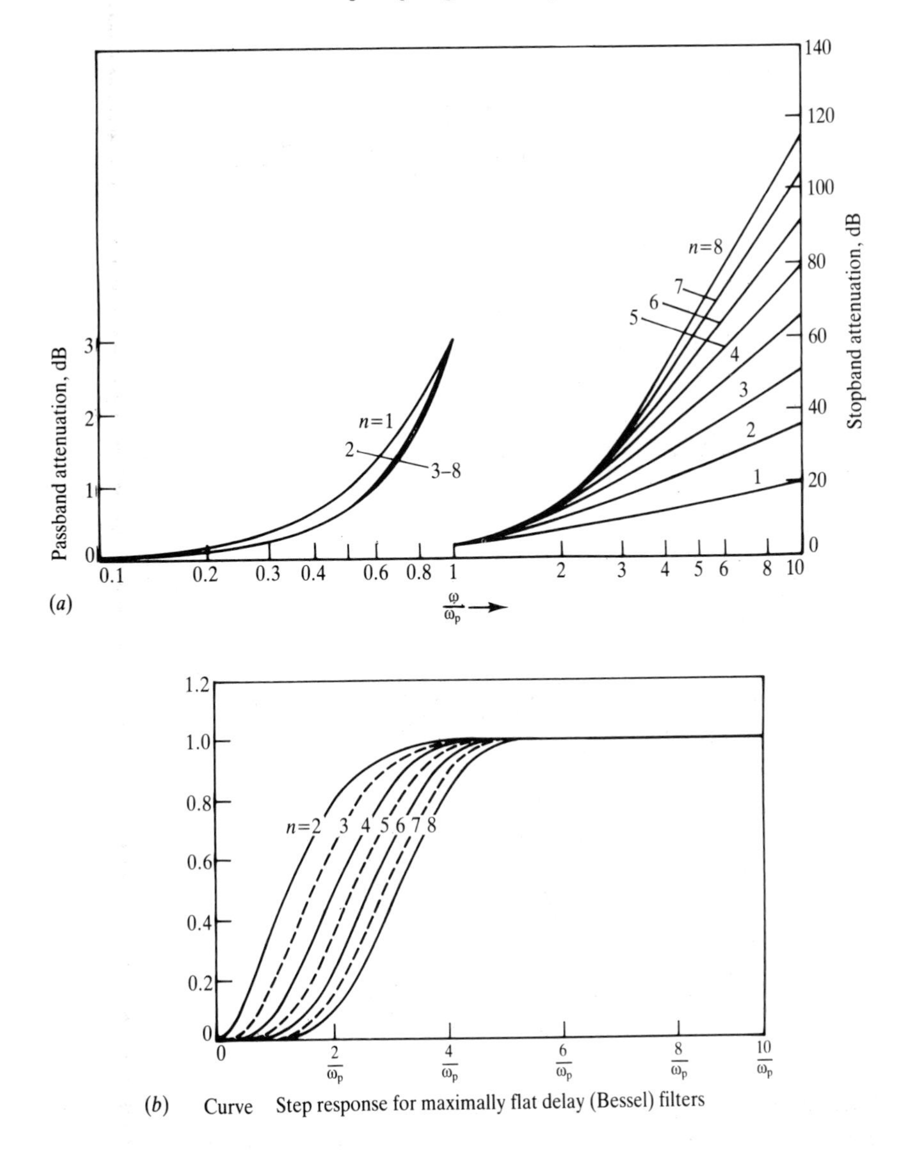

Fig. 6.48. Bessel filter response. (*a*) Gain–frequency response. (*b*) Step response.

Other filter groups

Butterworth, Chebyshev and Bessel are three of the most commonly-used filter groups. Four further filter groups are described below: the Butterworth–Bessel, the Legendre, the Inverse Chebyshev and the Elliptic groups of filters.

Butterworth–Bessel These filters are a blend of the maximally flat response of the Butterworth with the good linear phase response of the Bessel to give a filter with reasonably flat gain response and a reasonable phase linearity.

Legendre (or optimally monotonic) These filters combine properties from both Butterworth and Chebyshev filters. It has a gain response which is not as flat as the Butterworth but which has no ripples in its gain response as with the Chebyshev. It is optimally monotonic and has a better cut-off than the Butterworth, at the expense of the flat gain response.

Inverse Chebyshev This group of filters is the inverse of the normal Chebyshev in that it is monotonic over its pass band but contains equi-ripples in the stop band as shown in Fig. 6.49(*a*). The Inverse Chebyshev is used in applications where a non-infinite stop band attenuation can be tolerated but where the pass band gain must be as flat as possible. The ripples in the stop band gain are achieved by using zeros in the low pass transfer function.

Elliptic (Chebyshev–Cauer) This group of filters has ripples in both pass band and stop band to obtain a very sharp cut-off between pass band and stop band as shown in Fig. 6.49(*b*).

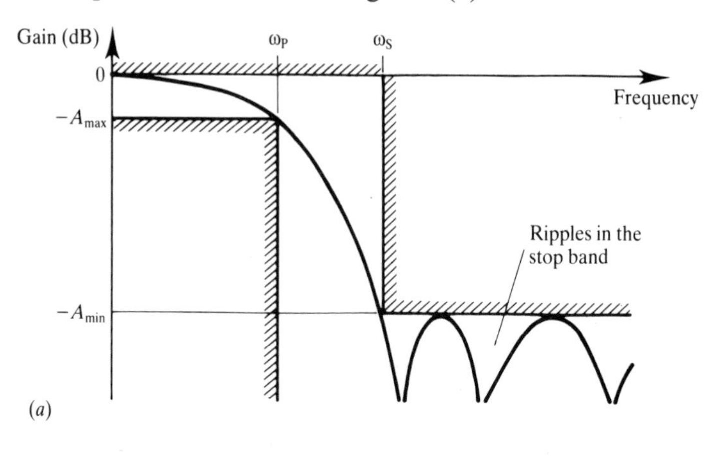

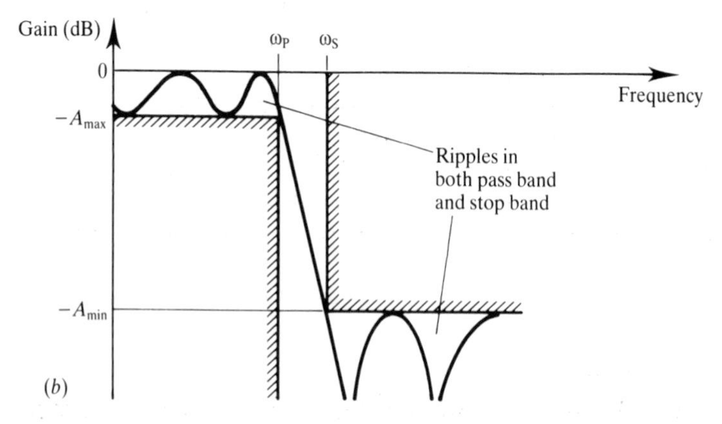

Fig. 6.49. Inverse Chebyshev and elliptic low pass filter responses. (*a*) Inverse Chebyshev. (*b*) Elliptic.

The properties of the various filter groups are summarized in Table 6.4.

Table 6.4. *Properties of filter groups*

	Elliptic	Inverse Chebyshev	Chebyshev	Legendre	Butterworth	Butterworth Bessel	Bessel
Gain response	Ripples in the pass and stop band	Ripples in the stop band	Ripples in the pass band	Optimally monotonic	Optimally flat	Monotonic	Monotonic
Cut-off	Very sharp	Sharp	Sharp		Moderately sharp		Less sharp
Phase linearity and pulse transmission	Very poor	Poor	Poor		Medium		Good (very small overshoot to step inputs)
Low pass transfer function	Poles and zeros	Poles and zeros	All pole	All pole	All pole	All pole	All pole (Low pass only for linear phase)

Table 6.5. *Butterworth transfer functions*

Butterworth Polynomials $T(s) = \dfrac{N(s)}{D(s)}$

$N(s) = 1$
$D(s) =$ (from the table)

Order n	$D(s)$
1	$(s + 1)$
2	$(s^2 + 1.41421s + 1)$
3	$(s + 1)(s^2 + s + 1)$
4	$(s^2 + 1.84776s + 1)(s^2 + 0.76537s + 1)$
5	$(s + 1)(s^2 + 1.61803s + 1)(s^2 + 0.61803s + 1)$
6	$(s^2 + 1.93185s + 1)(s^2 + 1.41421s + 1)$ $(s^2 + 0.51764s + 1)$
7	$(s + 1)(s^2 + 1.80194s + 1)(s^2 + 1.24698s + 1)$ $(s^2 + 0.44504s + 1)$
8	$(s^2 + 1.96157s + 1)(s^2 + 1.66294s + 1)$ $(s^2 + 1.11114s + 1)(s^2 + 0.39018s + 1)$

Transfer functions

Table 6.5 lists Butterworth transfer functions up to eighth order. For a particular specification, given A_{min}, A_{max}, ω_S and ω_P, the order of the filter is given by

$$n = \frac{\log\left[\dfrac{10^{0.1A_{\min}} - 1}{10^{0.1A_{\max}} - 1}\right]}{2\log\left[\dfrac{\omega_S}{\omega_P}\right]}$$

Note that since $A_{\min}$ and $A_{\max}$ are attenuations in dB they are positive values. Also, since n is an integer value it should always be rounded up from the above expression. The Butterworth polynomials listed in Table 6.5 are normalized to give a 3 dB bandwidth at 1 rad/s and a dc gain of unity (0 dB).

Table 6.6(*a–c*) lists Chebyshev transfer functions for pass band ripples of 0.5 dB, 1 dB and 3 dB respectively. These are normalized so that the maximum pass band gain is unity (i.e. 0 dB) and the attenuation of the transfer function $T(s) = N(s)/D(s)$ is $A_{\max}$ at 1 rad/s. A Chebyshev filter of order n will have n half cycles of ripple in its pass band. The order of Chebyshev filters required to satisfy a particular specification can be determined from the expression

$$n = \frac{\cosh^{-1}\left[\dfrac{10^{0.1A_{\min}} - 1}{10^{0.1A_{\max}} - 1}\right]}{\cosh^{-1}\left[\dfrac{\omega_S}{\omega_P}\right]}$$

where $A_{\min}$ and $A_{\max}$ are attenuations in dB and are positive. Note, $A_{\min}$ is also the amount of pass band ripple. For a given filter specification, a Chebyshev will usually be a lower order than a Butterworth. From the above equation, n must be rounded *up* to the nearest integer value. The transfer function can then be taken from the tables.

The denominator polynomial, $D(s)$, for an n order Bessel filter can easily be determined from the two previous polynomials, $D_{(n-1)}(s)$ and $D_{(n-2)}(s)$.

$$D_{(n)}(s) = (2n - 1)D_{(n-1)}(s) + s^2 D_{(n-2)}(s)$$

$$D_0(s) = 1$$

$$D_1(s) = s + 1$$

$$D_2(s) = s^2 + 3s + 3$$

etc.

The transfer function is $T(s) = N/D(s)$ for 0 dB gain at dc. The factorized denominator polynomials of a Bessel filter are given in

Table 6.6. (*a*) *Chebyshev transfer functions (0.5 dB pass band ripple)*

Order	$N(s)$	$D(s)$
1	2.863	$(s + 2.863)$
2	1.431	$(s^2 + 1.426s + 1.516)$
3	0.716	$(s + 0.626)(s^2 + 0.626s + 1.142)$
4	0.358	$(s^2 + 0.351s + 1.064)(s^2 + 0.847s + 0.356)$
5	0.1789	$(s + 0.362)(s^2 + 0.224s + 1.036)$
		$(s^2 + 0.586s + 0.477)$
6	0.0895	$(s^2 + 0.155s + 1.023)(s^2 + 0.424s + 0.590)$
		$(s^2 + 0.580s + 0.157)$
7	0.0447	$(s + 0.256)(s^2 + 0.114s + 1.016)$
		$(s^2 + 0.319s + 0.677)(s^2 + 0.462s + 0.254)$
8	0.0224	$(s^2 + 0.0872s + 1.012)(s^2 + 0.248s + 0.741)$
		$(s^2 + 0.372s + 0.359)(s^2 + 0.439s + 0.088)$

(*b*) *Chebyshev transfer function (1 dB pass band ripple)*

Order	$N(s)$	$D(s)$
1	1.965	$(s + 1.965)$
2	0.983	$(s^2 + 1.098s + 1.103)$
3	0.491	$(s + 0.494)(s^2 + 0.494s + 0.994)$
4	0.246	$(s^2 + 0.674s + 0.279)(s^2 + 0.279s + 0.987)$
5	0.123	$(s + 0.289)(s^2 + 0.468s + 0.429)$
		$(s^2 + 0.179s + 0.988)$
6	0.0614	$(s^2 + 0.124s + 0.991)$
		$(s^2 + 0.340s + 0.558)(s^2 + 0.464s + 0.125)$
7	0.0307	$(s + 0.205)(s^2 + 0.0914s + 0.993)$
		$(s^2 + 0.256s + 0.653)(s^2 + 0.370s + 0.230)$
8	0.0154	$(s^2 + 0.0700s + 0.994)(s^2 + 0.199s + 0.724)$
		$(s^2 + 0.298s + 0.341)(s^2 + 0.352s + 0.0703)$

(*c*) *Chebyshev transfer functions (3 dB pass band ripple)*

Order	$N(s)$	$D(s)$
1	1	$(s + 1)$
2	0.500	$(s^2 + 0.644s + 0.707)$
3	0.250	$(s + 0.298)(s^2 + 0.298s + 0.839)$
4	0.125	$(s^2 + 0.170s + 0.903)(s^2 + 0.410s + 0.196)$
5	0.0625	$(s + 0.177)(s^2 + 0.110s + 0.936)$
		$(s^2 + 0.287s + 0.377)$
6	0.0313	$(s^2 + 0.0763s + 0.955)(s^2 + 0.209s + 0.522)$
		$(s^2 + 0.285s + 0.0887)$
7	0.0156	$(s + 0.126)(s^2 + 0.0562s + 0.966)$
		$(s^2 + 0.157s + 0.627)(s^2 + 0.228s + 0.204)$
8	0.0781	$(s^2 + 0.0431s + 0.974)(s^2 + 0.123s + 0.704)$
		$(s^2 + 0.184s + 0.321)(s^2 + 0.217s + 0.0503)$

Table 6.7. *Bessel transfer functions* $T(s) = \frac{N}{D(s)}$

Order	3 dB freq (rad/s)	t_D (s)	N	$D(s)$
1	1.000	0.693	1	$(s+1)$
2	1.362	0.900	3	(s^2+3s+3)
3	1.756	0.958	15	$(s+2.322)(s^2+3.678s+6.459)$
4	2.115	0.979	105	$(s^2+5.792s+9.14)(s^2+4.208s+11.49)$
5	2.427	0.989	945	$(s+3.647)(s^2+6.704s+14.27)$ $(s^2+4.649s+18.16)$
6	2.703	0.994	10 395	$(s^2+5.032s+26.51)$ $(s^2+7.471s+20.85)$ $(s^2+8.497s+18.80)$
7	2.952	0.997	135 135	$(s+4.972)(s^2+5.371s+36.60)$ $(s^2+8.140s+28.94)$ $(s^2+9.517s+25.67)$
8	3.179	0.998	2 027 025	$(s^2+5.678s+48.43)$ $(s^2+8.737s+38.57)$ $(s^2+10.41s+33.93)$ $(s^2+11.18s+31.98)$

Table 6.7 where the polynomials are not normalized for any particular frequency or delay time. The −3 dB gain values are given with the time taken to reach 50% of final value in response to a step input. Normalized values are given in the tables which must be scaled to the desired frequency values. Increasing the frequency scale inversely decreases the time delay, i.e. scaling the frequency by a factor of k reduces the time delay by a factor of $1/k$.

Transformation and scaling

Most of the data in the previous section have been normalized to give a pass band gain of 0 dB and a cut-off frequency of 1 rad/s. In practice, this will virtually never be the case.

Gain scaling Gain scaling is a trivial matter of multiplying the transfer function by the required pass band gain.

Frequency scaling Frequency scaling is achieved by substituting s/ω_{scale} for s, where ω_{scale} is the desired frequency in the transfer function. The transfer function is therefore magnified such that what previously happened at for example $\omega = 1$ rad/s now happens when $\omega/\omega_{\text{scale}} = 1$ rad/s (i.e. when $\omega = \omega_{\text{scale}}$).

Transformation Until now, only the design of low pass filters has been considered. Other types of filters, such as bandpass, high pass or notch, do not have to be considered separately since they can be transformed into low pass equivalents. The design problem then becomes a low pass design problem. The resulting transfer function is then transformed back again to the required filter type as shown in Fig. 6.50.

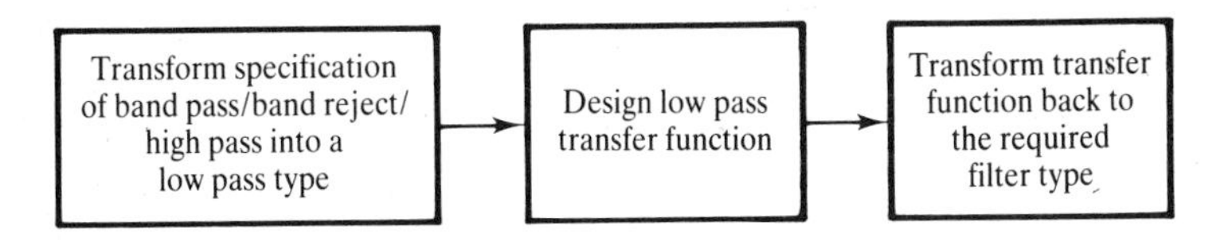

Fig. 6.50. Frequency transformation procedure.

(a) High pass filter transformations
The high pass transfer function can be obtained from the low pass by substituting

$$s/\omega_{\mathrm{LP}} \rightarrow \text{(Low pass to high pass)} \rightarrow \omega_{\mathrm{HP}}/s$$

in the transfer function. This will give a transfer function with a high pass response having the same gain at ω_{HP} as the low pass does at ω_{LP}. To avoid confusion, it is usual to normalize the high pass specification and then use the transformation as shown in Fig. 6.51.

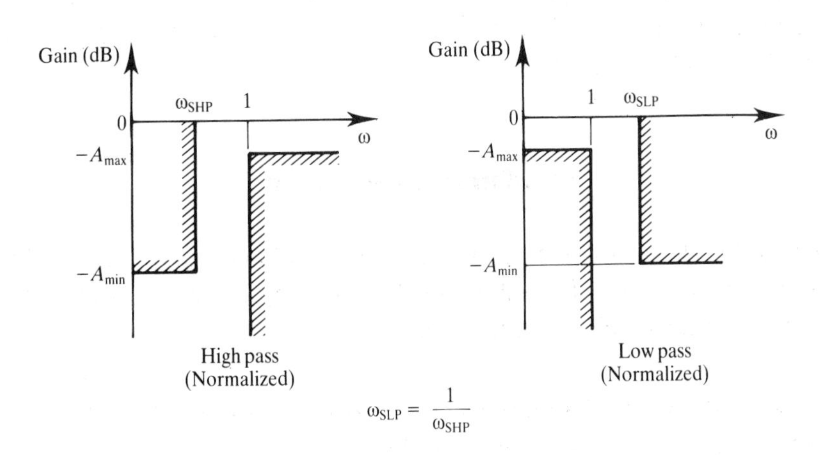

Fig. 6.51. High pass to low pass transformation.

(b) Bandpass filter transformations
A bandpass transfer function can be obtained from a low pass function by using the transformation and substituting this into the lowpass

transfer function.

$$s/\omega_{LP} \rightarrow \text{(Low pass to bandpass)} \rightarrow (s^2 + \omega_{BP}^2)/B_{BP} \cdot s$$

where ω_{BP} is the centre frequency of the pass band. B_{BP} is the bandwidth of the pass band gain given by $B_{BP} = \omega_{P2} - \omega_{P1}$ and $\omega_{BP} = \sqrt{\omega_{P1}\omega_{P2}}$ that is, ω_{P1} and ω_{P2} are positioned geometrically symmetrical with respect to ω_{BP}.

Note: transformations from low pass to bandpass doubles the filter order.

The bandpass transfer function will have the same gain at frequencies ω_{P2} and ω_{P1} as does the low pass transfer function at ω_{LP}. To avoid confusion, it is usual to normalize bandpass specifications so that $\omega_{BP} = 1$ rad/s. Also, before making the transformation, the bandpass specification must be geometrically symmetrical (Fig. 6.52) so that

$$\omega_{BP} = \sqrt{\omega_{P1} \cdot \omega_{P2}} = \sqrt{\omega_{S1} \cdot \omega_{S2}}$$

If a specification is asymmetrical it must be made symmetrical as shown in Fig. 6.53.

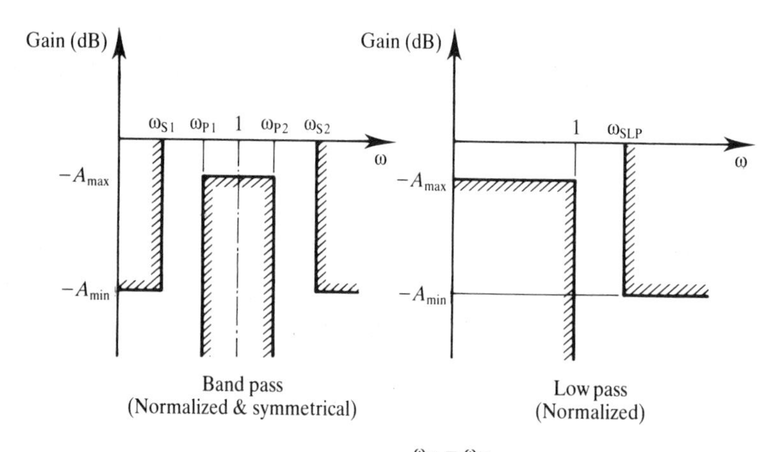

Fig. 6.52. Bandpass to low pass transformation (1).

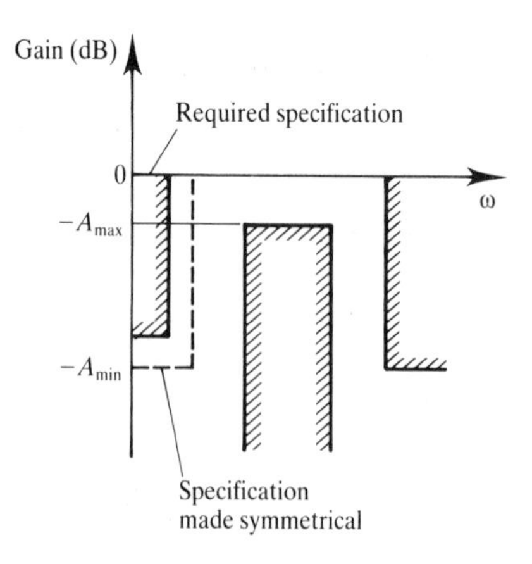

Fig. 6.53. Bandpass to low pass transformation (2).

(c) Bandstop filter transformations
A bandstop filter transfer function can be obtained from a low pass transfer function by making the substitution

$$s/\omega_{LP} \rightarrow \text{(Low pass to bandstop)} \rightarrow B_{BS} \cdot s/(s^2 + \omega_{BS}{}^2)$$

where ω_{BS} is the centre frequency of the stopband

$$\omega_{BS} = \sqrt{\omega_{P1}\omega_{P2}}$$

and B_{BS} is the bandwidth of the stopband

$$B_{BS} = \omega_{P2} - \omega_{P1}$$

As with bandpass filter transformations, the transformation from low pass to bandstop doubles the filter order.

The bandstop transfer function will have the same gain at frequencies ω_{P1} and ω_{P2} as will the low pass transfer function at ω_0. Again, to avoid confusion, it is usual to normalize the bandstop specification so that $\omega_{BS} = 1$ rad/s. The specification for the bandstop filter must be symmetrical so that

$$\omega_{BS} = \sqrt{\omega_{P1} \cdot \omega_{P2}} = \sqrt{\omega_{S1} \cdot \omega_{S2}}$$

If the specification is not symmetrical, then it must be made so in the same way as described for bandpass filters. Fig. 6.54 shows the details of the bandstop to low pass transformation.

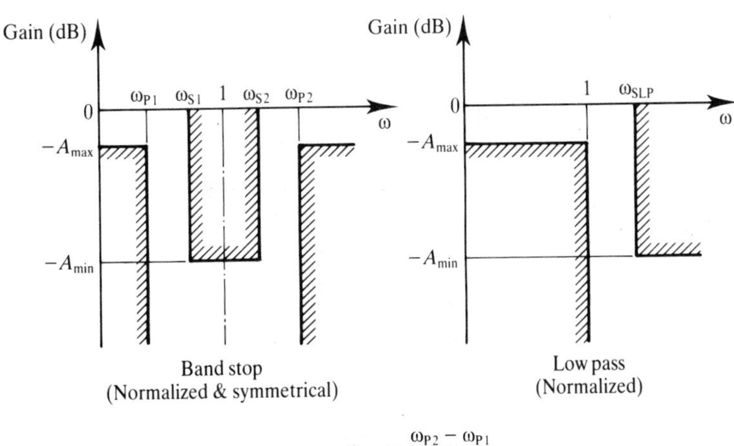

Fig. 6.54. Bandstop to low pass transformation.

From transfer function to circuitry

So far in this section we have mainly discussed filter transfer functions and how to establish a suitable transfer function for a particular application. After having determined the transfer function, the final stage is to design an active *RC* filter circuit to implement the function.

Two main techniques are used in circuit design to realize the transfer function. The first is called cascade filter design and the second technique involves the simulation of *R–L–C* ladder filters. These are compared below.

R–L–C ladder simulation

This technique involves simulating *R–L–C* ladder networks with active circuits. The *R–L–C* ladders can be simulated by direct replacement of inductors with active inductor simulator circuits or by using multi-loop simulation of the overall ladder network.

Advantages (1) the frequency response of *L–C* ladder networks is relatively insensitive to component tolerances so that accurate realization of the required frequency response can be achieved

Disadvantages (1) the design technique is more involved
(2) more op amps are required
(3) tuning is more difficult since the components interact with each other

The technique is best used where large quantities of filters are to be produced which require an accurate frequency response. Unfortunately as the design procedure is more involved, there is not sufficient space to describe this technique here.

Cascade filter design

This technique involves factorizing the transfer function into first and second order factors. These factors can then be realized separately in one- or two-pole segments in such a way that each segment does not interfere with other segments.

e.g. $$T(s) = \frac{b_m s^m + b_{(m-1)} s^{m-1} + \cdots + b_2 s^2 + b_1 s + b_0}{d_n s^n + d_{(n-1)} s^{n-1} + \cdots + d_2 s^2 + d_1 s + d_0}$$

which can be factorized into linear (first order) and quadratic (second order) factors i.e.

$$T(s) = K \underbrace{\frac{\left(s^2 + \frac{\omega_{011}}{Q_{F11}} s + {\omega_{011}}^2\right)}{\left(s^2 + \frac{\omega_{021}}{Q_{F21}} s + {\omega_{021}}^2\right)}}_{\text{1st factor}} \cdot \underbrace{\frac{\left(s^2 + \frac{\omega_{012}}{Q_{F12}} s + {\omega_{012}}^2\right)}{\left(s^2 + \frac{\omega_{022}}{Q_{F22}} s + {\omega_{022}}^2\right)}}_{\text{2nd factor}} \cdot \underbrace{\frac{(s + \omega_{013})}{(s + \omega_{023})} \cdot \frac{(\cdots)}{(\cdots)}}_{\text{etc.}}$$

Advantages
(1) design simplicity
(2) easy to tune since segments can be separately adjusted
(3) low power consumption since each segment can be designed with a minimum number of op amps

Disadvantages
(1) difficult to get very close tolerances in frequency response since the errors from each segment are additive.

There is usually an optimum combination of the way in which poles and zeros can be paired since any pole can be paired with any zero. The optimum combination will depend upon the particular application but the following factors will usually be required for most applications.

(i) maximizing dynamic range, i.e. ensuring that one stage does not saturate before other stages
(ii) minimizing sensitivity to op amp performance
(iii) a relatively simple tuning procedure

As a general rule of thumb, the dynamic range of a filter can be maximized by keeping the gain response of each segment as flat as possible. This can be done by combining high-Q poles with zeros which are as close to the poles as possible.

As with pole–zero pairing, there is an optimum sequence of filter segments which is dependent upon the application.

(i) To maximize dynamic range, it is usually better to have the pole Q of the cascade networks increasing from the input to the output
(ii) With large values of high frequency interference, it may be better to have a low pass stage at the input to avoid introducing slew rate limiting errors
(iii) It may be best to have a high pass or bandpass stage in the last stage of the filter so that output dc offset has only one stage contributing to it (band pass and high pass filters only).

7
Integrators and differentiators

7.1 Integration

Integration is one of the most basic mathematical operations and an electronic analogy of this operation involves making a circuit whose output rate of change is proportional to the input voltage. In graphical terms, the output voltage is proportional to the area under the input voltage curve. Many analog systems incorporate some form of integrator circuit. They are commonly used in active filters and in control systems to integrate dc errors. An integrator can be considered as a single pole low pass filter operating on the -20 dB/decade slope of the graph of frequency against amplitude. Two elementary integrator circuits are presented in Fig. 7.1.

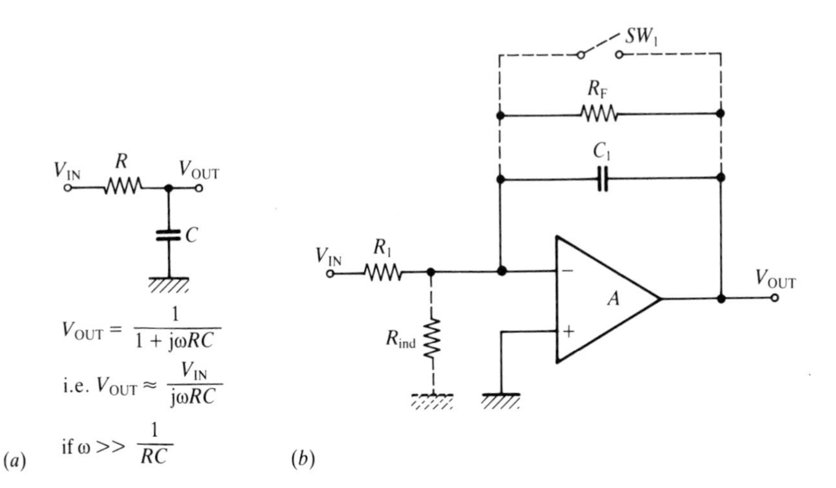

Fig. 7.1. Basic integrator circuits. (*a*) Simple *R–C* integrator. (*b*) Op amp based integrator.

The simple *RC* integrator shown in Fig. 7.1(*a*) has two severe drawbacks. Firstly, it significantly attenuates the input signal. Secondly, it has a high output impedance. As a result, it is rarely used in practice. The standard op amp based integrator shown in Fig. 7.1(*b*) consists of R_1 and C_1 connected around the op amp *A*. The input current to the inverting input is determined by R_1. The very high gain of the op amp forces the inverting input to be a virtual earth. As a result, the input current is defined by the input voltage and by resistor R_1. Nearly all the input current, therefore, flows into C_1, charging it up, thereby giving the integrating action.

Transfer equation: $$\frac{V_{OUT}}{V_{IN}} = \frac{-\dfrac{1}{C_1 R_1}}{s + \dfrac{1}{C_1 R_1 A} + \dfrac{1}{R_F C_1}}$$

$$\approx -\frac{1}{C_1 R_1 s}$$

Frequency limits of operation:

Low frequency $f_1 = \dfrac{1}{2\pi A R_1 C_1} + \dfrac{1}{2\pi R_F C_1}$

High frequency $f_2 = f_A$

[where A is the op amp gain and f_A is the gain–bandwidth product]

Input resistance = R_1

Output drift rate (worst case):

due to input offset voltage V_{IO} and bias

current I_B of the op amp $\Rightarrow \dfrac{V_{IO}}{CR} + \dfrac{I_B}{C}$

due to leakage resistance $R_F \Rightarrow \dfrac{V_{OUT}}{R_F C}$

due to leakage resistance $R_{IND} \Rightarrow \dfrac{V_{OUT}}{A_V R_{IND} C}$

Final output offset $= \left(\dfrac{R_F}{R_1} + 1\right) V_{IO} + I_B R_F$

The main problem with analog integrators is output drift due to the feedback capacitor C_1 charging up due to the unwanted leakage and bias currents and input offset voltage of A. In effect, the integrator is integrating the dc error voltages. If nothing is done to cure this problem, then the output will wander off in one direction and either a large and non-constant final output offset will occur or the op amp output will saturate. Three approaches can be tried to effect a cure.

(i) If the op amp is part of a larger feedback circuit or system, such as the state variable filter circuit described in Chapter 6, then the integrator drift may not be a problem since the overall feedback will limit its value.

(ii) If the signal to be integrated has no DC content, then you can include a resistor R_F in the feedback loop of the op amp as shown in Fig. 7.1. R_F thereby provides an alternative path for bias currents to flow rather than the path through C_1. This remedy is only useful in practice for input signals which have a frequency content greater than around 1 Hz, since at lower frequencies too large a value of R_F is required to be effective. The value chosen for R_F must be small enough to reduce the final offset to a tolerable amount, but yet large enough for the circuit to remain operating as an integrator over the entire frequency range of the input signal.

(iii) If DC signals are to be integrated, then a reset switch SW_1 may be added to the feedback loop so that the capacitor C_1 can be discharged at regular intervals.

As an example of the limits of drift performance which can be achieved, if a chopper stabilised CMOS op amp is used with $C_1 = 10\ \mu F$ and $R_1 = 10\ M\Omega$, taking $V_{IO} = 1\ \mu V$ and $I_B = 1\ pA$ as typical values for the op amp, then the output drift rate will be 0.4 mV/hr. For low drift applications, you will need to take extra care in the layout and construction of the integrator, otherwise extra leakage current will be added to the bias current lost from the inverting node of the integrator. It is usual practice to place a guard ring around the inverting input on both sides of the circuit board. The board should also be carefully cleaned. For ultra-low leakage currents, a PTFE stand-off could be used to insulate the inverting input of the op amp.

If an analog switch is used to discharge the capacitor, ensure that the leakage current introduced by the switch is lower than the bias current of the op amp. For very low leakage currents, FETs or analog switches can be used in series.

The ideal integrator would have a frequency response with a constant -20 dB/decade roll of across its spectrum. The practical integrator deviates from this ideal as shown in Fig. 7.2 for small signal inputs. The low frequency end of the response is limited by either the finite gain of the op amp or by the finite value of R_F. The integrator may well be unusable at these low frequencies due to output drift. At high frequencies, the integrator is limited by the finite gain–bandwidth product of the op amp. For the circuit to provide an integrating action, the input signal must have a frequency content safely within the upper and lower frequency limits (i.e. at least 10 times greater than the lower limit and less than around 1/10th of the upper limit).

The upper frequency end of the integrator response is limited by the

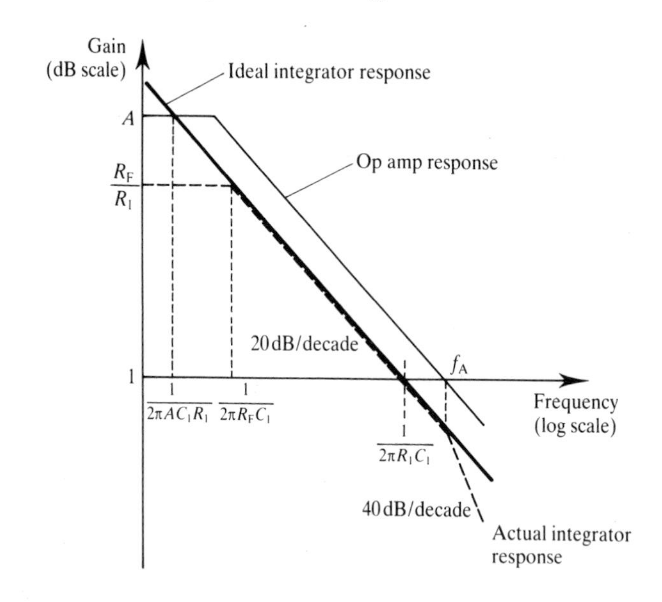

Fig. 7.2. Integrator small signal frequency response.

finite bandwidth of the op amp which introduces an extra pole at a frequency of approximately f_A, where f_A is the gain–bandwidth product of the op amp. This extra pole causes a phase and gain error in the output of the integrator at high frequencies. One method of compensating for this error involves adding a small capacitor C_{COMP} in parallel with R_1 to cancel the high frequency pole. Assuming that $1/2\pi R_1C_1 \ll f_A$, choose C_{COMP} such that $f_A = 1/2\pi(R_1C_{COMP})$. It is difficult to achieve exact cancellation since f_A is not known accurately; improvements approaching an order of magnitude can nevertheless be achieved by this method but be careful since if C_{COMP} is too large your circuit may oscillate.

For larger signals, the output may suffer considerable distortion from slew rate limiting of the op amp. Make sure that you check that the maximum desired output rate of change does not exceed the maximum slew rate possible from the op amp and that the maximum rate of change of the op amp output is not limited by the speed with which the op amp charges the capacitive load. This is particularly the case with large values of C_1 requiring fast changes. Remember that the maximum output rate of change $= I_{out_{max}}/(C_1 + C_L)$ where $I_{out_{max}}$ is the maximum output current and C_L is the total capacitance on the output.

If large values for integrator time constants (i.e. C_1R_1) are required, this means that the values for R_1 and C_1 must be large, i.e. C_1 in μF and R_1 in MΩ. This can present problems since large value capacitors, above around 1 μF, are not only physically much larger but also suffer more from parasitic effects such as high leakage resistance, increased dielectric absorption and higher dissipation factor. Large value resistors,

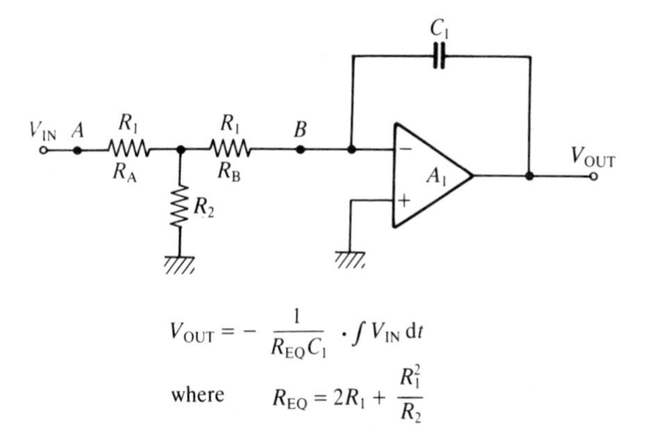

Fig. 7.3. Using a T-network of resistors.

by contrast, above around 1 MΩ, suffer from the effects of parasitic leakage resistance and parasitic stray capacitance; in addition such resistors cannot be obtained with very low tolerances and tend also to be more expensive. To avoid these problems, the resistor R_1 can be replaced with a T-network of resistors as shown in Fig. 7.3. It is still important to avoid parasitic capacitances and leakage resistances across the T-network (AB) since these will be in parallel with R_{EQ}. This involves laying the circuit out so that A and B are well separated with the possible need to use guard tracks. The effects of leakage resistance and capacitances across R_A and R_B have a much reduced effect since both R_A and R_B can be much lower in value and hence the usefulness of this network. Note that a T-network can also be used to achieve large values for R_F.

The basic integrator circuit is easily modified to integrate the sum of several signals applied to the inverting input as shown in Fig. 7.4. The largest number of signals that can be treated in this way is limited by the effective value of input resistance R_{1EQ}

where

$$R_{1EQ} = R_1 // R_2 // R_3 // \cdots // R_N$$

This value replaces R_1 in the design equations for drift and obviously reduces with more signals causing increased drift.

To integrate the difference between two signals, the circuit in Fig. 7.5 can be used. This circuit is very similar to the differential amplifier configuration but with two capacitors replacing two resistors. This circuit requires careful matching of the resistors and the capacitors otherwise a poor common mode rejection ratio will result. The CMRR due to mismatch in component values is given by

$$\text{CMRR} \simeq \frac{1 + sCR}{s\,\Delta(RC)}$$

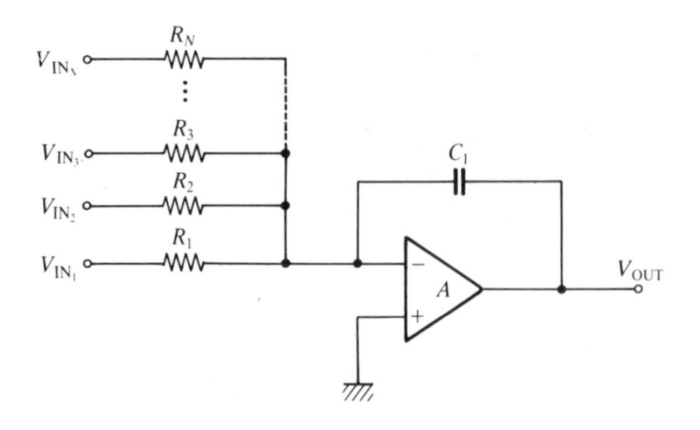

Fig. 7.4. Integration and summation of several input signals.

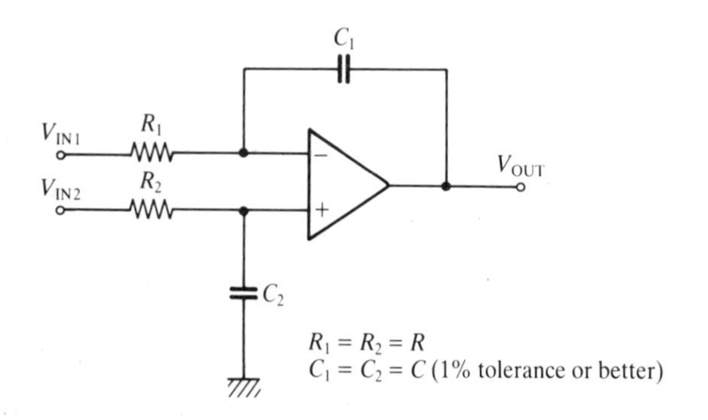

Fig. 7.5. Integrating the difference between two input signals.

where $\Delta(RC)$ is the difference in time constants $(R_1C_1 - R_2C_2)$ and output drift is given by

$$V_{IO} + \frac{V_{IO}t}{RC} + \frac{I_{OS}t}{C}$$

If you need a differential integrator with higher CMRR, an extra op amp is added as an inverter with a summing integrator as shown in Fig. 7.6. In this way, better common mode rejection is achieved since the CMRR is now dependent entirely on resistor matching with no requirement for capacitor matching.

If you want a non-inverting integrator you can either ground the inverting input (V_{IN1} input) of the differential integrator as shown in Fig. 7.5 or add an inverting stage after the integrator. The inverter is best placed after the integrator to preserve dynamic range since the integrator stage attenuates the higher frequency signals.

The basic integrator can be converted into a current integrator (see Chapter 3 on charge amplifiers) by simply omitting the input resistor as shown in Fig. 7.7(*a*). Alternatively, a differential current integrator can be formed as shown in Fig. 7.7(*b*). The differential current integrator has several serious disadvantages including the need to carefully and

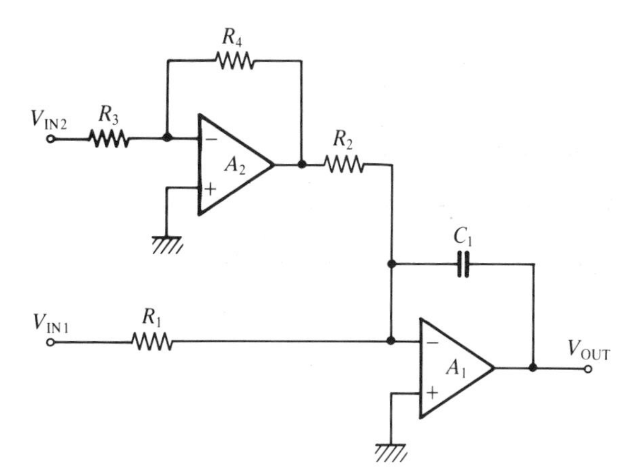

Fig. 7.6. A differential integrator with higher CMRR.

accurately match the capacitors and the requirement for a current source with a high impedance and wide output voltage compliance. To avoid these problems, another op amp can be used, with one op amp now being used as a current integrator and the other as a current mirror as shown in Fig. 7.7(*c*).

As a further variation on a theme, the two circuits shown in Fig. 7.8, for inverting and non-inverting configurations, are capable of adding the integral of the input signal to itself. Remember that the rate of output drift with these circuits is the same as with the basic integrator.

If a double integrator is needed, as for example in double integrating the output of an accelerometer to measure displacement, one approach worth considering is using the −40 dB per decade cut-off rate of a two-pole low pass filter instead of using two integrators. A circuit which realizes this approach is shown in Fig. 7.9.

The response of this circuit is:

$$-\frac{(1 + sR_3(C_2 + C_3))}{s^2C_2C_3\left(1 + \dfrac{sR_1R_2C_1}{(R_1 + R_2)}\right)(R_1 + R_2)}$$

Select components such that $C_1 = C_2 = C_3 = C$ and $R_1 = R_2 = R$ and $R_3 = R/4$ (then the poles and zeros will cancel)

therefore:

$$\frac{V_{OUT}}{V_{IN}} = -\frac{2}{s^2C^2R^2}$$

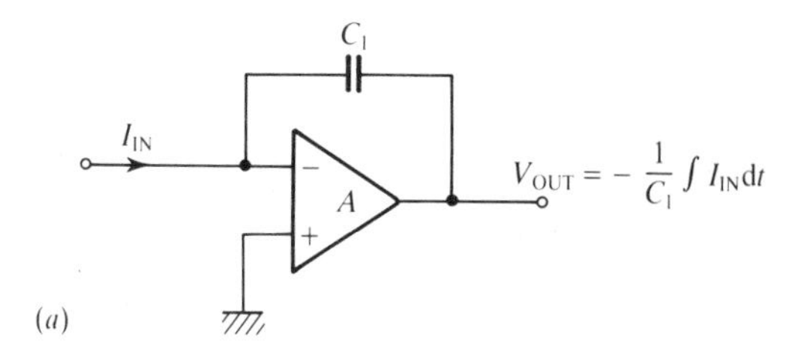

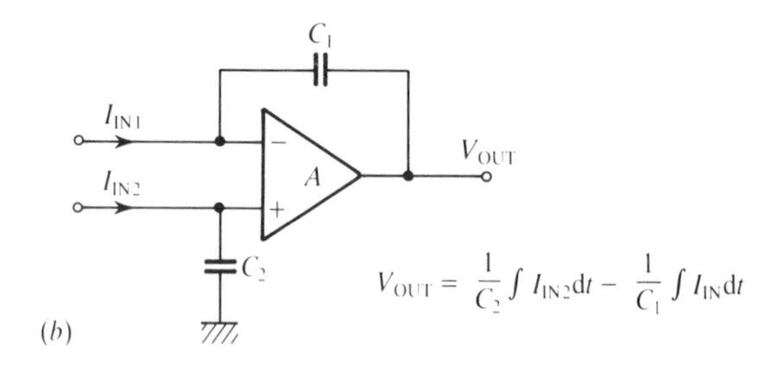

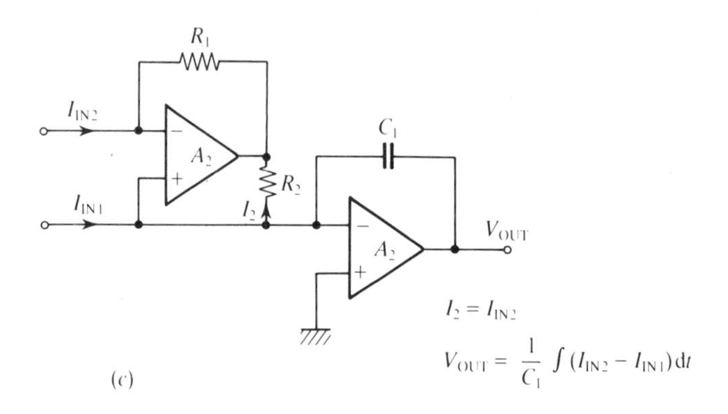

Fig. 7.7. Current integrators. (*a*) Simple virtual earth. (*b*) Differential. (*c*) Differential virtual earth.

Note that the pole and zero cancellation occurs at a frequency which is usually close to the middle of the operating frequency range. Very close component matching is needed for good pole-zero cancellation.

The error due to output drift is

$$\left[\frac{V_{IO}}{R^2C^2} + \frac{2I_{OS}}{RC^2}\right]t^2 + \left[\frac{2I_B}{C}\right]t + [V_{IO} + 2I_B R]$$

An alternative approach to integrating an analog signal is shown in Fig. 7.10 which uses a semi-digital technique. In this approach the input signal is converted into a frequency using a voltage-to-frequency converter (VFC). The integral of the input signal is then generated simply by counting the VFC output frequency with a binary counter. The integral value can then be converted back into an analog form if required with a DAC. The big 'plus' of this approach is that the integral value is stored as a digital value in the counter, not as a charge stored on a capacitor, and so may be free from drift.

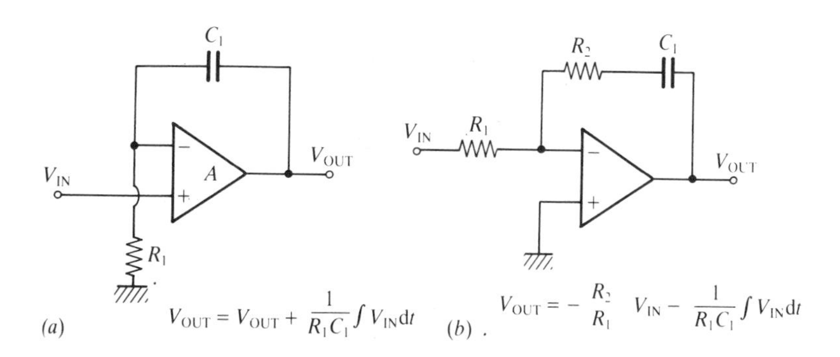

Fig. 7.8. Adding the integral of the input signal to itself. (*a*) Non-inverting. (*b*) Inverting.

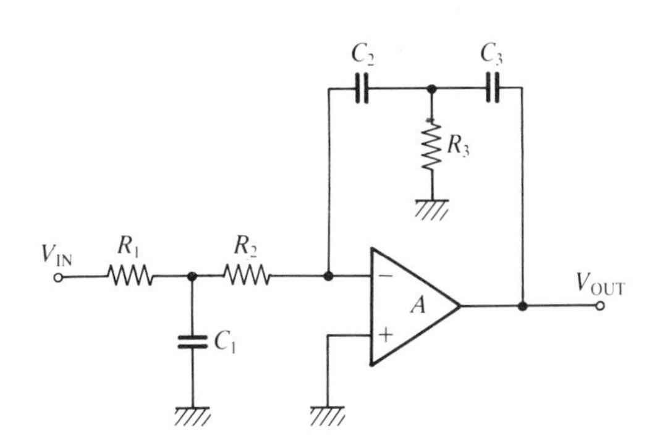

Fig. 7.9. Using a low pass filter as a double integrator.

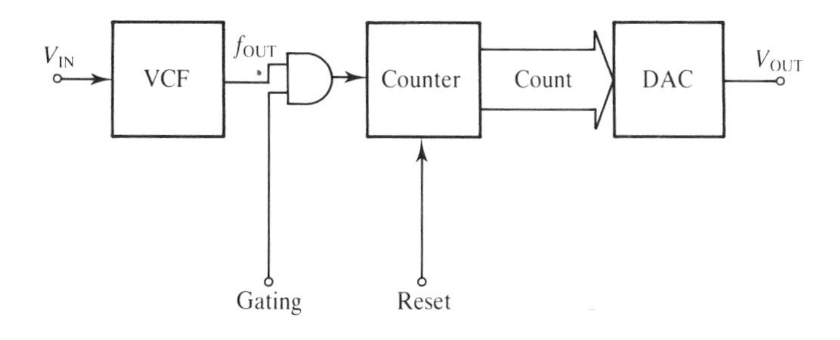

Fig. 7.10. A semi-digital integrator.

Notes on component selection

The op amps For very long duration integrators, approaching 1000 seconds, an op amp is required with a very low leakage current and a very low input offset voltage. Chopper stabilized amplifiers (with MOS inputs) are commonly used for these applications. For shorter durations, up to about 100 seconds, use standard FET input op amps. For fast integrators with time durations of a fraction of a second, the requirement for very low input bias currents to prevent output drift can be relaxed and bipolar input op amps can be used. Also, choose a fully compensated op amp which is stable for unity gain feedback conditions.

The capacitors For very long duration integrators, insulation resistance is one of the most important parameters and so polystyrene or PTFE capacitors are usually selected in this case, since they have an insulation resistance around 10^6 MΩ·μF. These capacitors can usually only be obtained in values up to 1 μF. For higher values, polycarbonate capacitors are usually selected but these have lower insulation resistances from 10^4 to 10^5 MΩ·μF. For shorter duration integrators, insulation resistance is not so important and other capacitor types, such as mica or ceramic, can be used. For very high speed applications, capacitor dissipation factor may become important. For long duration integrators, where charge is quickly switched in and out of the capacitor, in between its charging and discharging, then dielectric absorption may also be important. Polypropylene is good for low dielectric absorption.

7.2 Differentiators

A differentiator is a circuit whose output is proportional to the rate of change of the voltage at the input. A differentiator can be considered as a single pole high pass filter where the 20 dB/decade part of the response curve is used. The main problems with integrators were drift and low frequency response. With differentiators, the main problems are concerned with noise, stability and response at the higher frequencies.

The simple *RC* differentiator shown in Fig. 7.11(*a*) is very crude and has two main drawbacks: it attenuates the input signal and has a high output impedance. The differentiator circuit shown in Fig. 7.11(*b*) consists of C_1 and R_1 placed as shown around the op amp. Changes in the input voltage forces current to flow through C_1; this current must also flow through R_1. The high gain of the amplifier forces the inverting input to be a virtual earth and so the output voltage is proportional to

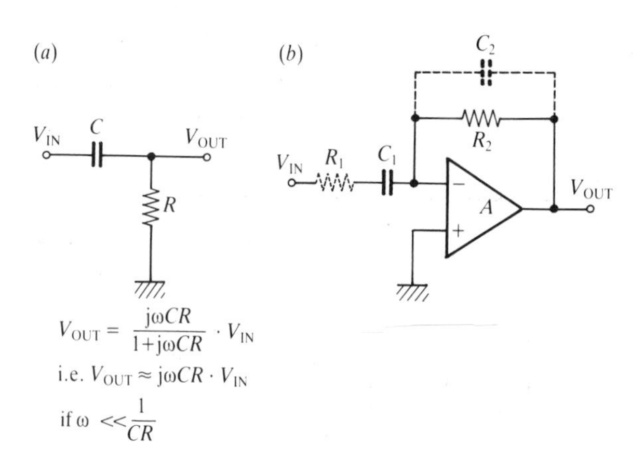

Fig. 7.11. Basic differentiator circuits. (*a*) Simple *R–C* differentiator. (*b*) Op amp based differentiator.

the rate of change of the input voltage. The circuit consisting of C_1, R_1 and the op amp is inherently unstable and tends to oscillate at higher frequencies. To avoid instability, either R_1 or C_2, or both, are added as shown. Further modifications are described below.

Transfer function Ideal: $-sC_1R_2$

Actual: $$-\frac{sC_1R_2}{(1 + sC_2R_2)(1 + sC_1R_1)}$$

the pole frequencies are $1/2\pi C_2R_2$ and $1/2\pi C_1R_1$

Input impedance $= R_1 + 1/j\omega C_1$

Output offset $= V_{IO} + I_B R_2$

where V_{IO} = the input offset voltage of A_1 and I_B = the input bias current of A_1

The inherent instability of the simple differentiator is highlighted in the response diagram in Fig. 7.12 which shows the way in which the open loop gain and the $1/\beta$ curves intersect (β being the feedback factor). The intersect occurs with a closure rate of 40 dB/decade part and so the circuit is conditionally unstable.

$$\beta = \frac{1}{1 + j\omega C_1R_2}$$

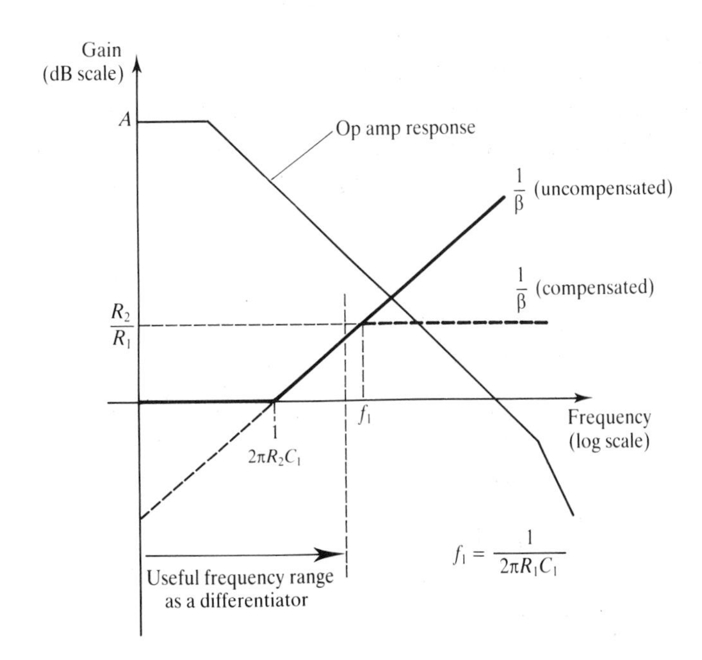

Fig. 7.12. Frequency instability of the simple differentiator.

Another way of viewing the relationships shown in the above response curve is by considering that the op amp introduces a 90° phase lag due to its internal frequency compensation. In addition, the feedback resistor and capacitor, also introduce a 90° phase shift. The total phase shift around the closed loop, including the 180° phase shift from the usual inverting operation of the op amp, is now 360° and so the system is unstable. At the best, the output response will show some ringing, at the worst the circuit will burst into oscillation. To stabilize the response, a pole is brought into the $1/\beta$ response by adding resistor R_1. This makes sure that the closure point, between the op amp response and $1/\beta$, occurs on the 20 dB/decade part of the response curve. This ensures stability for the circuit. The frequency of this pole $(1/(2\pi R_1 C_1))$ must be safely less than the frequency at which intersection occurs, otherwise the pole will have little effect. The response of the differentiator, due to the effect from R_1, is also shown in Fig. 7.12.

This response also shows up another problem. The gain at high frequencies is R_2/R_1 which is significantly greater than the gain at the lower frequencies. This can create problems with higher frequency noise from the op amp for example being much amplified. To avoid this problem, you can insert an extra pole, using C_2, to limit the high frequency response of the circuit as shown in Fig. 7.13. If C_2 is not present, the high frequency response is either limited by stray capacitances across R_2 or by the op amp as shown. C_2 may also help to stabilize the circuit. For large signals, the high frequency response of the differentiator may be limited by the slew rate of the op amp.

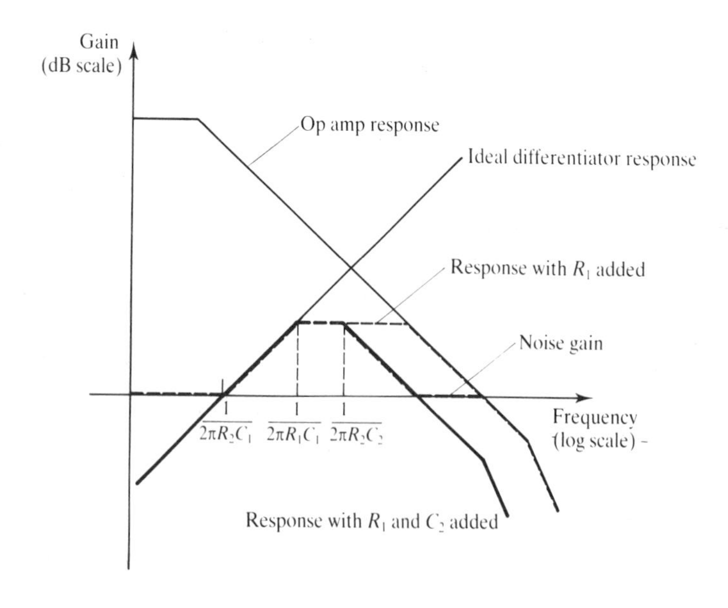

Fig. 7.13. Differentiator frequency response.

Equations for noise in integral form are given below.

$$\text{Total output noise} = V_{n\cdot\mathrm{TOT}}{}^2 = V_{n1}{}^2 + V_{n2}{}^2 \quad (\text{volts rms})^2$$

where

$$V_{n1}{}^2 = \int_0^\infty \frac{e_n{}^2}{|\beta(\mathrm{j}\omega)|^2}\cdot \mathrm{d}\omega \approx \frac{R_2{}^2 e_n{}^2}{4R_1{}^2}\left(\frac{1}{R_2C_2} - \frac{1}{R_1C_1}\right)$$

and

$$V_{n2}{}^2 = \int_0^\infty |Z_2|^2\cdot i_n{}^2\cdot \mathrm{d}\omega \approx \frac{i_n{}^2 R_2}{4C_2}$$

with

$$\beta(\mathrm{j}\omega) = \frac{Z_1}{Z_1 + Z_2} \qquad Z_1 = R_1 + \frac{1}{\mathrm{j}\omega C_1} \qquad Z_2 = \frac{R_2}{1 + \mathrm{j}\omega R_2 C_2}$$

These are usually evaluated approximately, sometimes using graphs or input equivalent noise voltage density e_{n} ($V/\sqrt{\mathrm{Hz}}$) and current density i_{n} ($A/\sqrt{\mathrm{Hz}}$) to determine the overall rms noise. Approximate expressions for noise are given, but these assume e_n and i_n are constant, the op amp has an infinite bandwidth and $1/R_2C_2 > 1/R_1C_1$. Generally, the output noise is dominated by e_n, amplified by R_2/R_1 over the high frequency region $1/R_1C_1$ to $1/R_2C_2$. A very important point to note concerning noise with differentiators is that the input signal should be band-limited, i.e. should be filtered so that it does not contain any higher frequency noise component. Be warned: high frequency noise on the input signal can drive the output of your differentiator berserk.

With very high gain or high accuracy differentiators, offset errors caused by the input offset voltage and input bias currents of the op amp become important. To reduce the output offset voltage caused by the flow of input bias current you can follow the standard approach of connecting a resistor equal in value to R_2 between the non-inverting input of the op amp and ground. This resistor is decoupled using a large value capacitor to avoid introducing stray feedback at the non-inverting input.

If there is a dc voltage at the input, then the effects of leakage current through the capacitor, C_1 should also be considered. Also, with very high gain differentiators, for applications with very slowly changing input signals, dielectric absorption in the capacitor C_1 may cause problems where there is a dc voltage on the input. The problem is caused at switch-on, since the charge absorbed by the dielectric will have the same effect as a changing input signal and the differentiator will take time to settle down.

If you need a large gain differentiator, the time constant C_1R_1 must be made large. This would involve using a large value resistor for R_2 (above 1 MΩ). Resistors of this size suffer from the parasitic effects of leakage resistance and from stray capacitance and cannot be obtained to close tolerances. Using a large value resistor for R_2 can be avoided by employing a T-network of smaller value resistors (low value resistors are less sensitive to stray capacitances and leakage resistances). Once again, take extra care with component layout to keep stray capacitances and leakage resistance across the network to a minimum. Guard tracks are well worth using. The modification is shown in Fig. 7.14.

If you are still unable to get a sufficiently high time constant then extra gain stages will have to be used. It is better to put as much gain before the differentiator as possible, in this case. This will avoid offset and offset drift problems later in the system as the differentiator is effectively ac-coupled and offsets from earlier stages can, consequently, be conveniently blocked by capacitor C_1. Putting as much gain as possible before the differentiator also reduces its relative noise contribution which can be a big advantage as the differentiator is usually a noisy circuit.

The lower limit of time constant is limited by the smallest value of R_2 and C_1 which can be practically obtained. R_2, for example, cannot be reduced below a certain limit because of its loading effects on the output of the op amp. The minimum value of C_1 is limited by parasitic stray capacitances and leakage resistances. The T-network shown in Fig. 7.15 allows larger value capacitors to be used to produce a lower effective capacitance. Careful layout and mounting is once again

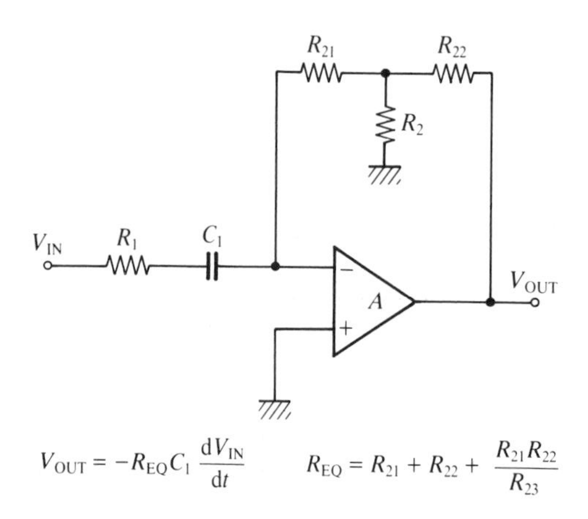

Fig. 7.14. Using a T-network of resistors in the feedback loop.

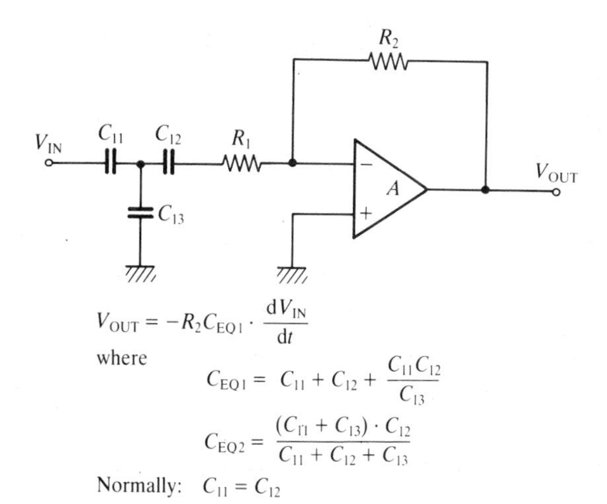

Fig. 7.15. Using a T-network of capacitors at the op amp input.

required. Note that C_{EQ2} should be used to determine the output noise and stability of the circuit.

A circuit which adds the derivatives of several signals can be made quite easily by adding extra components to the inverting input of the op amp, as shown in Fig. 7.16, where the inverting input acts as a current summing point.

Each input must have a stabilizing resistor so that the pole which it introduces is at a low enough frequency to stabilize the circuit. The number of signals that can be summed in this manner is limited due to noise since the greater the number of summed signals, the greater the effective high frequency gain of the circuit given by $-R_F/R_{TOT}$ where $R_{TOT} = R_1//R_2//\cdots R_N$ and the greater the noise, therefore, at the output.

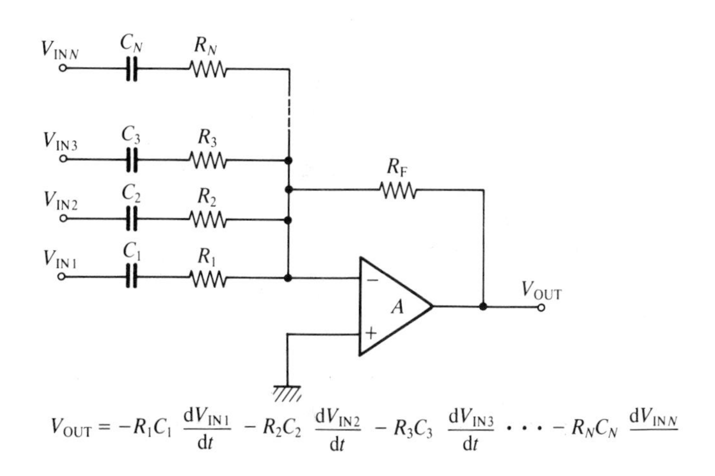

Fig. 7.16. Adding signals at the input to the differentiator.

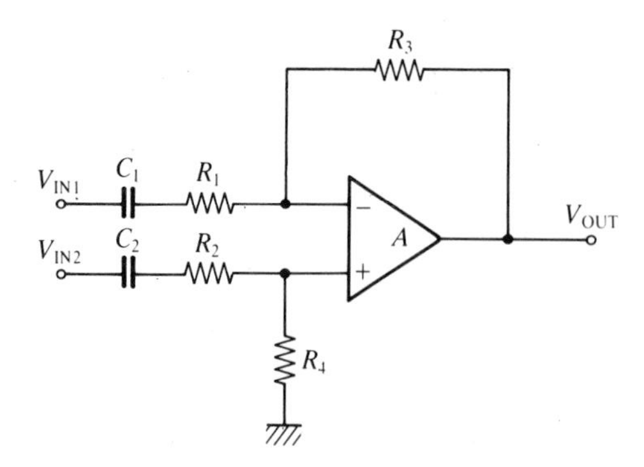

Fig. 7.17. A differential differentiator.

A differential differentiator can be created using the circuit in Fig. 7.17 which is similar to the standard difference amplifier described in Chapter 1. The response of this circuit can be shown to be

$$V_{\mathrm{OUT}} = \frac{1}{(1 + sC_1R_1)}\left[\frac{(1 + sC_1(R_1 + R_3))}{(1 + sC_2(R_2 + R_4))}\,sC_2R_4V_{\mathrm{IN2}} - sC_1R_3V_{\mathrm{IN1}}\right]$$

R_1 and R_2 are used, as before, to introduce a high frequency pole to stabilize the circuit. Good matching between capacitor and resistor values is required to avoid poor common mode rejection ratio. Assuming the components are selected such that $R_1 = R_2 = R$; $R_3 \simeq R_4 \simeq R_{\mathrm{F}}$ and $C_1 \simeq C_2 \simeq C$, and that signal frequencies are kept well below $1/(RC)$, then

$$V_{\mathrm{OUT}} \simeq R_{\mathrm{F}}C\left(\frac{\mathrm{d}V_{\mathrm{IN2}}}{\mathrm{d}t} - \frac{\mathrm{d}V_{\mathrm{IN1}}}{\mathrm{d}t}\right)$$

CMRR due to component mismatch can be shown to be

$$\mathrm{CMRR} \simeq \frac{1 + \mathrm{j}\omega CR_{\mathrm{F}}}{R_{\mathrm{F}}C\,\Delta(R_{\mathrm{F}}C)}$$

where

$$\Delta(R_{\mathrm{F}}C) = R_4C_2 - R_3C_1$$

(The op amp will also introduce a small CMRR error.)

In practice it may be difficult to achieve good performance with this circuit due to the difficulty in obtaining closely matched capacitors. So, it is often better to use a separate differential amplifier before the differentiator.

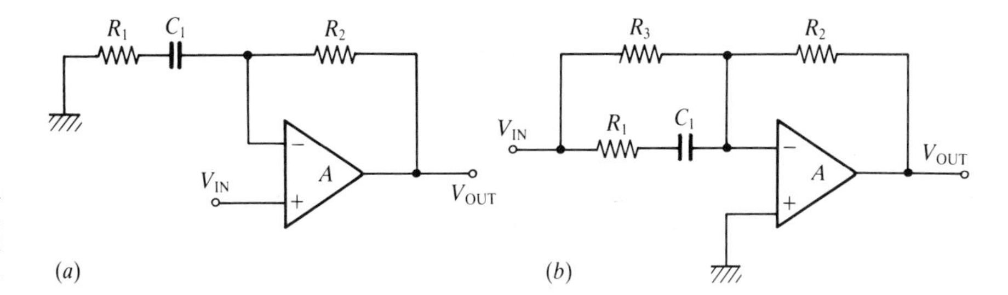

Fig. 7.18. Adding a differentiated signal to itself. (*a*) Non-inverting and (*b*) Inverting.

If the differential of a signal is to be added to itself, then either one of the two circuits shown in Fig. 7.18 could be used.
Non-inverting output

$$V_{OUT} = V_{IN} + \frac{sC_1R_2}{(1 + sC_1R_1)} \cdot V_{IN}$$

$$\approx V_{IN} + C_1R_2 \frac{dV_{IN}}{dt} \qquad \text{for } \omega \ll \frac{1}{R_1C_1}$$

Inverting output

$$V_{OUT} = \frac{(1 + sC_1(R_3 + R_1))R_2}{R_3(1 + sC_1R_1)} \cdot V_{IN}$$

$$\approx -\frac{R_2}{R_3} V_{IN} - C_1R_2 \frac{dV_{IN}}{dt}$$

Once again, provided signal frequencies are safely below $1/2\pi(R_1C_1)$, $R_1 \ll R_2$ and $R_1 \ll R_3$, then

$$V_{OUT} \simeq V_{IN} + C_1R_2 \frac{dV_{IN}}{dt} \quad \text{(non-inverting)}$$

or

$$V_{OUT} \simeq -\frac{R_2}{R_3} V_{IN} - C_1R_2 \frac{dV_{IN}}{dt} \quad \text{(inverting)}$$

Notes on component selection

The op amp This must have a high enough bandwidth and low enough higher frequency noise. For high accuracy amplification, or detecting very small rates of change, then an op amp with low values of offset voltage, offset drift and bias current is also desirable.

The capacitor C_1 The value of differentiator gain (i.e. time constant R_2C_1) is usually dictated by the application. For a particular value of R_2C_1 choose C_1 as large as possible yet preferably lower than 1 μF since capacitors above this value tend to be larger and more expensive. Ensure that C_1 is not so large as to require a small value for R_2 (such as less than 1 kΩ for example), as this lower value will excessively load the op amp.

The capacitor C_2 The type of capacitor chosen for C_2 must be stable enough over the temperature range of the application. Common types which are used include ceramic (in particular multi-layer NPO), mica, polystyrene and polycarbonate).

8

Log and antilog converters

Log and antilog converters use the exponential property of a forward biased p–n junction, using either a diode or a bipolar transistor, to provide the necessary log or antilog function. These special types of converters are widely used as a building block in circuits carrying out various mathematical functions. Log converters are also used to compress signals which have a very wide dynamic range, such as speech signals. Some of the converters described in this chapter can be designed to operate over a seven decade range of input signal. Custom-built log and antilog converters are described along with a selection of commercially available devices.

8.1 Log converters

The circuit of Fig. 8.1 is extremely simple but it has many major drawbacks including a very poor log conformity and drifting of the output due to temperature variations. Although this circuit is very rarely used, it does provide a good starting point from which more practical log converter designs can easily be developed.

The equation for a diode can be approximated by $I = I_0(e^{-\frac{qV}{kT}} - 1)$ where

I is the current flowing through the diode

V is the voltage across the diode

k is Boltzmann's constant

q is the electron charge

I_0 is the reverse diode leakage current

T is the temperature in K

Applying this equation to the above circuit gives

$$I_{IN} = V_{IN}/R_1 = I_0(e^{-\frac{qV_{out}}{kT}} - 1)$$

$$\therefore\ V_{OUT} = -\frac{kT}{q}\cdot\log_e\left(\frac{V_{IN}}{R_1 I_0} + 1\right)$$

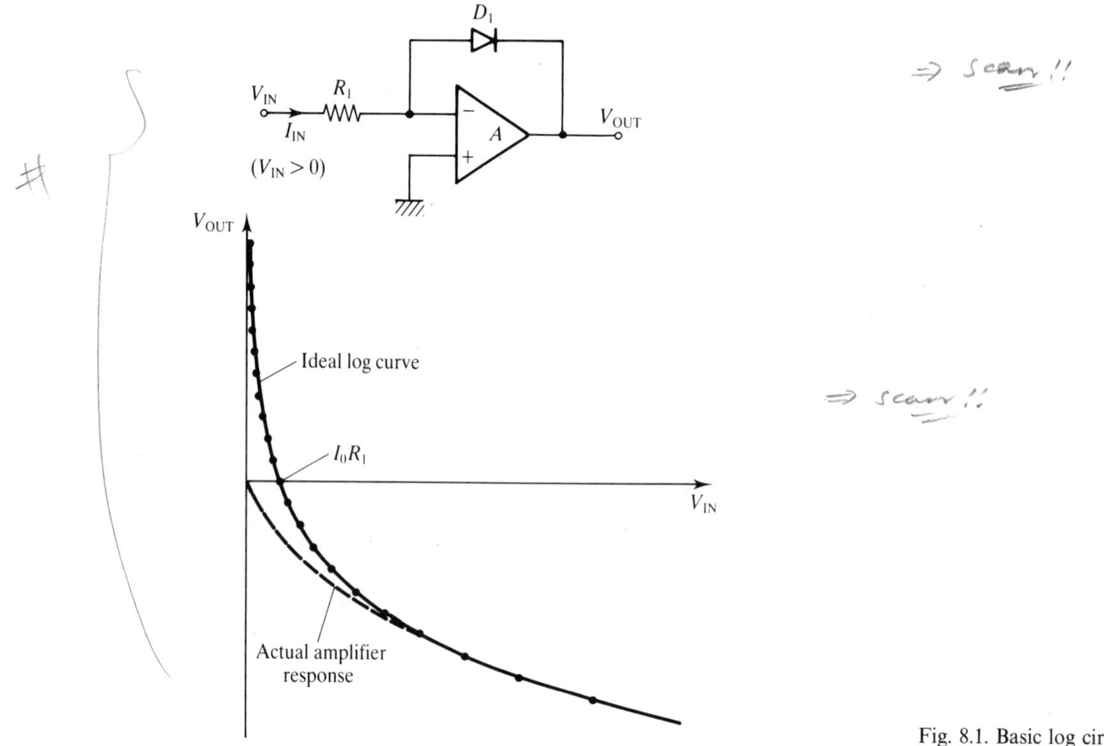

Fig. 8.1. Basic log circuit.

To operate as a log converter, $V_{IN}/R_1 \gg I_0$

so,

$$V_{OUT} = -\frac{kT}{q}\log_e\left(\frac{V_{IN}}{R_1 I_0}\right) = -\frac{kT}{q}\log_e\left(\frac{I_{IN}}{I_0}\right)$$

Note that I_0 is typically 10^{-9} A for a silicon diode and $kT/q = 25$ mV at room temperature.

The gain of a log amp is usually specified in terms of volts output per decade change of input. A 3 decade logger, for example, would allow a signal input ranging from 1 mV to 1 V. A 7 decade logger, on the other hand, would allow signal inputs from 1 μV to 10 V. Note that by omitting R_1, you can convert the basic log amp into a current input log amp.

Effect of the diode bulk resistance

$$V_{OUT} = -\frac{kT}{q}\log_e\left(\frac{I_{IN}}{I_0}\right) - I_{IN}R_B$$

where R_B is typically around 10 Ω

Op amp offset error effects

for a voltage input $$V_{OUT} = -\frac{kT}{q}\log_e\left(\frac{\frac{V_{IN}-V_{IO}}{R_1}-I_B}{I_0}\right)+V_{IO}$$

for a current input $$V_{OUT} = -\frac{kT}{q}\log_e\left(\frac{I_{IN}-I_B}{I_0}\right)+V_{IO}$$

where V_{IO} is the op amp input offset voltage and I_B is the input bias current of the op amp

#6 (The simple log amplifier is rarely used since it has two severe limitations. Firstly, its response is very temperature sensitive (note the T and I_0 terms in the transfer equation). Secondly, diodes do not provide a good log conformity, which means that the relationship between their forward voltage and their current does not accurately follow a logarithmic form.) In most cases, it is very difficult to make log amplifiers from general purpose silicon diodes which allow more than 3 decades of signal range.

#7 (Transistors, however, provide a far better log conformity than diodes. Many general purpose transistors configured as diodes can provide good logging of up to seven decades of magnitude of input current. The bipolar transistor offers this better performance due to the fact that its conduction is based upon a majority carrier, either electrons or holes, whereas the diode conducts with both holes and electrons.) The relationship between the collector current and the base–emitter voltage, with the base–collector voltage at 0 V, is given in Fig. 8.2. I_S is the reverse collector

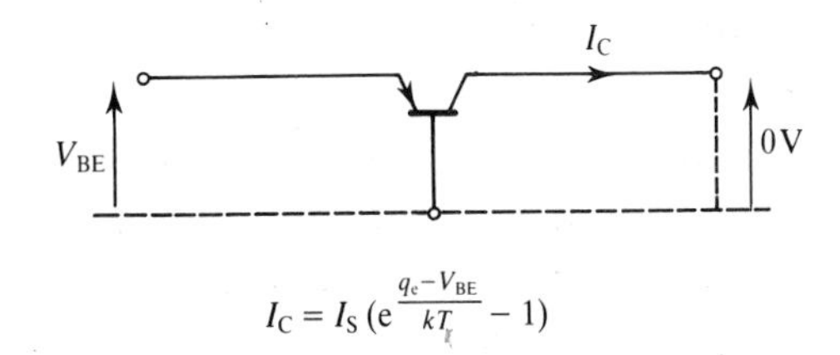

$$I_C = I_S\left(e^{\frac{q_e - V_{BE}}{kT}} - 1\right)$$

Fig. 8.2. Log relationship of a transistor.

saturation current of the transistor. I_S is typically around 0.1 pA for general purpose low power bipolar transistors and is temperature dependent. # #8 (Two popular configurations for replacing the diode, using a bipolar transistor, are termed the *transdiode* configuration and the *diode connected transistor* configuration. Both of these configurations are shown in Fig. 8.3.)

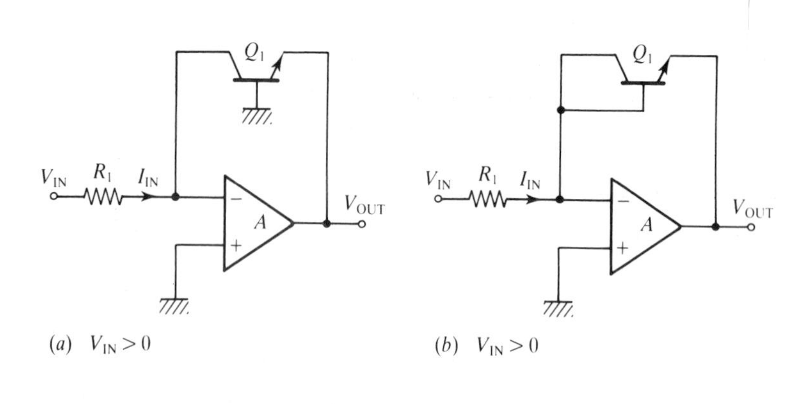

Fig. 8.3. Using bipolar transistors for logging purposes. (a) Transdiode configuration.
(b) Diode-connected transistor configuration.

$$I_{IN} \gg I_S \quad \text{and} \quad V_{OUT} = \frac{kT}{q} \log_e\left(\frac{V_{IN}}{R_1 I_S}\right) = -\frac{kT}{q \log_{10}(e)} \cdot \log_{10}\left(\frac{V_{IN}}{R_1 I_S}\right)$$

Note that with both configurations the base–collector voltage is almost at 0 V. With the diode connected transistor approach, the base and the collector are connected together and consequently both the base and the collector currents of the transistor flow into the virtual earth. With the transdiode approach, the collector is connected to the virtual earth of the op amp. Table 8.1 summarizes the properties of each configuration.

The output voltage of the two circuits shown in Fig. 8.3 ranges from 0 V to −0.7 V, i.e. one forward diode voltage drop. This output can be increased by adding two resistors, R_2 and R_3 as in Fig. 8.4. R_2 and R_3 must be chosen small enough so that the current flowing through the transistor has a negligible effect on the voltage divider action of R_2 and R_3 which controls the gain of the log amp.

You will notice that the log converter inverts the polarity of the input. To obtain a positive polarity output from negative polarity input signals, as shown in Fig. 8.4, simply connect the transistor the other way around for the diode connected transistor configuration; with the transdiode configuration a p–n–p transistor must replace the n–p–n, or alternatively add an inverter stage at the input.

Inputs of the wrong polarity can destroy the transistor since the op amp will saturate causing almost the full negative supply voltage to be placed across the reverse biased base–emitter junction (the base–emitter junction is usually sensitive to large reverse voltages). Protection diodes are commonly used, therefore, to prevent damage to the transistor. Three different arrangements are shown in Fig. 8.5 for the transdiode configuration. The salient points concerning each of these protection arrangements are given below.

Table 8.1. *Summary of two main logging approaches*

Transdiode	Diode connected transistor
Very wide and accurate logging range	The logging range is limited to 4 or 5 decades of input voltage since the base current is added to the collector current of the transistor
Possible frequency stability problems which can be avoided by adding an extra resistor and capacitor to the circuit	Usually a stable dynamic response
Slow response at low input currents due to the additional capacitor used for stability	Faster response
Transistor cannot be reversed, the emitter must be connected to the op amp output. For negative input voltages, a p–n–p transistor must be used, or an inverter stage	Polarity of the circuit is easily changed by reversing the transistor
	Use a transistor with a high β (h_{FE}) for a good log conformity

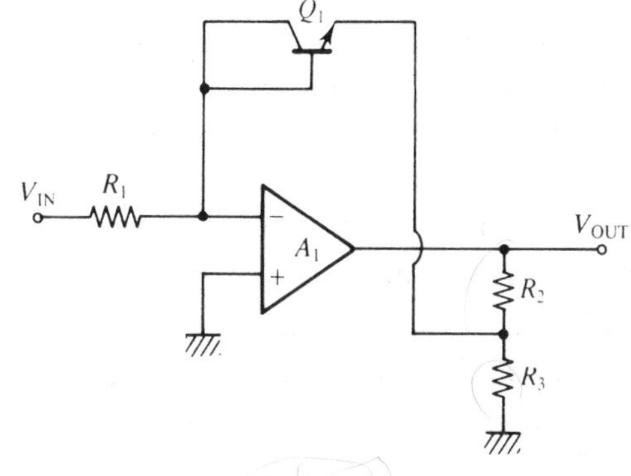

$$V_{OUT} = -\left(1 + \frac{R_2}{R_3}\right)\frac{kT}{q_e} \cdot \log_e\left(\frac{V_{IN}}{I_S R_1}\right)$$

$$= -59\,\text{mV}\left(1 + \frac{R_2}{R_3}\right)\log_{10}\left(\frac{V_{IN}}{I_S R_1}\right)$$

Fig. 8.4. Increasing the output voltage range.

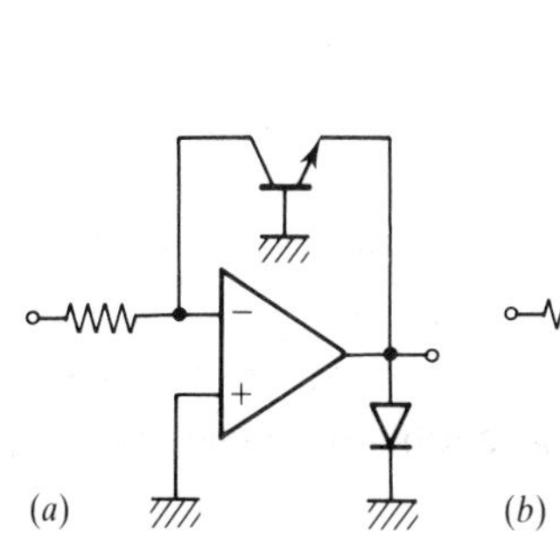

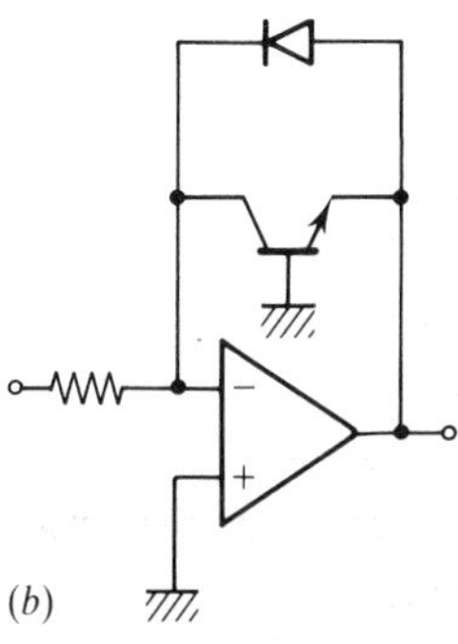

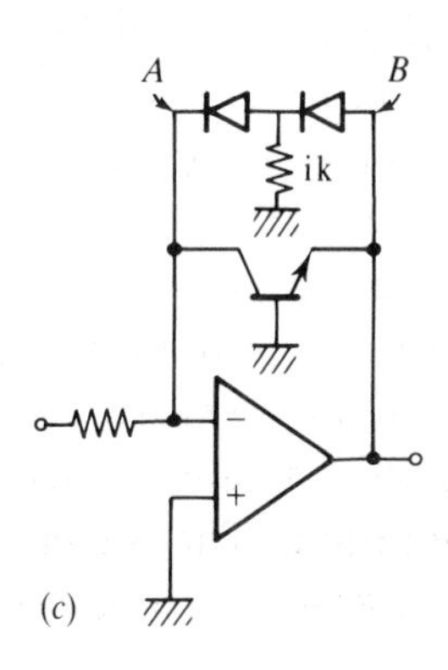

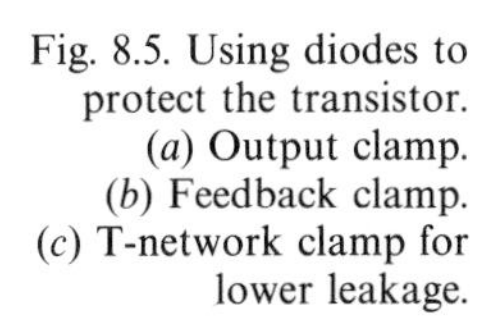

Fig. 8.5. Using diodes to protect the transistor. (*a*) Output clamp. (*b*) Feedback clamp. (*c*) T-network clamp for lower leakage.

(i) Clamping the op amp output. The op amp must have an output current limit. This approach has the advantage that the diode will not cause an error during normal operation, however, the op amp must recover from an output limit condition after a reverse input transient has occurred and so may recover slowly. Also, the op amp will heat up during a reverse input transient which may result in increased offset drift.

(ii) Feedback clamp. The clamping diode in this case introduces additional error due to the flow of reverse leakage current during normal operation.

(iii) Clamping the transistor with a T-network. This provides very low leakage current during normal operation, without imposing a hard limit on the output of the op amp, and a fast recovery from reverse transients.

From the transistor equation, the lower limit of currents that the transistor can log is I_S ($\simeq 0.1$ pA). However, with practical log converters, the lowest input which the converter can log is usually determined by the input offset voltage and the input bias currents of the op amp. You should use FET input op amps for small input systems due to their very low bias currents. To allow small input signals to be logged, and so achieve the required wide logging range, the input offset voltage and the bias currents can be nulled. With very accurate log converters, the input offset voltage and the bias currents can be nulled independently.

The largest input current that the transistor can log is limited by the error due to the bulk resistance R_{BULK} of the transistor. This bulk resistance is composed mainly of the semiconductor resistance and the resistance of the terminals connected to the semiconductor with their connecting leads. For a small signal transistor, the bulk resistance is typically around 10 Ω and the errors from R_{BULK} can limit the input currents to less than 0.1 mA. However, you can compensate for R_{BULK} as shown in Fig. 8.6. R_2 and R_3 must be chosen so that $R_2/R_3 = R_{BULK}/R_1$. Normally, R_2 is small (10 Ω–100 Ω) with R_3 very much greater in value.

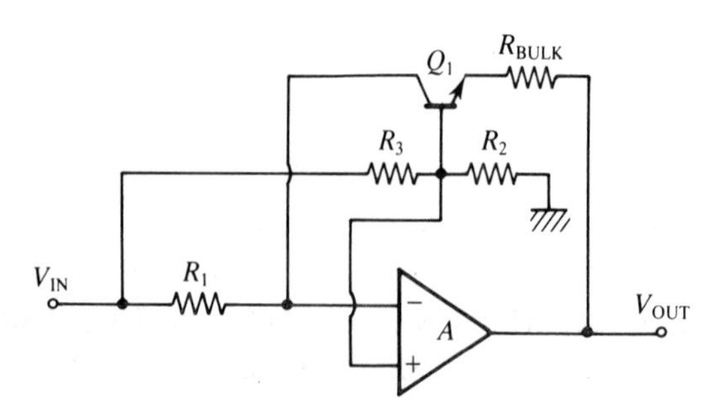

Fig. 8.6. Compensating for bulk resistance.

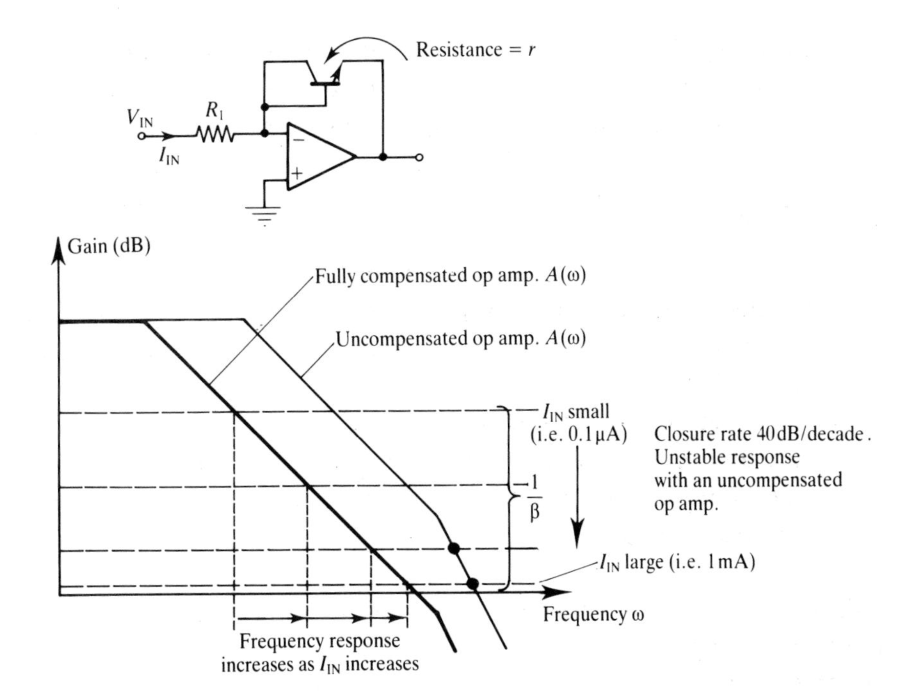

Fig. 8.7. Diode connected transistor Bode plot.

The frequency response of the diode connected transistor configuration is shown in Fig. 8.7 in which $A(j\omega)$ is the open loop gain of the op amp and $1/\beta$ is the feedback factor where

$$1/\beta = 1 + r_e/R_1$$

r_e being the small signal dynamic collector–emitter resistance of the transistor.

Note: $$r_e = kT/q_e I_{IN}$$

$$= 25\ \text{mV}/I_{IN} \text{ at room temperature}$$

You will notice that as I_{IN} increases, r_e decreases, as shown below.

I_{IN}	r_e
100 nA	250 kΩ
1 μA	25 kΩ
10 μA	2–5 kΩ
100 μA	250 Ω
1 mA	25 Ω

As the input current reduces, the bandwidth of the log converter will also reduce. This decreasing frequency response as the input signal decreases is a property which is common to most log amplifier designs. You may also remember from the fundamentals of frequency response analysis that if $1/\beta$ and $A(\omega)$ cross at 40 dB/decade or greater, then the feedback loop will be unstable.

In general, the diode connected transistor configuration will have a stable frequency response providing that a fully compensated op amp is used. Where the op amp is not fully compensated, the log converter may be prone to ringing or it may become unstable at higher frequencies (i.e. approaching the unity gain crossover frequency of the op amp as shown in Fig. 8.7) when there are high input currents.

The transdiode configuration can present severe frequency stability problems unless it is compensated. The frequency instability occurs since this configuration introduces additional gain and so the overall gain around the feedback loop is also increased. One of the most common causes of instability is additional phase shift introduced by C_1, the input capacitance and the collector–base capacitance of the transistor, as shown in Fig. 8.8.

From Fig. 8.8 the feedback factor $1/\beta$ is given by

$$\beta = V_1/V_{\mathrm{OUT}}$$

so

$$1/\beta = V_{\mathrm{OUT}}/V_1 = (r_e/R_1)//C_1 = r_e/R_1 \cdot (1 + j\omega R_1 C_1)$$

where

$$r_e = kT/q_e I_{\mathrm{IN}}$$

$$= 25\ \mathrm{mV}/I_{\mathrm{IN}} \text{ at room temperature}$$

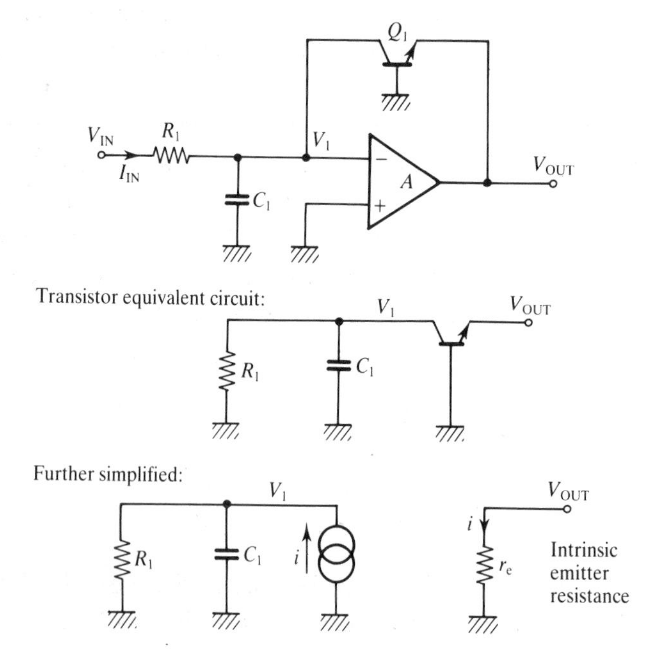

Fig. 8.8. Input capacitance of the transdiode configuration.

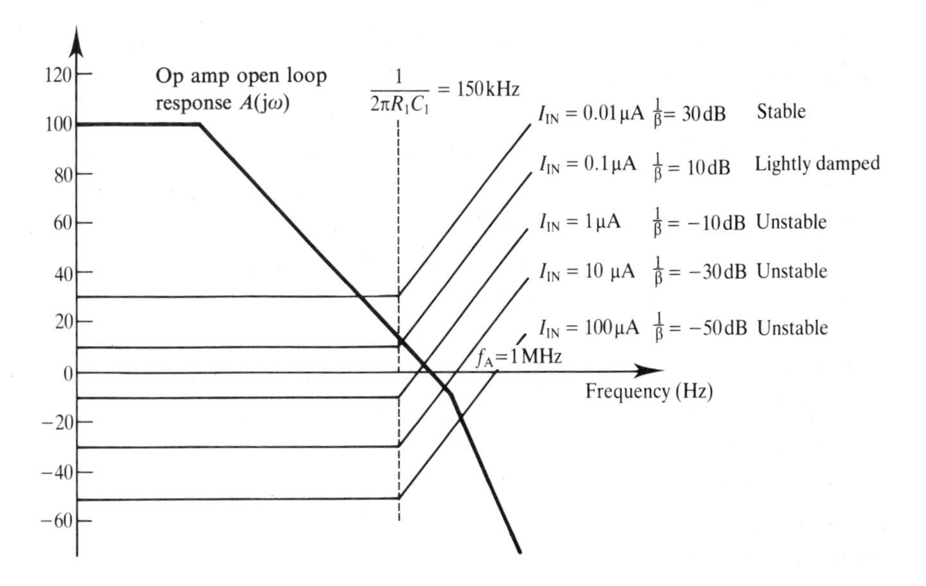

Fig. 8.9. Transdiode Bode plot.

A Bode plot for the transdiode configuration is shown in Fig. 8.9 which is based on some typical values. In this case, the op amp is fully compensated with a unity gain frequency of 1 MHz, $R_1 = 50\ k\Omega$, $C_1 = 20$ pF and $V_{IN} = 0.5$ mV to 5 V (i.e. I_{IN} from 10 nA to 0.1 mA). This Bode plot illustrates the instability of the transdiode configuration where $1/\beta$ and $A(j\omega)$ cross with closure rates greater than 40 db/decade for $I_{IN} > 100$ nA. You can eliminate this instability by adding a capacitor C_2 between the inverting input and output of the op amp which will introduce a pole into the $1/\beta$ response and limit the bandwidth of the log amp to a frequency of $1/(2\pi r_e C_2)$. However, you often need a large value for C_2 since r_e can be quite small at the higher input currents ($C_2 \simeq 1$ nF for a stable response). This large value severely restricts the slew rate and the bandwidth at low input currents. The slew rate is limited since C_2 must be charged up with the input current I_{IN} and the slew rate at $I_{IN} = 0.01\ \mu A = I_{IN}/C = 0.01\ \mu A/1\ nF \simeq 10$ V/second which is extremely slow! The bandwidth at $I_{IN} = 0.01\ \mu A = 1/(2\pi r_e C_2) = 64$ Hz. So, to avoid the use of such a large value capacitor, the effective value of r_e can be increased by adding an extra resistor R_2 as shown in Fig. 8.10. The effective value of r_e can be changed to $(R_2 + r_e)$ in the above expressions.

$1/\beta(j\omega)$ is now given by

$$\frac{R_2 + r_e}{R_1} \cdot \frac{(1 + j\omega(C_1 + C_2)R_1)}{(1 + j\omega(R_2 + r_e)C_2)}$$

The effects of using R_2 and C_2 can be summarized on the Bode plot shown in Fig. 8.11. R_2 must be selected so that it is as large as possible

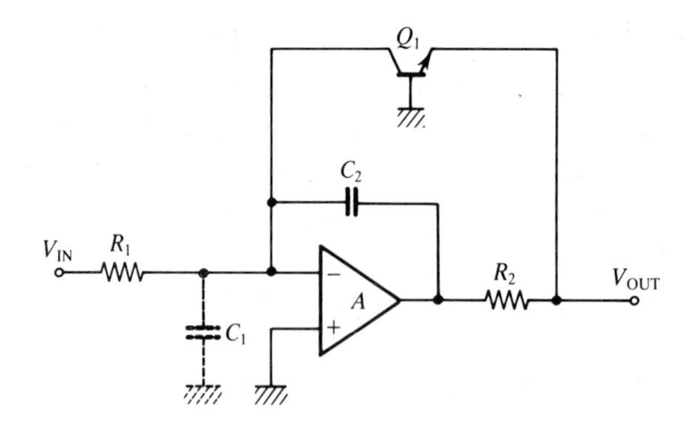

Fig. 8.10. A frequency compensated log converter.

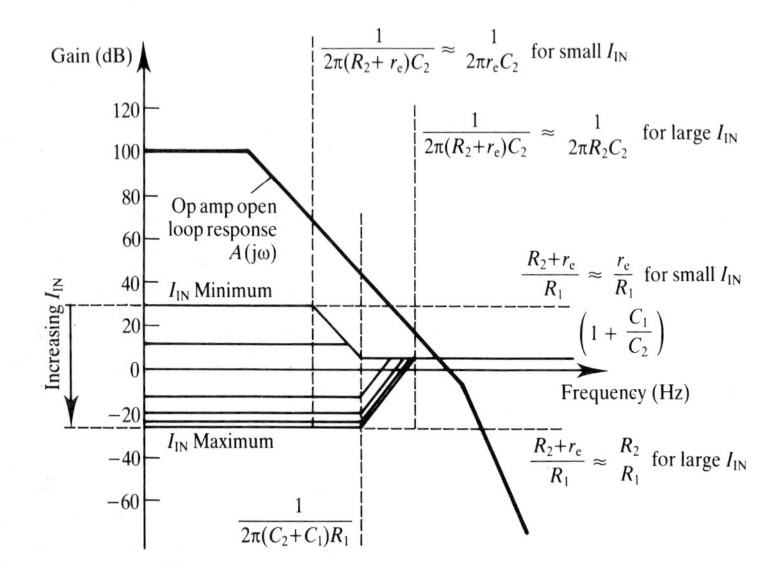

Fig. 8.11. Effects of R_2 and C_2 on frequency response.

without the output saturating for maximum value of input current. The maximum value of R_2 is given by

$$\frac{(V_{SAT} - V_{OUTMAX})}{(I_{INMAX} + I_{LMAX})}$$

where

V_{SAT} is the saturation voltage of the op amp
V_{OUTMAX} is the maximum output of the log converter circuit
I_{INMAX} is the maximum input current
I_{LMAX} is the maximum load current

In this case, consider the op amp supplied with ± 15 V supplies. A typical op amp will saturate, say, around ± 12 V. V_{OUTMAX} is one diode drop (i.e. 0.7 V) and there is an external load which draws a maximum current of 1 mA. The maximum input current is 100 μA. So, R_2 must be chosen to be less than a maximum $\simeq 11\ \text{k}\Omega$. With R_2 chosen at approximately half this value, i.e. 4.7 kΩ, there is a good safety margin and also a reduction in the maximum op amp output to 6 V.

For stability at the maximum input current, C_2 must introduce a pole safely less than the op amp zero crossover frequency f_A (=1 MHz) so choose 500 kHz. Then, $1/(2\pi R_2 + r_e) = 500$ kHz (at $I_{IN} = 0.1$ mA and $r_e = 250\ \Omega$). Which gives a minimum value for C_2 of 64 pF so choose 100 pF.

The bandwidth at minimum input current

$$\simeq 1/(2\pi(R_2 + r_e)C_2) = 640 \text{ Hz} \quad (r_e = 2.5\ \text{M}\Omega \text{ at } 0.01\ \mu\text{A})$$

To summarize the procedure for evaluating R_2 and C_2.

(i) Determine a value of R_2 consistent with non-saturation of the op amp over the entire operating conditions.
(ii) Determine C_2 so that $1/((2\pi R_2 + r_e)C_2)$ is safely less than the unity gain crossover frequency of the op amp. If the log converter is unstable with maximum input signal, then increase C_2.
(iii) Check that the bandwidth of the log converter is high enough with the smallest input signal. If not, use a fast op amp and recalculate C_2 (with the value being made smaller).

One technique which allows a large effective value of R_2 at high input currents without saturating the op amp is shown in Fig. 8.12. The R_3's and D_2 arrangement presents a large slope resistance for high input signals but only a small slope resistance for small input signals. Thus the effective value of R_2 is only large at high input signals and so is without the risk of saturating the op amp, which would happen if a large value resistor was chosen for R_2. Thus a much smaller value capacitor

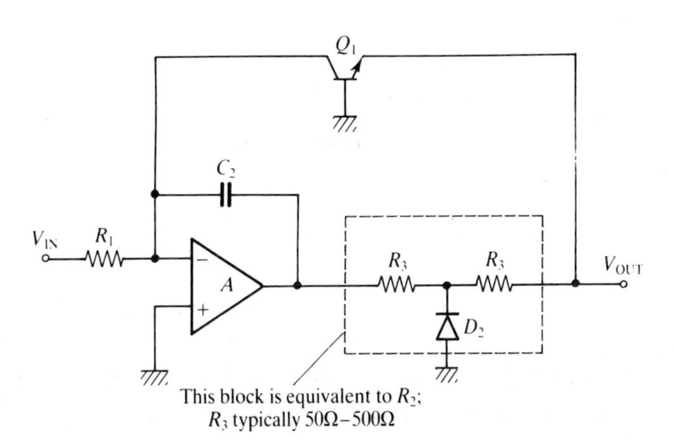

Fig. 8.12. Using an active compensation network.

may be used for C_2 so giving a faster response. R_3 is typically between 50 Ω and 500 Ω. You would obtain the best results if D_2 were replaced with a diode connected transistor which is matched (i.e. one from a transistor array on the same package) with the logging transistor.

As described earlier, as the input signal reduces, the bandwidth of the log converter also reduces. Consequently, the log converter will respond slowly to step responses in small inputs and quickly to step responses in large inputs. With the previous circuit values, for a small step in inputs of $\pm 5\%$:

V_{IN}	Output response (time to 95% of final value)
0.5 mV	750 μs (limited by the compensating network for small inputs $\simeq 3C_2(R_2 + r_e)$
7.0 V	1 μs (limited by the op amp with large inputs, output may ring slightly)

In addition, the step response of the log converter will be non-linear. It will have a faster response for increasing signals than decreasing signals (i.e. a faster rise time and a slower fall time). This is because as the signal increases in size its time constant gets smaller and as the signal size decreases its time constant gets longer.

The frequency response of a log converter can be tested by adding a sine wave to a dc level and then applying this composite signal to the input of the log converter. The pk–pk value ac component should not be more than 5% of the dc level. In this way, the frequency response of the converter can be measured over the entire range of dc inputs. Similarly, the step response at various dc level inputs can be measured by using a small amplitude square wave added to a dc level.

The main problem with the simple configurations shown previously is temperature variations in the I_S and kT/q terms. The following practical circuits use the transistor op amp configuration as the basic logging circuit. Variations in I_S are eliminated by using matched transistors. Temperature changes in the kT/q term may be eliminated by using a temperature sensitive resistor. To keep the circuits simple, circuit additions such as protection diodes, offset voltage and bias current nulls and bias current compensation have not been included. These refinements can be simply added onto the practical circuit as required.

By adding an extra op amp and a transistor such that Q_1 and Q_2 are matched, the I_S's can be cancelled. The circuit shown in Fig. 8.13 uses this approach and is found in some commercial log converters.

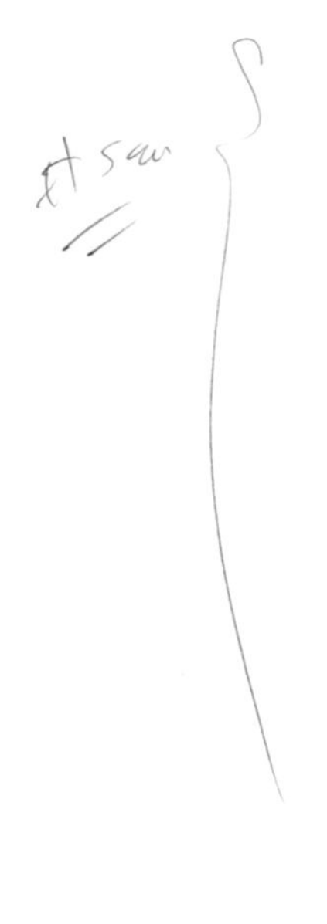

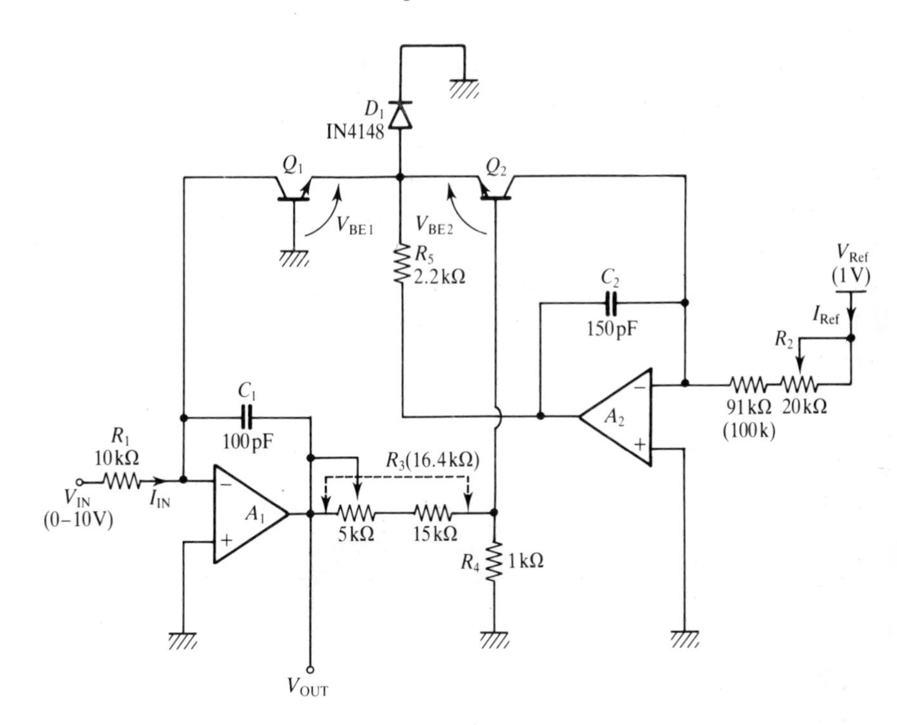

Fig. 8.13. Practical log converter circuit (1).

A_1, A_2 are fully compensated FET input op amps
Q_1, Q_2 are a matched pair (e.g. LM 394).

This circuit operates as follows:

$$\frac{R_4}{R_3 + R_4} V_{OUT} = V_{BE1} - V_{BE2}$$

Now,

$$V_{BE1} = \frac{kT}{q} \log_e\left(\frac{I_{IN}}{I_{S1}}\right) \qquad V_{BE2} = \frac{kT}{q} \log_e\left(\frac{I_{Ref}}{I_{S2}}\right)$$

so

$$\frac{R_4}{R_3 + R_4} V_{OUT} = -\frac{kT}{q} \log_e\left(\frac{I_{IN}}{I_{S1}} \cdot \frac{I_{S2}}{I_{Ref}}\right)$$

but Q_1 and Q_2 are matched, so that $I_{S1} = I_{S2}$. Then

$$V_{OUT} = -\left(1 + \frac{R_3}{R_4}\right) \frac{kT}{q} \cdot \log_e\left(\frac{I_{IN}}{I_{Ref}}\right)$$

$$V_{OUT} = -\left(1 + \frac{R_3}{R_4}\right) \frac{kT}{q \cdot \log_{10}(e)} \cdot \log_{10}\left(\frac{V_{IN} \cdot R_2}{V_{Ref} R_1}\right)$$

In the circuit, R_2 trims the reference current (I_{Ref}) and R_3 trims the gain. For the values shown, $V_{OUT} = 0$ when $V_{IN} = 100$ mV and the gain is 1 V/decade.

Note that the output of this circuit is negative if $I_{IN} > I_{Ref}$ zero if $I_{IN} = I_{Ref}$ and positive if $I_{IN} < I_{Ref}$. Remember that I_{IN} and I_{Ref} *must* be positive.

The logging accuracy for small input signals is limited by the input offset and bias current of A_1. The largest input signal is limited by the bulk resistance of transistor Q_1. Refer to the earlier notes on how these limits can be extended. Log conformity is limited mainly by transistor Q_2, since the output voltage varies its base–collector voltage. This causes V_{BE2} to vary slightly (normally less than 1%). Log conformity can be improved by connecting the non-inverting input of A_2 to the base of Q_2 instead of ground. A current source must then be used to supply I_{Ref}, instead of simply R_2.

C_2 and R_5 are used to frequency compensate Q_2 and A_2 as described previously. With op amp A_1 it is more difficult to calculate the optimum compensation values since the feedback loop includes Q_1, Q_2 and A_2. A_2 is compensated using C_1, R_3 and R_4. The usual procedure is to calculate the value of C_1 as described earlier, then increase C_1 to compensate for the effects of Q_2 and A_2. The optimum value for C_1 may be determined practically on a trial and error basis.

The gain of the log converter is given by

$$\frac{R_3 + R_4}{R_4} \cdot \frac{kT}{q}$$

which has a temperature coefficient due to its linearity dependency on T of 3300 ppm/°C at room temperature. Often, R_4 is made temperature sensitive with a change of +3300 ppm/°C to cancel the effect from T. Also, $R_4 > R_3$ (usually around 10 times bigger) to provide a good output swing of ±several volts. R_3 can be used to vary the gain. If R_4 is a temperature sensing resistor, it should be placed alongside Q_1 and Q_2. Temperature compensation in this manner can improve gain drift by an order of magnitude.

Another configuration similar to the previous one is shown in Fig. 8.14. Once again, this circuit is found in commercial log converters. In this circuit, Q_1 does the logging and Q_2 is used as a reference. Q_2 is diode connected as this is quite adequate since it only provides a reference. Once again, R_4 can be made temperature sensitive. The component values shown give this log converter a similar gain to the circuit in Fig. 8.13.

If the log ratio of two signals is required (not just a signal and a reference) then the two previous circuits are unsuitable. Note though, that their reference input could be used as a signal input: this approach does not give a good log conformity. For logging the ratio of two input

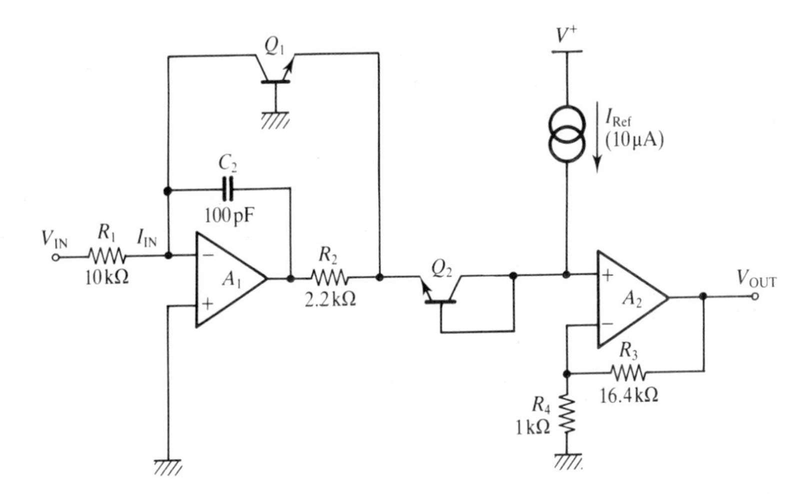

Fig. 8.14. Practical log converter circuit (2).

$$V_{OUT} = -\left(1 + \frac{R_3}{R_4}\right)\frac{kT}{q}\cdot\log_e\left(\frac{I_{IN}}{I_{Ref}}\right) \quad \text{if } Q_1 \text{ and } Q_2 \text{ are matched.}$$

signals, the circuit shown in Fig. 8.15 could be used. This circuit is basically two simple logging converters built around A_1 and A_2 feeding their outputs into the input terminals of op amp A_3 configured as a differential amplifier (see Chapter 1 for further details on differential amplifiers). The differential amplifier subtracts one log from the other. Note that the resistors around A_3 must be very closely matched. An extra op amp A_4 is shown with either R_5 or R_6 using temperature-sensitive resistors to compensate for any temperature drift in the log converter gain.

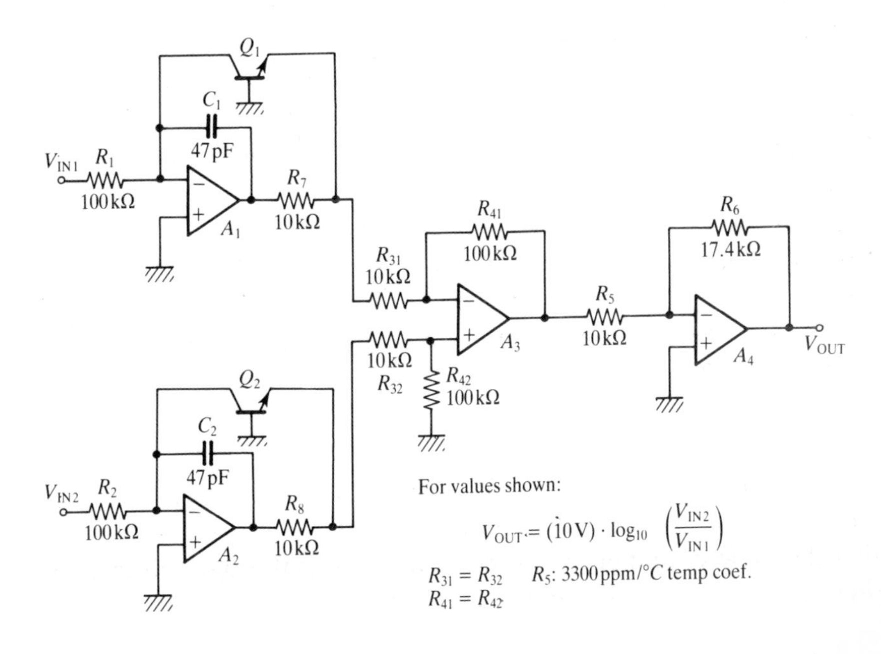

Fig. 8.15. A log ratio amplifier.

The response of a log amp is typically of the form

$$V_{OUT} = -V_K \log_{10}\left(\frac{V_{IN} + V_{IO}}{V_{Ref}}\right)$$

The offset cannot be adjusted for zero output, as with a normal amplifier, since the log of zero equals minus infinity. So, with log converters a slightly more complex nulling procedure must be followed as detailed below.

(i) set V_{Ref} to a very small value
(ii) short the input to ground (i.e. $V_{IN} = 0$ V) so that $V_{OUT} = -V_K \cdot \log_{10}(V_{IO}/V_{Ref})$
(iii) trim V_{IO} until V_{OUT} is a large enough positive output

For example, if $V_K = 1$ V (therefore output changes 1 V/decade change at the input). Setting $V_{Ref} = 0.1$ V

then, if $V_{OUT} = -2$ V then $V_{IO} = 1$ mV

if $V_{OUT} = -3$ V then $V_{IO} = 100\ \mu$V

and if $V_{OUT} = -4$ V then $V_{IO} = 10\ \mu$V.

Consequently, nulling V_{IO} until V_{OUT} is greater than -4 V ensures that $V_{IO} < 10\ \mu$V. If V_{Ref} is also a signal input, any offsets in V_{Ref} may also be nulled using a similar technique.

The most suitable reference voltage V_{Ref} is the geometric mean of the smallest signals $V_{IN\,MIN}$ and the largest signals $V_{IN\,MAX}$ so that $V_{Ref} = (V_{IN\,MIN} \cdot V_{IN\,MAX})^{1/2}$. Using this value will allow equal positive and negative output swings. If this value is not used, the output swing will not be symmetrical about zero.

To set the reference voltage, set $V_{IN} = V_{Ref}$ (i.e. the desired reference voltage) and then tune V_{Ref} until $V_{OUT} = 0$ V. In this case,

$$V_{OUT} = V_K \log_{10}\left(\frac{V_{IN}}{V_{Ref}}\right) = 0\text{ V} \qquad \text{when } V_{IN} = V_{Ref}.$$

Notes on component selection

The transistors A super matched transistor pair such as the LM304 or MAT-01 for example can be used for discrete designs.

The op amp The op amp limits the smallest input signal that can be used due to its input offset voltage and input bias current. For a current input, the input error is the input bias current of the op amp and so a FET input op amp should be chosen. For voltage inputs, the input error is given by $V_{IO} + I_B R_1$. If V_{IO} is nulled, then the bias offset current may cause a significant error and FET input op amps should, once again, be chosen. For very low V_{IO}, try a CMOS input chopper stabilized op amp. If high frequency signals are to be logged, a fast op amp will need to be used.

The resistors Temperature sensitive resistors can be obtained with temperature coefficients of 3000 ppm to provide temperature compensation in the kT/q term.

8.2 Antilog converters

This basic circuit (Fig. 8.16) is very similar to the elementary log converter in Fig. 8.1 except that the diode and the resistor are swopped

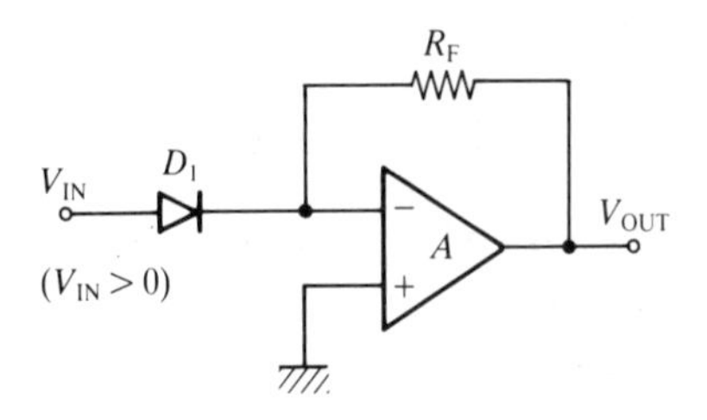

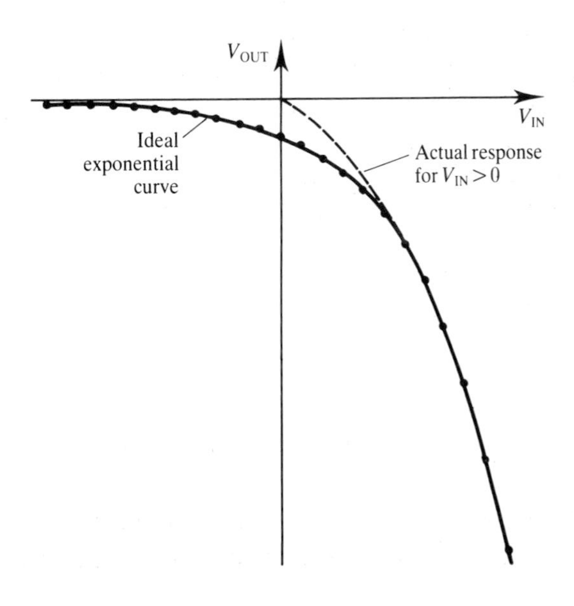

Fig. 8.16. Basic antilog circuit.

around. The antilog function is achieved using the voltage and current log relationship of the p–n junction as described previously for the log converter. Once again, this circuit is very crude and would rarely be used in practice due to its many limitations; it does provide a starting point, though, from which to develop a practical antilog converter. The design of antilog converters is very similar to log converter design so that most of the previous design notes for log converters are relevant and should be consulted by a reader unfamiliar with the design of log or antilog converters.

Transfer equation

$$I_{IN} = I_O(e^{\frac{qV_{IN}}{kT}} - 1)$$

$$I_{IN} = -\frac{V_{OUT}}{R_F}$$

$$\therefore\ V_{OUT} = -I_O R_F(e^{\frac{qV_{IN}}{kT}} - 1) = I_O R_F e^{\frac{qV_{IN}}{kT}} \quad (\text{as } e^{\frac{qV_{IN}}{kT}} \gg 1)$$

Offsets

$$V_{OUT} = -I_O R_F(e^{\frac{q}{kT}(V_{IN} - V_{IO})}) - I_B R_F + V_{IO}$$

where I_B is the input bias current and V_{IO} is the input offset voltage.

The main problems with this simple circuit are that I_0 and kT/q are temperature sensitive and that the diode does not provide a very accurate logging function. It is usual to replace the diode with a transistor which does provide more accurate antilogging (to within 0.5% of the ideal case) over a much wider range, up to seven decades in some cases. The transistor is usually connected as shown in Fig. 8.17.

$$V_{OUT} = I_S R_F\, e^{(\frac{q}{kT} \cdot V_{IN})}$$

where I_S is the reverse collector saturation current ($\simeq$0.1 pA) and is a temperature-dependent parameter.

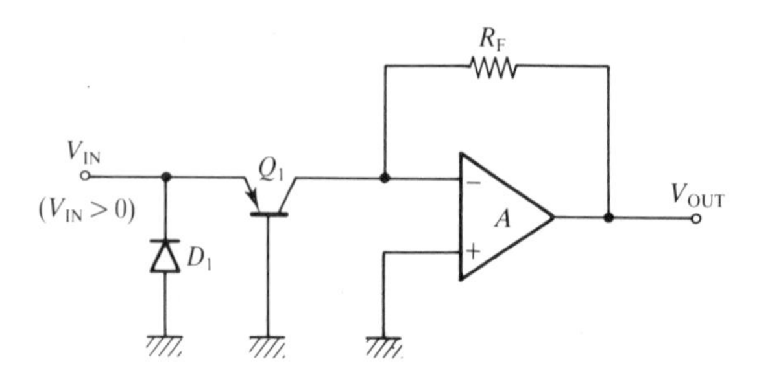

Fig. 8.17. Basic transdiode antilog converter.

The transdiode configuration shown in Fig. 8.17 is generally used in preference to the diode connected transistor arrangement for antilog converters because closer exponential conformity can be obtained. Remember that transistors are particularly vulnerable to large reverse base–emitter voltages and so a protection diode has been added at the input in Fig. 8.17 to clamp negative input signals. The input voltage must also be less than approximately 1 V otherwise a large current would flow and possibly damage the transistor. Of course, a voltage divider could be added at the input to increase the voltage range. However, the resistor values may have to be prohibitively small to prevent a loading error from the transistor. To convert negative input voltages, an n–p–n transistor can be used in Fig. 8.17 instead of a p–n–p.

The smallest input voltage that can be converted with any accuracy is limited by the transistor equation:

$$I = I_S e^{\left(\frac{q}{kT} \cdot V_{IN}\right)} - 1)$$

where $V_{IN} \gg kT/q$ (=25 mV at room temperature) and the smallest input V_{IN} must be greater than 25 mV.

The input offset voltage and input bias current of the op amp are not as critical as with the log converter since the gain for small signals at the input is very small with the antilog circuit (the logging circuit described previously had a very high gain for small input signals and so keeping the errors of the op amp small was then important). The maximum input signal is limited by the errors caused by the bulk resistance of the transistor. This error can be reduced by up to an order of magnitude by the circuit shown in Fig. 8.18.

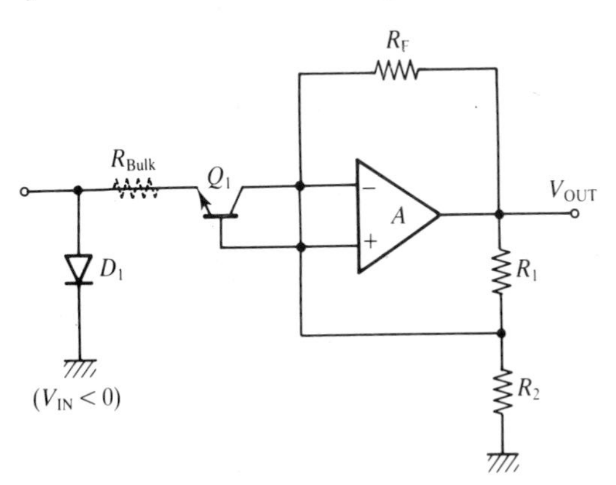

Fig. 8.18. Reducing transistor bulk resistance errors.

$R_{BULK}/R_F = R_2/R_1$ where, usually, $R_2 \ll R_1$.

$$V_{OUT} = -\left(1 + \frac{R_2}{R_1}\right) I_S R_F \cdot e^{\left(\frac{q}{kT} \cdot V_{IN}\right)}$$

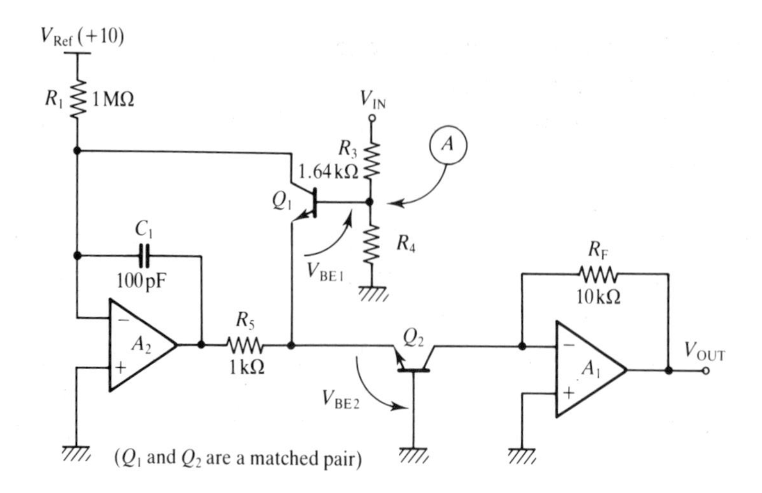

Fig. 8.19. A practical anti-log converter.

The main problem with the basic antilog circuit is the temperature variations in the kT/q and I_S terms. The practical circuit in Fig. 8.19 uses the previous basic circuit as the foundation for the antilogging function. Temperature variations in I_S are eliminated by using matched transistors. Temperature variations in the kT/q term are reduced by using temperature sensitive resistors. This circuit is somewhat simplified in that some of the additions, such as use of protection diodes, are omitted.

R_3 and R_4 must be chosen so that the base current of Q_1 can be neglected, i.e.

$$\frac{R_4 \cdot V_{IN}}{R_3 + R_4} = V_{BE1} - V_{BE2}$$

But

$$V_{BE1} = \frac{kT}{q} \log_e\left(\frac{I_{Ref}}{I_{S1}}\right) \quad \text{and} \quad V_{BE2} = \frac{kT}{q} \log_e\left(\frac{I_{OUT}}{I_{S2}}\right)$$

so,

$$\frac{R_4}{R_3 + R_4} \cdot V_{IN} = \frac{kT}{q} \log_e\left(\frac{I_{Ref} \cdot I_{S2}}{I_{S1} \cdot I_{OUT}}\right) = \frac{kT}{q} \log_e\left(\frac{V_{Ref} \cdot R_F \cdot I_{S2}}{V_{OUT} \cdot R_1 \cdot I_{S1}}\right)$$

Since Q_1 and Q_2 are supermatched transistor pairs, $I_{S1} = I_{S2}$

so,

$$V_{OUT} = \frac{R_F}{R_1} \cdot V_{Ref} \cdot e^{-\left[\frac{q}{kT} \cdot \frac{R_4}{R_2 + R_4} \cdot V_{IN}\right]}$$

or,

$$V_{OUT} = \frac{R_F}{R_1} \cdot V_{Ref} \cdot 10^{-\left[\frac{q}{kT} \frac{R_4 \log_{10} e}{(R_3 + R_4)}\right]}$$

Note that Q_1 and A_1 provide a logged reference voltage, V_{BE1} (see previous design notes on log converters).

For this circuit to operate adequately the following conditions should be observed.

V_{Ref} must be positive, otherwise the voltage at point A will not be sufficiently positive to forward bias the base–collector junction of Q_1 and cause Q_1 to cut-off.

The maximum input should also be less than $(1 + R_3/R_4)V_{BE1}$, otherwise the base-emitter voltage of Q_2 (i.e. V_{BE2}) will not be sufficiently greater than kT/q_e and Q_2 will not antilog accurately.

Obtaining the ideal antilog function is limited mainly because the input signal varies the base–collector voltage of transistor Q_1. This causes the logged reference voltage V_{BE1} to vary with the input signal. Better accuracy can be obtained by connecting the non-inverting input of A_2 to point A in Fig. 8.19 instead of ground and by using a current source to supply I_{Ref} instead of resistor R_1.

The temperature term T may be cancelled by making R_4 temperature sensitive with a temperature coefficient equal to +3300 ppm at room temperature.

As a final point, the trimming procedure for an antilog converter is described below, remembering that the equation for a typical antilog converter is of the form $V_{OUT} = -V_K \exp(V_{IN}/V_{Ref}) + V_{OO}$.

(i) To null the output offset V_{OO}, set V_{IN} to maximum negative input. Adjust V_{OO} until $V_{OUT} = 0$ V.

(ii) To adjust V_K, set V_{IN} equal to 0 V. Adjust the gain until V_{OUT} is the required value for V_K.

(iii) To adjust V_{Ref}, set V_{IN} to some nominal value and calculate the required V_{OUT} at this value. Then, adjust V_{Ref} to make V_{OUT} equal to this calculated value.

Notes on component selection

The transistors A supermatched transistor pair such as the LM394 or MAT-01 can be used for discrete designs.

The op amp The requirements are not as stringent as for the log converter and for most applications any general purpose op amp will

Table 8.2. *A selection of commercial log and antilog converters*

	SSM2100	759P	LOG 100	4127JG	ICL8048BC	ICL8049BC
Transfer						
Type	Log, Antilog, Log ratio	Log, Antilog, Log ratio	Log, Antilog, Log ratio	Log, Antilog, Log ratio	Log, Log ratio	Antilog
Function	$K \log_e \frac{I_1}{I_2}$	$K \log_{10} \frac{I_1}{I_2}$	$K \log_{10} \frac{I_1}{I_2}$	$K \log_{10} \frac{I_1}{I_2}$	$K \log_{10} \frac{I_1}{I_2}$	$V_{Ref} \cdot 10^{-\left(\frac{V_{IN}}{K}\right)}$
Scale factor (K)	–	2, 1, $\frac{2}{3}$ V/dec.	1, 3, 5 V/dec.	–	–	–
Input range						
Voltage	–	−1 mV to −10 V	–	1 mV to 10 V	1 mV to 10 V	10 mV to 10 V[C]
Current	10 nA to 1 mA	−1 nA to −1 mA	1 nA to 1 mA	1 nA to 1 mA	1 nA to 1 mA	–
Errors						
Log conformity	0.4%	–	0.15%	0.5%[B]	0.5%[B]	10 mV[B]
Scale factor error	–	1%[A]	0.3%	–	–	–
Input offset voltage	4 mV	2 mV[A]	0.7 mV	–	15 mV	–
Input offset drift	–	10 μV/°C	80 μV/°C	–	–	–
Bandwidth (small signal 3 dB)						
I_{IN} = 1 nA	–	250 Hz	110 Hz	–	100 Hz	–
I_{IN} = 10 nA	–	–	–	80 Hz	500 Hz	–
I_{IN} = 100 nA	–	–	–	250 Hz	2 kHz	–
I_{IN} = 1 μA	–	100 kHz	27 kHz	5 kHz	10 kHz	–
I_{IN} = 10 μA	–	200 kHz	38 kHz	50 kHz	40 kHz	–
I_{IN} = 100 μA	–	–	–	90 kHz	50 kHz	–
I_{IN} = 1 mA	–	200 kHz	45 kHz	–	60 kHz	–
Comments	Scale factor K set by two external resistors. Temperature stabilized with an internal heater. Contains a +5 V reference.	One of three scale factors set by a pin connection. Current input or voltage input. 759P used for positive inputs and 759P for negative.	Scale factor 1, 3, 5 selected by pin connection. Current input.	Scale factor and reference current set by external resistors. Contains a reference current source and an uncommitted op amp.	Monolithic log amp. Scale factor set by external resistors.	Monolithic antilog amp. Scale factor and reference set by external resistors.

A Max B Total error (max) C Output range

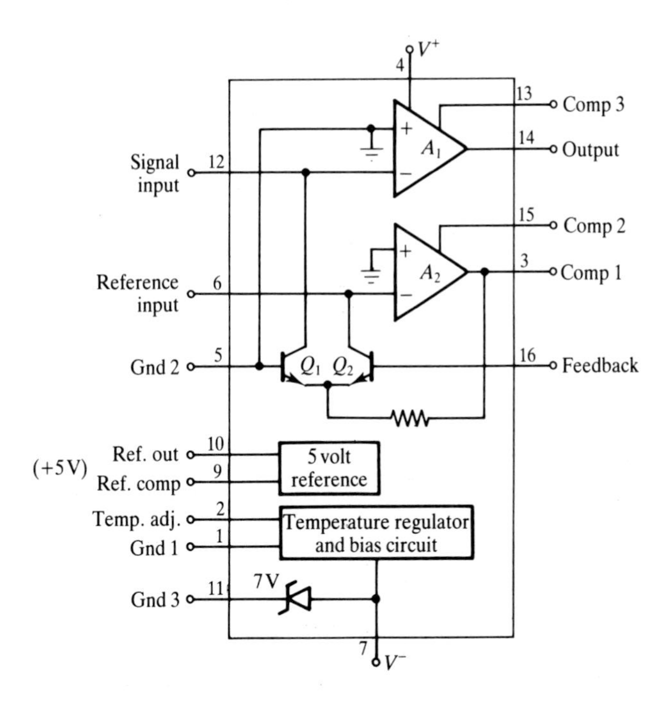

Fig. 8.20. SSM-2100 log converter.

be suitable if it has low enough offsets and bias currents and is fast enough.

The resistors Temperature sensitive resistors can be used to compensate for the kT/q term which have temperature coefficients of 3000 ppm.

8.3 Commercial log and antilog converters

There are several commercially available log and antilog converters, a selection of such devices is given in Table 8.2. Some units are dedicated log converters, others are dedicated antilog converters and some can be configured as either log or antilog converters. Fig. 8.20 shows the schematic diagram of a popular antilog and log converter ic, the SSM-2100. Table 8.2 compares and contrasts a selection of commercially available log and antilog converters.

9
Arithmetical operations

The arithmetical operations covered in this chapter have been divided into addition/subtraction and multiplication/division. The addition/subtraction operations can be realized using a single op amp and resistors in a straightforward manner. The multiplication/division operations are much more complicated. Generally, very accurate multiplication of analog signals is expensively difficult and should be avoided if at all possible.

9.1 Addition and subtraction

The circuit shown in Fig. 9.1 is, in effect, an extension of the single op amp differential amplifier (see Chapter 1 on instrumentation amplifiers) in which the op amp amplifies the difference in the voltages present on the inverting and non-inverting inputs. The gain, as usual, is determined by the ratio between the feedback resistor R_F and the input resistors.

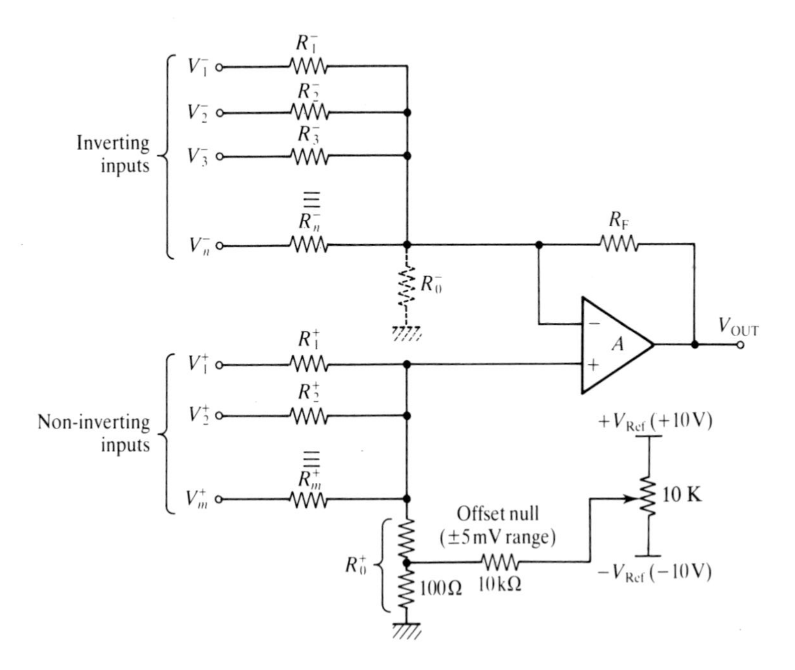

Fig. 9.1. Addition and subtraction circuit.

Output voltage

$$V_{OUT} = A_1^+V_1^+ + A_2^+V_2^+ + \cdots + A_m^+V_m^+ - A_1^-V_1^- - A_2^-V_2^- - \cdots - A_n^-V_n^-$$

where

$$A_1^- = \frac{R_F}{R_1^-}$$

$$A_2^- = \frac{R_F}{R_2^-}$$

$$\cdots \qquad \cdots$$

$$A_n^- = \frac{R_F}{R_n^-}$$

and

$$A_1^+ = \left(1 + \frac{R_F}{R_{Tot}^-}\right)\cdot\frac{R_{Tot}^+}{R_1^+}$$

$$A_2^+ = \left(1 + \frac{R_F}{R_{Tot}^-}\right)\cdot\frac{R_{Tot}^+}{R_2^+}$$

$$\cdots \qquad \cdots$$

$$A_m^+ = \left(1 + \frac{R_F}{R_{Tot}^-}\right)\cdot\frac{R_{Tot}^+}{R_m^+}$$

where

R_{Tot}^- and R_{Tot}^+ are given by:

$$R_{Tot}^- = R_0^-//R_1^-//R_2^-//\cdots//R_m^-$$

$$R_{Tot}^+ = R_0^+//R_1^+//R_2^+//\cdots//R_m^+$$

Offsets

$$\text{Output offset} = \left(1 + \frac{R_F}{R_{Tot}^-}\right)V_{IO} + I_B^-R_F - I_B^+R_{Tot}^+\left(1 + \frac{R_F}{R_{Tot}^-}\right)$$

$$= \left(1 + \frac{R_F}{R_{Tot}^-}\right)V_{IO} + I_{OS}R_F$$

if

$$\frac{1}{R_{Tot}^+} = \frac{1}{R_F} + \frac{1}{R_{Tot}^-}$$

(i.e. the resistors are matched to balance offsets)

where V_{IO} is the input offset voltage of the op amp, I_B^- and I_B^+ are the input bias currents of the inverting and non-inverting op amp inputs and I_{OS} is the input offset bias current of the op amp.

You should calculate resistor values such that the desired gains $A_1^+, A_2^+ \cdots A_m^+$ and $A_1^-, A_2^- \cdots A_m^-$ are the correct values and that the resistance seen at the inverting and non-inverting inputs are equal, so that the effects of the input bias currents of the op amp are minimized. The steps outlined below allow you to calculate resistor values which satisfy these requirements.

(a) Select a value for R_F, usually between 1 kΩ and 1 MΩ. If R_F is large, i.e. 1 MΩ, the input resistance will be relatively high and operating current relatively low. The bandwidth, however, may be reduced and larger offsets will occur due to the flow of bias currents, particularly if the op amp has a bipolar input. If R_F is low, around 1 kΩ, the response may be faster since the effects of stray capacitances are reduced, the input resistance may, though, be too low and operating current relatively high. The value of R_F, therefore, is something of a compromise depending upon the application. Typically, R_F is chosen between 10 kΩ and 100 kΩ.

(b) Calculate all the inverting input resistor values from the inverting input gains, i.e.

$$A_1^- = \frac{R_F}{R_1^-} \Rightarrow R_1^- = \frac{R_F}{A_1^-}$$

$$A_2^- = \frac{R_F}{R_2^-} \Rightarrow R_2^- = \frac{R_F}{A_2^-}$$

etc.

(c) Calculate all the non-inverting input resistor values from the non-inverting gains.

$$A_1^+ = \left(1 + \frac{R_F}{R_{Tot}^-}\right) \cdot \frac{R_{Tot}^+}{R_1^+} = \frac{R_F}{R_1^+}$$

if the resistors are matched for minimum offsets, so,

$$R_1^+ = \frac{R_F}{A_1^+}$$

$$R_2^+ = \frac{R_F}{A_2^+}$$

etc.

(d) Now that all the values of R_1^+ to R_m^+ and R_1^- to R_m^-

and R_F are known, R_0^+ to R_0^- need to be determined. It is usual to omit either one of these resistors, or both, depending upon the application.

Calculate

$$\frac{1}{R_1^-} + \frac{1}{R_2^-} + \frac{1}{R_3^-} + \cdots + \frac{1}{R_n^-} + \frac{1}{R_F} = \frac{1}{R^-}$$

$$\frac{1}{R_1^+} + \frac{1}{R_2^+} + \frac{1}{R_3^+} + \cdots + \frac{1}{R_n^+} = \frac{1}{R^+}$$

There are three possible situations at this stage:

(i) $R^- = R^+$ then omit both R_0^+ to R_0^-

(ii) $1/R^- > 1/R^+$ then omit R_0^- and determine R_0^+ from

$$\frac{1}{R_0^+} = \frac{1}{R^-} - \frac{1}{R^+}$$

(iii) $1/R^+ > 1/R^-$ then omit R_0^+ and calculate R_0^- from

$$\frac{1}{R_0^-} = \frac{1}{R^+} - \frac{1}{R^-}$$

The values for R_1^- to R_m^- and R_1^+ to R_m^+ must include source resistances. Ensure that these values are very much greater than their corresponding values of source impedance for good accuracy (by a factor of at least 100). Using resistors in the same package will decrease gain errors due to temperature variations since resistor value changes will tend to cancel. If large values are necessary for the resistors (i.e. above 10 kΩ–100 kΩ), then a FET input op amp should be used. Otherwise, a large output offset may result due to the flow of amplifier input bias currents.

If large resistor values are used for all resistors (above 100 kΩ) then frequency stability problems may arise, particularly when using a fast op amp. The circuit may be stabilized by adding a small value capacitor (of up to 100 pF, determined by trial and error) placed across R_0^+ and R_F. If the bandwidth is limited by C_F, then the 3 dB frequency will be given by $1/2\pi R_F C_F$ (note: the C_F due to strays is around 1 pF typically). If the bandwidth is limited by the op amp, the maximum frequency is given at the point where the $1/\beta$ crosses the op amp open loop frequency response. Note that β is the feedback factor and $1/\beta = 1 + R_F/R_{Tot}^-$ and for a fully compensated op amp this bandwidth is given by approximately βf_A, where f_A is the gain–bandwidth product of the op amp.

The maximum number of inputs to this circuit is usually limited by the circuit errors which can be tolerated. An increase in the number of inputs must generally be accompanied by a reduction in the feedback

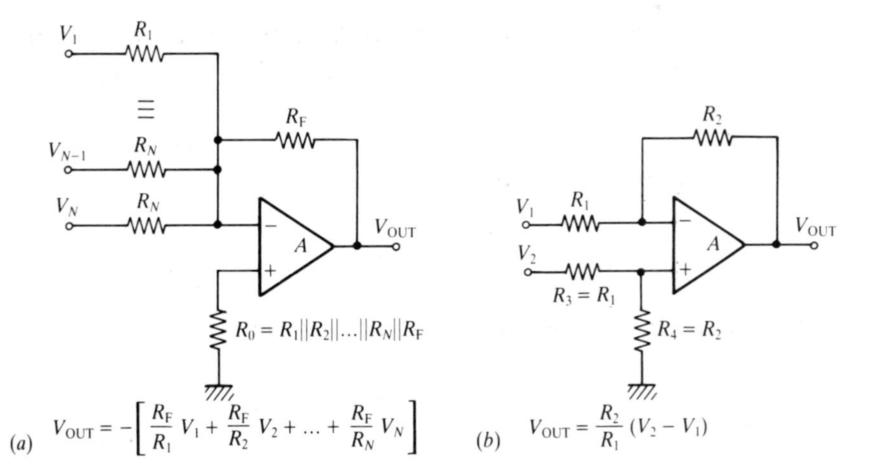

Fig. 9.2. Simplified addition/subtraction circuits. (*a*) Addition. (*b*) Subtraction.

factor β. Consequently the offset error and drifts will increase, the bandwidth may be reduced and non-linearities due to the finite op amp gain will increase. If a large number of input signals are to be added or subtracted, then several op amp stages should be used.

When choosing the op amp, bear in mind the following points.

If there are many inputs or a large gain is required, the op amp should have a very high open loop gain.

If both inverting and non-inverting inputs are used, then the op amp should also have a large common mode rejection ratio (CMRR).

If resistor values are large, use a FET input op amp.

Ensure that the output offset voltage is low enough.

For a fast response, use a fast op amp of course and keep resistor values low.

The circuit shown in Fig. 9.1 is often used in simplified configurations, Fig. 9.2(*a*) shows an inverting adder and Fig. 9.2(*b*) a differential amplifier.

9.2 Multipliers

Analog multipliers are one of the most commonly used functional circuits in present day analog systems and have many applications beyond the straightforward multiplication of two signals. These other applications include squarers, square roots, power measurement, voltage controlled circuits such as amplifiers or filters, oscillator amplitude control, RMS circuits and linearizing circuits. The following sections deal with subjects which are common to most multiplier circuits.

Multiplier operation and errors (Fig. 9.3)

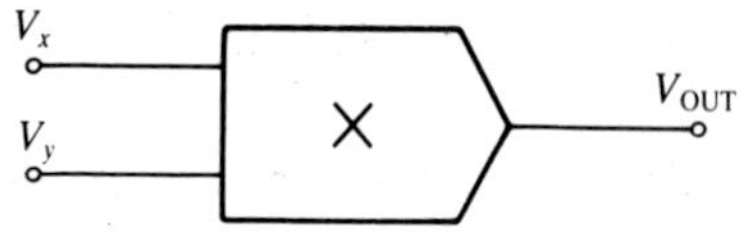

$$V_{OUT} = k_m V_x V_y = \frac{V_x V_y}{E_m}$$

where $E_m = \frac{1}{k_m}$

Fig. 9.3. An analog multiplying element.

Typical range of values for V_x, V_y and V_{OUT} are ± 10 V and so k is typically equal to 0.1 Volt^{-1}. This will give a 10 V output for inputs of V_x and V_y which also equal 10 V.

An ideal multiplier should operate in all four quadrants (i.e. any polarity combinations of x and y at the inputs) as shown in Fig. 9.4. Some practical multipliers offer only one quadrant operation (e.g. both x and y must be positive) or two quadrant operation (i.e. one input must be unipolar).

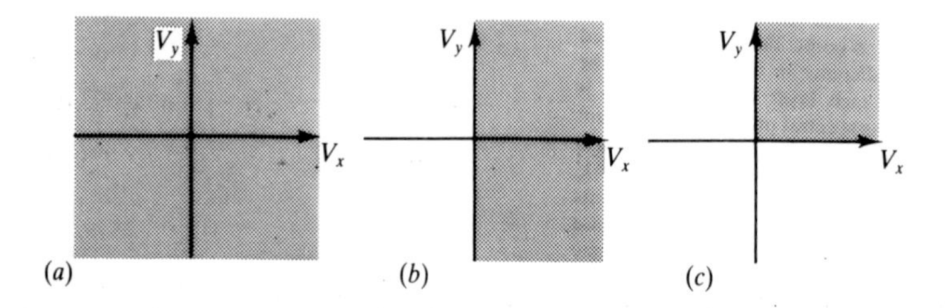

Fig. 9.4. Quadrant operation of multipliers. (*a*) Four quadrant. (*b*) Two quadrant. (*c*) Single quadrant.

A practical multiplier never behaves in an ideal manner and has a number of basic limitations. The inputs, for example, have a finite differential and common mode voltage range with finite impedance and the outputs have a finite current output capacity and finite output impedance. The limitations which are specific to multipliers are discussed below for a four quadrant multiplier.

The actual output of a multiplier may more accurately be expressed as

$$V_{OUT} = (k_m + \Delta k_m)(V_x + V_{xIO})(V_y + V_{yIO}) + V_{OO} + V_{xf} + V_{yf} + V_n(x, y)$$

where

Δk_m is the scale factor error
V_{xIO} is the input offset voltage of x-input
V_{yIO} is the input offset voltage of y-input
V_{OO} is the input offset voltage
V_{xf} is the non-linear feedthrough from the x-input to the output
V_{yf} is the non-linear feedthrough from the y-input to the output
and $V_n(x, y)$ is the non-linearity in gain response.

Rearranging this expression gives

$$V_{\mathrm{OUT}} = \underbrace{k_m V_x V_y}_{\text{ideal output}} + \underbrace{\Delta k_m V_x V_y}_{\text{gain error}} + \underbrace{[(k_m + \Delta k_m) V_x V_{y\mathrm{IO}} + V_{xf}]}_{\text{total } x\text{-input feedthrough}}$$

$$+ \underbrace{[(k_m \, \Delta k_m) V_y V_{x\mathrm{IO}} + V_{yf}]}_{\text{total } y\text{-input feedthrough}} + \underbrace{V_{\mathrm{OO}}}_{\text{output offset}} + \underbrace{V_n(x, y)}_{\text{non-linear output}}$$

The multiplier may have four external adjustments:

(i) the x-input offset null (for $V_{x\mathrm{IO}}$)
(ii) the y-input offset null (for $V_{y\mathrm{IO}}$)
(iii) the output offset null (for V_{OO})
(iv) a gain adjustment (for $V_n(x, y)$)

(a) Feedthrough and offsets
An ideal multiplier would have zero output when either input is zero. In practice, this is not the case since there are three errors operating:

(i) output offset voltage
(ii) x-input feedthrough, which is a small error signal feeding through to the output from the x-input when the y-input signal is zero
(iii) y-input feedthrough, which is a small error signal feeding through to the output from the y-input when the x-input signal is zero.

Output offset error is not usually a major problem since it is readily nulled. Output offset drift may cause a small error since this cannot be distinguished from the signal unless the system is auto-zeroing. A typical value for output offset is around 5 mV.

The x-input feedthrough has two components (similar remarks apply to y-input feedthrough). The first component is due to the x-input signal being multiplied by the finite y-input offset voltage. This component may be minimized by offset nulling the y-input. The second feedthrough component varies non-linearly with the x-input signal and cannot be nulled. So, by nulling the input offset voltages of both inputs, the feedthrough may be minimized. Input offset temperature drifts will, however, cause the feedthrough error to increase unless the system has a self-nulling facility. Feedthrough is also frequency dependent and may increase considerably with frequency. Some manufacturers of multiplier ics specify feedthrough values with the input offset voltages externally nulled to zero. Note that there may be a considerable difference between the x-input and y-input feedthrough errors. This difference is often greater than a factor of ten. Feedthrough is expressed as a peak-to-peak output voltage or as a percentage of maximum output with one input at zero and the other input with a full scale (20 V peak-to-peak), low frequency (50 Hz) sine wave applied. Typical values of feedthrough are around 50 mV (0.5%) or less.

The offset nulling procedure for a multiplier consists of the following steps.

(i) Connect the x-input to ground. Apply a low frequency sine wave of maximum amplitude to the y-input (e.g. 20 V p-to-p at 50 Hz). Null the x-offset by adjusting until the ac output is minimum.
(ii) Connect the y-input to ground. Apply a low frequency sine wave of maximum amplitude to the x-input. Null the y-offset.
(iii) Connect both x-input and y-input to ground. Null output offset.

(b) Gain error
For most multipliers, the gain or scale factor can easily be trimmed to the correct value by adding, for example, a small trim pot. However, you should watch out for drifts in scale factor, particularly those due to temperature changes as these errors cannot be so easily compensated and present a more fundamental limitation.

(c) Non-linearity error ($V_n(x, y)$)
Non-linearity error is the maximum deviation of the input/output response from the ideal straight line. This error is usually determined with a maximum (+ or −) level applied to one input, e.g. +/− 10 V with a test signal applied to the other input, e.g. a 50 Hz, 20 V p-to-p sine wave. Non-linearity introduces distortion and unfortunately is difficult to compensate for. Typically, commercial multiplier ics have non-linearity errors between 0.01% and 0.5% and it is worth noting that one input may be significantly more linear (up to ten times better) than the other, which may determine how the multiplier inputs are assigned in your system.

(d) Total error
Total error is a kind of catch-all specification which includes scale factor error, input feedthrough error, output offset error and non-linearity error. It is a useful bench mark for comparison purposes when selecting a multiplier ic. Typical values are 0.1% to a few per cent.

(e) Dynamic errors

Most multiplier data sheets contain the following specifications: 3 dB bandwidth (with one input held at its maximum positive or negative level), output slew rate, output settling time and the bandwidth for which the vector output error is less than 1%. There are two dynamic effects you should watch out for: firstly, distortion usually increases at higher frequencies; secondly, the bandwidth may be dependent on the dc signal levels and may be small with low dc levels.

Converting a multiplier to a divider

A multiplier can be modified to make a divider by the addition of an op amp as shown in Fig. 9.5.

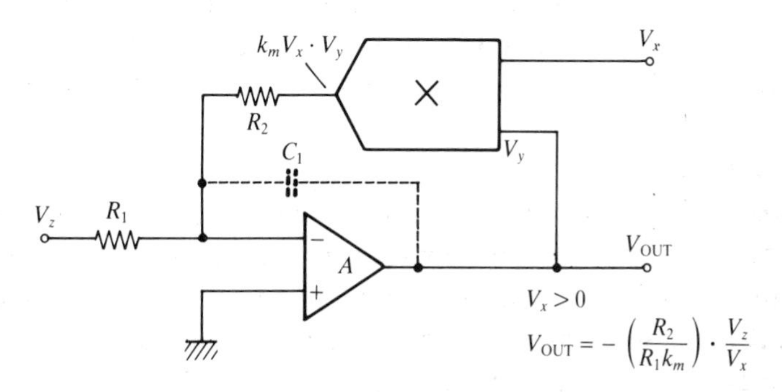

Fig. 9.5. Modifying a multiplier to make a divider.

Note that the input V_x must be positive so that the feedback around the op amp is negative. The feedback dynamics of this circuit are somewhat complicated by the addition of the multiplier in the feedback loop of A. Consequently, an extra capacitor C_1 may be needed to ensure that the circuit has a stable closed loop operation. The range of operation is limited by the error of the multiplier V_{OO} and the input offset voltage of the op amp V_{IO} as given by

$$V_{OUT} = -\left(\frac{R_2}{k_m R_1}\right) \cdot \frac{V_z}{V_x} + \left(1 + \frac{R_2}{R_1}\right) \cdot \frac{V_{IO}}{k_m V_x} + \frac{V_{OO}}{k_m V_x}$$

Clearly, as V_x reduces the error increases.

For typical values, $R_1 = R_2, k = 0.1\ \text{volt}^{-1}, V_{OO} = 10\ \text{mV}, V_{IO} = 5\ \text{mV}$

$$\text{Error} = 0.2V/V_x$$

$$= \frac{1}{0.1 V_x}[(1 + 1) \cdot 5\ \text{mV} + 10\ \text{mV}]$$

So, for a 0.1 V input, error = 2 V!!!

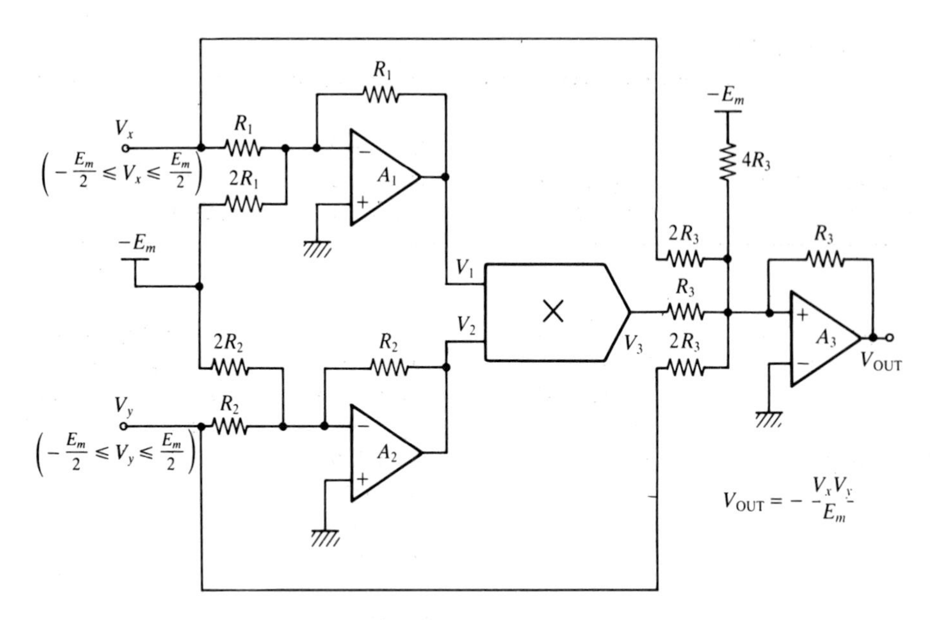

Fig. 9.6. Using a one quadrant multiplier for four quadrant operation.

The input offset voltage and error from the multiplier may limit the operation to only over one decade, typically between 1 V and 10 V.

A one quadrant multiplier can be converted into a four quadrant device by using an offsetting and scaling technique, at the inputs and the outputs, as shown in Fig. 9.6. In this arrangement, the output of the one quadrant multiplier is $V_3 = V_1 V_2 / E_m$, $0 \leqslant V_1 \leqslant E_m$, $0 \leqslant V_2 \leqslant E_m$.

It can easily be proved that the circuit output is $V_{OUT} = V_x V_y / E_m$.

To minimize feedthrough error and output offset, the resistor values must be in the ratios shown in the diagram. Also, the offsetting voltage, $-E_m$, must be equal in magnitude to the gain of the multiplier. Typically, $E_m = 10$ V.

A similar technique can be used to convert a two quadrant multiplier into a four quadrant multiplier or alternatively convert a four quadrant multiplier for two quadrant operation.

Summary of popular multiplier circuits

So far in this chapter, multipliers have been treated as a functional block. The remainder of the chapter gives a more detailed description of practical multiplier circuits and those circuits are reviewed in Table 9.1. Only a selection of multiplier circuits are dealt with in this way. The circuits described are a representative and popular selection from a larger number of possible configurations.

Table 9.1. *A selection of multiplier techniques*

FET controlled resistance multiplier	Variable trans-conductance multiplier	Log–antilog multiplier	Pulse-width pulse-height multiplier	Multiplying DAC
Technique				
Uses a FET as a voltage controlled resistor. This resistor is used to vary the gain of an amplifier	Involves controlling the trans-conductance of one or more bipolar transistor differential stages	Signals are logged, added and then anti-logged	One input varies the pulse width, the other varies the pulse height of a pulse train which is then low pass filtered	Involves using a multiplying DAC and/or an ADC
Total error				
Few %	0.1%–Few %	0.1%–1%	Can be as low as 0.01%	Depends on the no. of bits
Frequency response (dB)				
Up to a few MHz	Up to several 100 MHz	Approx. few 100 kHz	Slow, only a few 100 Hz	Depends on the DAC conversion time
Comments				
Very simple, low cost circuits requiring matched FETs. Useful in applications where performance is not critical. Generally, two quadrant operation.	Very widely used technique in commercial multiplier ics. Also has a divide input available in some circuits.	Very widely used technique in commercial multipliers. Generally, one quadrant but can be easily extended to 4-quadrant operation. A number of signals can be multiplied or divided.	Excellent dc accuracy but low bandwidth. Only two quadrant operation.	Useful for multiplying a binary number and analog signal together.

FET controlled resistor multiplier

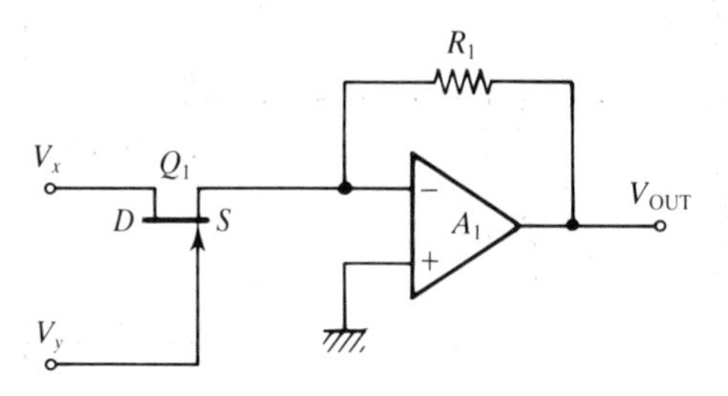

Fig. 9.7. A simple FET controlled resistor multiplier.

The circuit in Fig. 9.7 is drawn to highlight the principle involved, and as shown, is not a suitable practical circuit. The circuit in Fig. 9.8 is a more realistic form. For small source–drain voltages (i.e. well below the pinch-off voltage, typically 0.5 V), a FET acts as a controlled resistor, the resistance of which is inversely related to the gate–source voltage (see Chapter 5 and the section on controlled resistors). Thus the circuit may be considered as an amplifier with V_x as the input, whose gain is approximately proportional to V_y, hence V_x and V_y are multiplied together. The circuit has several major drawbacks including, for instance, very poor linearity since the gain between V_x and V_{OUT} is not linearly related to V_y. The properties of a FET are also very temperature sensitive causing large temperature dependent errors. Lastly, there is a restricted range of voltages for V_x and V_y inputs, i.e. $-0.5\text{ V} < V_x < +0.5\text{ V}$ and $0 \leqslant V_y \leqslant 0.5\text{ V}$. Even so, with careful design it is possible to achieve a total error of only a few % with this approach as shown in Fig. 9.8.

Variable transconductance multiplier

The variable transconductance approach is the basis of a large variety of analog multiplier ics and is one of the most important analog multiplication techniques in common use. The underlying principles of this approach are shown in idealized form in Fig. 9.9. Multiplication is achieved by varying the gain of a common emitter differential amplifier stage (Q_1 and Q_2) by varying its operating current I_1. In this example, I_1 is proportional to V_y through the use of a current mirror (Q_3 and Q_4). The output is given by

$$V_{OUT} = \frac{R_1}{r_e} \cdot V_x$$

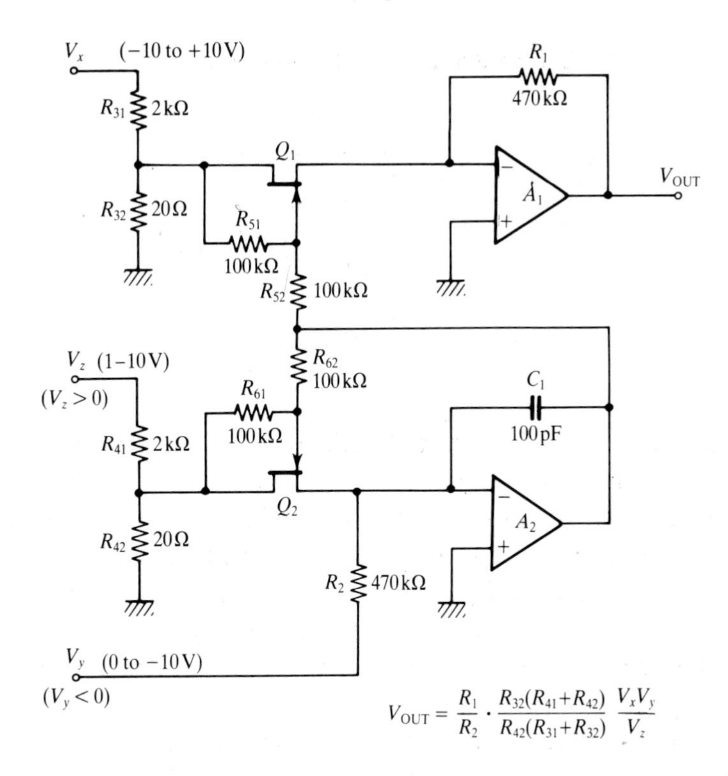

Fig. 9.8. An improved FET resistor multiplier.

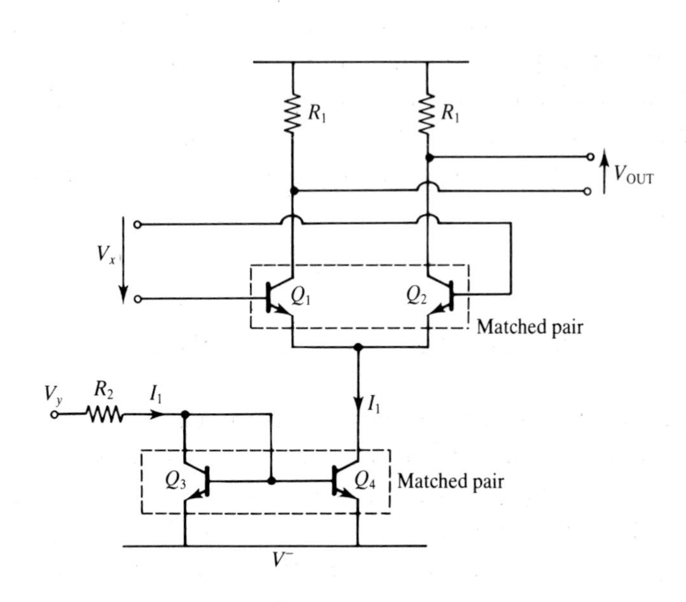

Fig. 9.9. A simple variable transconductance muliplier.

where r_e is the dynamic emitter resistance:

$$\frac{2kT}{qI_1}$$

k = Boltzmann's constant; T = absolute temperature; q_e = electron charge.

So,

$$V_{\text{OUT}} = \left(\frac{q_e R_c}{2kT}\right) V_x I_1$$

As shown here, this configuration has limited practical use as a multiplier because it has a number of major drawbacks. Firstly, the input voltage V_x must be small (tens of mV) otherwise the circuit will be highly non-linear. Secondly, the V_y input is referenced to the negative supply voltage V^- and not to ground. Finally, the gain of the multiplier is temperature sensitive. However, by incorporating more circuitry, the multiplying property of the common emitter differential stage can be more fully exploited.

Operational transconductance amplifier ics are popular analog building blocks which are based on the controllable transconductance property of a transistor differential stage. These devices can be configured to make a cheap and cheerful, low accuracy multiplier. Possible devices include CA3080 and the LM 13700.

Many high performance analog multiplier ics are also based on the variable transconductance property of the common emitter differential stage, however their internal circuitry is considerably more sophisticated than the simple circuit in Fig. 9.8. The core of these devices is a transconductance multiplier element which consists of several matched differential transistor stages which are arranged so that temperature drifts and non-linearities are compensated for as shown in Fig. 9.9. The multiplier element is often called a Gilbert Cell after one of the pioneers of this type of circuitry. The architecture of two popular high performance ics, the MPY-100 (Burr-Brown) and AD534 (Analog Devices) is shown in Fig. 9.10.

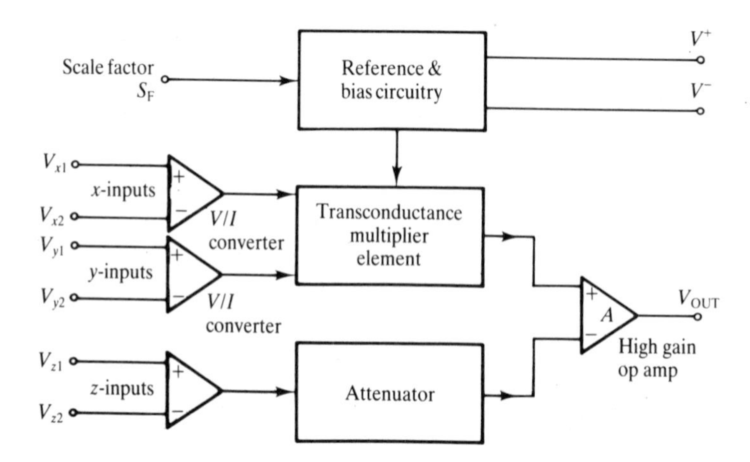

Fig. 9.10. MPY-100 and AD534 multiplier architecture.

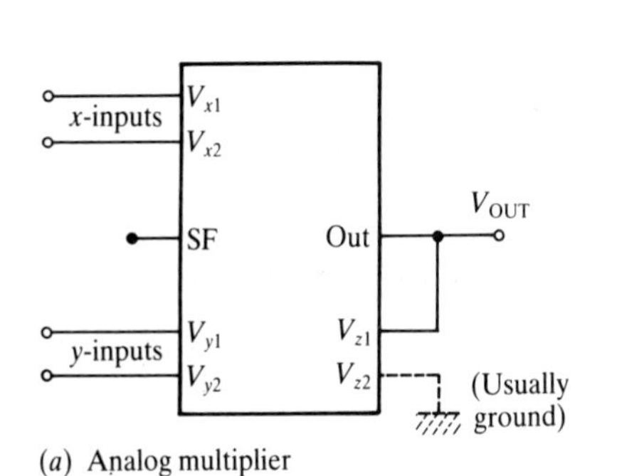

(*a*) Analog multiplier

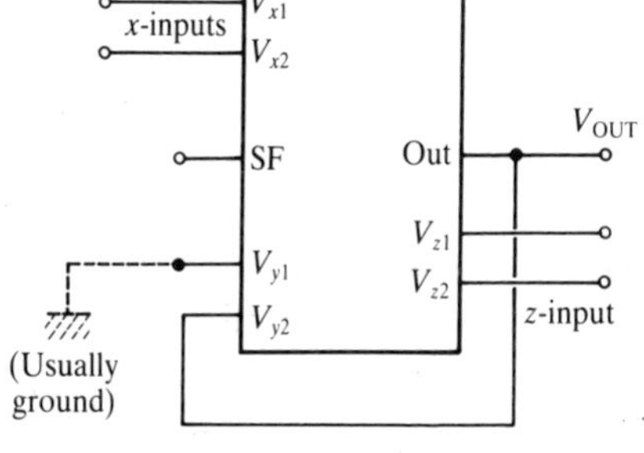

(*b*) Analog divider

Fig. 9.11. MPY-100 and AD534 configurations.

The transfer function of these ics is given by

$$V_{OUT} = A\left[\frac{(V_{x1} - V_{x2})(V_{y1} - V_{y2})}{SF} - (V_{z1} - V_{z2})\right]$$

These ics can readily be configured as a multiplier, as shown in Fig. 9.11(*a*), or as a divider as shown in Fig. 9.11(*b*).

The output of the multiplier

$$V_{OUT} = A\left[\frac{(V_{x1} - V_{x2})(V_{y1} - V_{y2})}{SF} - (V_O + V_{z2})\right]$$

so

$$V_{OUT} = \frac{A}{(1 + A)} \cdot \left[\frac{(V_{x1} - V_{x2})(V_{y1} - V_{y2})}{SF} + V_{z2}\right]$$

and

$$V_{OUT} = \frac{(V_{x1} - V_{x2})(V_{y1} - V_{y2})}{SF} + V_{z2} \qquad \text{when } A \gg 1$$

The output of the divider

$$V_{OUT} = A\left[\frac{(V_{x1} - V_{x2})(V_{y1} - V_{OUT})}{SF} - (V_{z1} - V_{z2})\right]$$

from which

$$V_{OUT} = \frac{(V_{z2} - V_{z1}) \cdot SF}{(V_{x1} - V_{x2})} + V_{y1}$$

if

$$\frac{(V_{x1} - V_{x2})A}{SF} \gg 1 \qquad \text{and} \qquad V_{x1} - V_{x2} > 0$$

Note that the bandwidth of the multiplier is proportional to the magnitude of $V_{x1} - V_{x2}$. Note also that $V_{x1} - V_{x2}$ must be positive to ensure that the overall feedback around the multiplier is negative. You can appreciate, now, that this combination of op amp output, multiplier core and differential input stage is very flexible. This type of multiplier is very widely used due to its properties of wide bandwidth, good linearity, differential inputs, four quadrant operation and relatively low cost.

Log–antilog multiplier (Fig. 9.12)

In this technique, multiplication is split into log conversion, addition and then antilog conversion since these operations are easy to achieve electronically using matched bipolar transistors and op amps. The technique is widely used in many multiplier ics since the method offers low cost, many signals can be multiplied and divided and functions of the form

$$V_{\mathrm{OUT}} = \left(\frac{V_x V_y}{V_z}\right)^m$$

can readily be produced. Although it offers only one quadrant multiplication, the basic log–antilog circuit can easily be converted into four quadrant operation by adding a few extra resistors. The following notes will describe practical circuits using this technique. The reader is recommended, also, to read the relevant parts of Chapter 8 dealing in further detail with log and antilog converters.

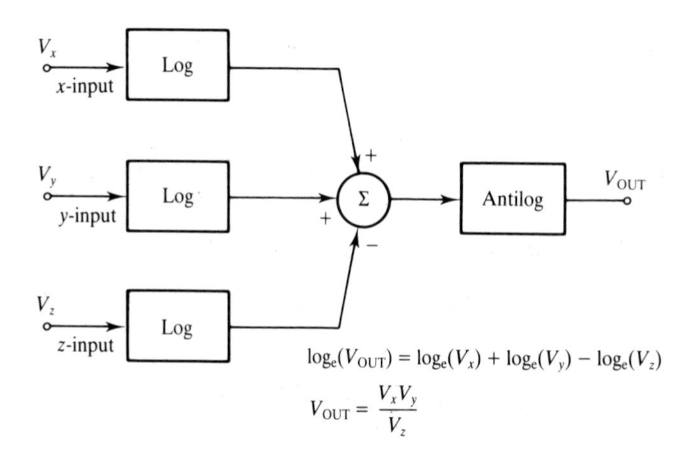

Fig. 9.12. Block diagram of a log–antilog multiplier.

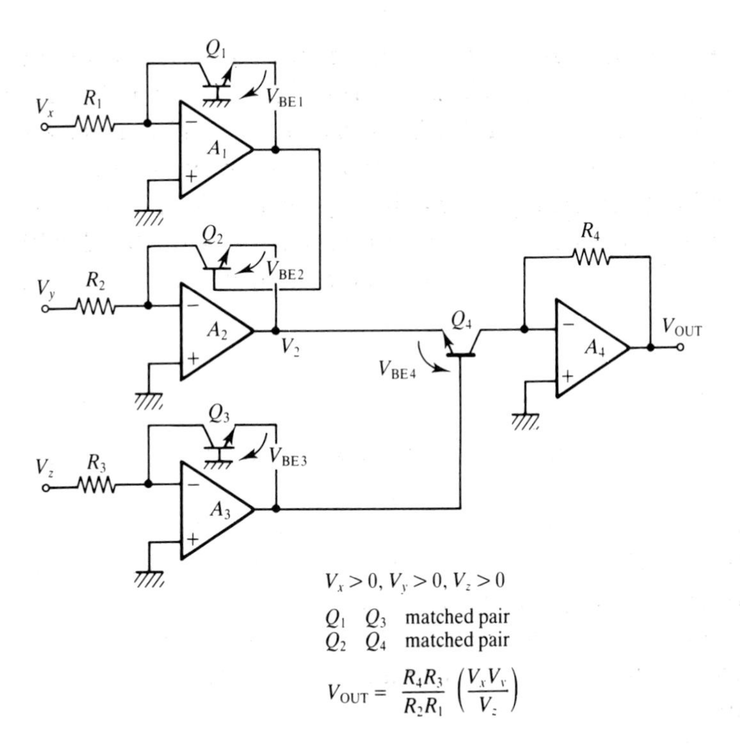

Fig. 9.13. A simplified log–antilog multiplying circuit.

A circuit configuration is shown in Fig. 9.13 in simplified form.

The operation of this circuit is dependent on the following simplified transistor equation used with log converters

$$V_{BE} \simeq V_T \log_e \frac{I_C}{I_S} \quad \text{and} \quad V_T = \frac{kT}{q}$$

From the circuit, transistors Q_1 and Q_2: $V_2 = V_{BE1} + V_{BE2}$

But from transistors Q_3 and Q_4: $V_2 = V_{BE3} + V_{BE4}$

So,

$$V_{BE1} + V_{BE2} = V_{BE3} + V_{BE4}$$

i.e.

$$V_{T1} \log_e\left(\frac{I_1}{I_{S1}}\right) + V_2 \log_e\left(\frac{I_2}{I_{S2}}\right) = V_{T3} \log_e\left(\frac{I_3}{I_{S3}}\right) + V_{T4} \log_e\left(\frac{I_4}{I_{S4}}\right)$$

Usually, all the transistors will be on the same ic so that

$$V_{T1} = V_{T2} = V_{T3} = V_{T4}$$

so,

$$\log_e\left(\frac{I_1 I_2}{I_3 I_4} \cdot \frac{I_{S3} I_{S4}}{I_{S1} I_{S2}}\right) = 0$$

Also, the transistors must be matched so that the leakage current term cancels. So, if matched transistors are used, Q_1 and Q_3 must be a matched pair as must Q_4 and Q_2. Then, $I_{S1} = I_{S3}$ and $I_{S2} = I_{S4}$, which means that the transistors cancel each other's temperature errors to leave

$$\log_e\left(\frac{I_1 I_2}{I_3 I_4}\right) = 0$$

so,

$$I_4 = \frac{I_1 I_2}{I_3}$$

Replacing these currents with their voltages:

$$I_4 = \frac{V_{OUT}}{R_4} : I_1 = \frac{V_x}{R_1} : I_2 = \frac{V_y}{R_2} : I_3 = \frac{V_z}{R_3}$$

So,

$$V_{OUT} = \frac{R_4 R_3}{R_1 R_2} \cdot \left(\frac{V_x V_y}{V_z}\right)$$

Usually, $R_1 = R_2 = R_3 = R_4$ and then $V_{OUT} = V_x V_y / V_z$.

The circuit in Fig. 9.13 is somewhat simplified and the one shown in Fig. 9.14 is a more practical design. This circuit also offers four quadrant multiplication.

The additions to this circuit are as follows.

Resistors and capacitors to stabilize the closed loop response of A_1, A_2 and A_3 (R_{11}, C_1, R_{21}, C_2, R_{31}, C_3 respectively).

Diodes D_1, D_2, D_3 and D_4 to protect all the base–emitter junctions of the transistors against reverse breakdown.

Resistors R_5 to R_9 are used to convert the one quadrant operation into a four quadrant operation. For accurate four quadrant operation, these resistors need to be closely matched.

If you are an enthusiastic customizer, then you can build this circuit using discrete components and a transistor network ic such as the CA3046. However, there are several commercial devices which employ this method.

There are a few points to note about the performance of this type of multiplier. Firstly, the bandwidth decreases as signal size decreases (see Chapter 8). Secondly, linearity is limited by the accuracy of the log

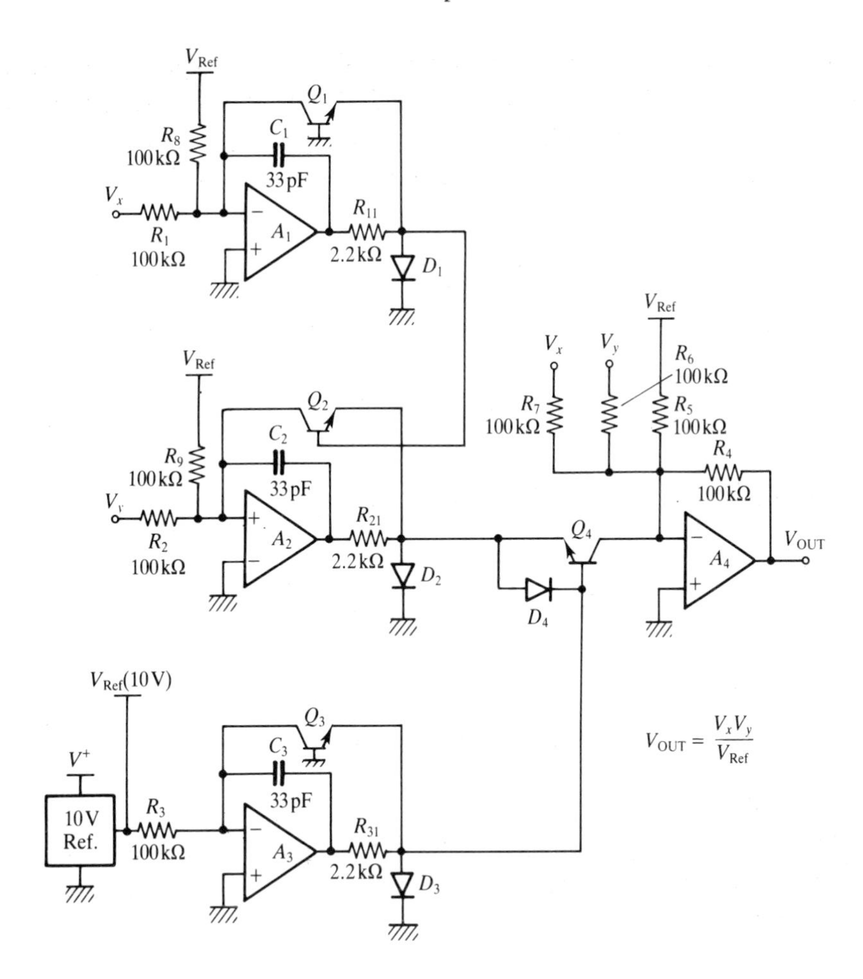

Fig. 9.14. An improved log–antilog multiplier.

converters. This is limited for small signals by the error (input offset voltage and input bias current) of the op amps and for large signals by the bulk emitter resistance of the transistors. The error due to the op amp can be limited by careful selection. Compensation can also be applied for the transistor bulk emitter resistance (see Chapter 8).

A pulse width/pulse height multiplier (Fig. 9.15)

This technique is worth a mention since it offers excellent static accuracy, approaching 0.01%. It has, however, a very low frequency response, not much greater than a few hundred Hz. With this technique, one signal varies the width of a pulse train, the other signal varies its height. This train of pulses is then filtered with a low pass filter whose cut-off is much lower than the clock frequency. The block diagram of a practical circuit for realizing the operation is shown in Fig. 9.16.

Table 9.2. *A selection of commercial multipliers*

	MPY600	MPY100A	MPY632A	MPY634A
Function				
Transfer function (V_{OUT})	$A\left[\frac{(x_1 - x_2)(y_1 - y_2)}{SF} - (z_1 - z_2)\right]$	as MPY600	$A\left[\frac{(x_1 - x_2)(y_1 - y_2)}{2V} - (z_1 - z_2)\right]$	as MPY600
Input range	±2 V	±10 V	±10 V	±10 V
Supplies (rated)	±5 V	±15 V	±15 V	±15 V
Error				
Total error	25 mV	2%	1%	1%[A]
Feedthrough				
x-input	−65 dB[B]	±0.5%	±0.15%	±0.3%
y-input	−70 dB[B]	±0.03%	±0.01%	±0.01%
Non-linearity				
x-input	−60 dB	0.08%	±0.08%	±0.4%
y-input	−65 dB	0.08%	±0.01%	±0.01%
Offset				
Output offset	–	50 mV	5 mV	5 mV
Output offset drift	–	0.7 mV/°C	200 μV/°C	200 μV/°C
Dynamic				
3 dB bandwidth	30 MHz	550 kHz	1 MHz	10 MHz
1% bandwidth	–	70 kHz	50 kHz	100 kHz
Comments	Operates as both multiplier and divider without external components. High speed multiplier/divider block. Provides also differential current outputs.	Operates as both multiplier and divider without external components. Three differential inputs, x, y and z. General purpose multiplier/divider block.	Multiplier/divider block with three differential inputs. Multiplier and dividers easy to configure without external components.	Fast, general purpose multiply and divide block. Can be used with external components. Three differential inputs, x, y and z.

A: Maximum
B: at 500 kHz
C: Minimum

Table 9.2. (*Continued*)

AD734A	SG1495	AD834J	LH0094	AD538AD
$A\left[\frac{(x_1 - x_2)(y_1 - y_2)}{v_1 - v_2} - (z_1 - z_2)\right]$	$\Delta I_{OUT} = \frac{(x_1 - x_2)(y_1 - y_2)}{SF}$	$\Delta I_{OUT} = \frac{xy}{(1\ V)^2} \cdot 4\ mA$	$y \cdot \left(\frac{z}{x}\right)^m$	$y\left(\frac{z}{x}\right)^m$
±10 V ±15 V	±10 V ±15 V	±1 V ±4 V–9 V	0 to +10 V ±15 V	0 to +10 V ±15 V
0.1%	–	0.5%	0.25%	0.5%
−85 dB −85 dB	– –	0.2% 0.1%	– –	– –
0.05% 0.025%	2%[A] 4%[A]	±0.5% ±0.5%	– –	– –
– –	– –	– –	– –	200 μV –
10 MHz –	3 MHz 30 kHz	500 MHz[C] –	10 kHz –	400 kHz –
Fast, precision multiply and divide block. Three differential inputs x, y and z, with extra v input to control the scaling voltage.	Low cost multiplier block. Two differential inputs. Provides a differential output current and requires several external resistors.	Very fast multiplier. Two differential inputs. Differential current output.	Multifunction converter, uses a log and antilog technique. The exponent m by two resistors, $0 < m < 10$. Internal resistors for $m = 0.5$, 1 and 2. Single quadrant operation, inputs must be positive.	Computational unit which uses a log and antilog technique. The exponent m is fixed by two external resistors $0.2 < m < 5$. Pin strap for $m = 1$. Mainly single quadrant operation. Also contains a voltage reference.

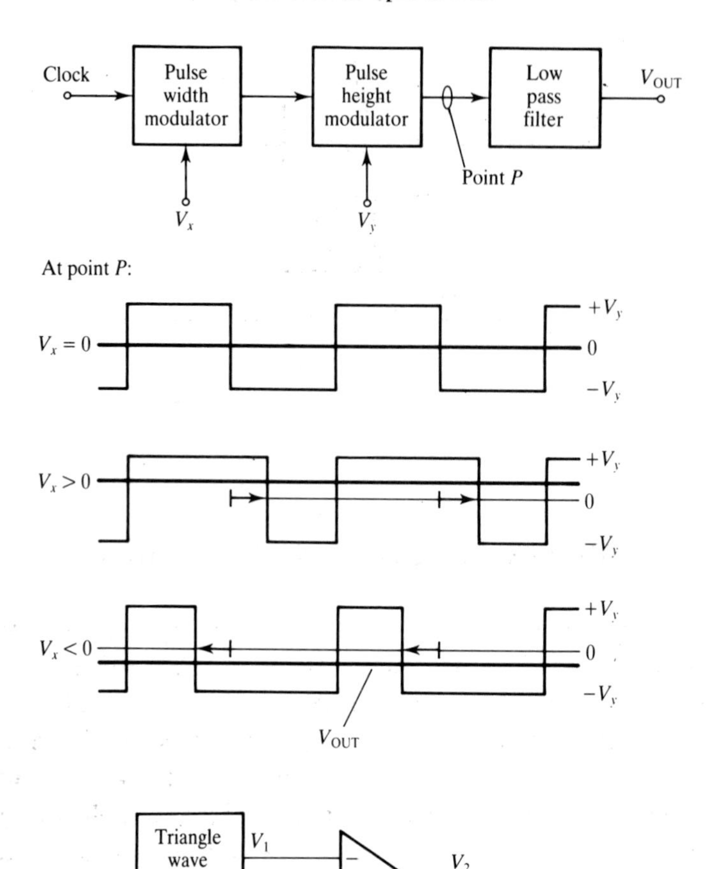

Fig. 9.15. A pulse width–pulse height multiplier.

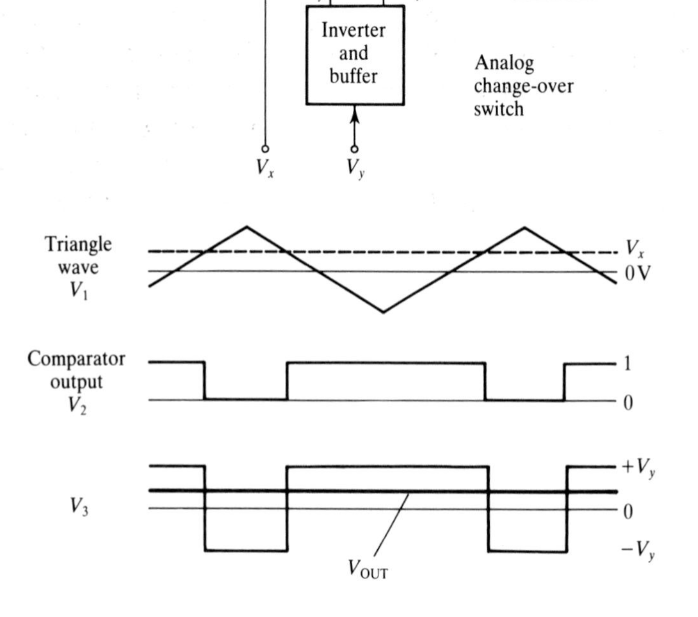

Fig. 9.16. Outline of a pulse width/pulse height multiplying circuit.

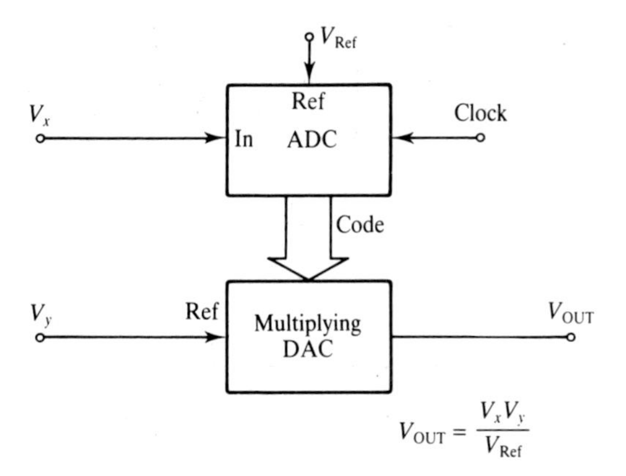

Fig. 9.17. Analog multiplication using D/A and A/D converters.

Multipliers using D/A and A/D converters

The technique shown in Fig. 9.17 is a somewhat sledgehammer approach to analog multiplication in terms of circuitry. The reducing cost and improved performance of many recent monolithic A/D and D/A converter ics, however, may make the approach feasible. There are also applications where it is necessary to multiply a digital code with an analog signal with the result being an analog signal; a multiplying DAC is ideal for this type of operation since this is exactly what it does do.

9.3 Commercial analog multipliers

Commercial multipliers display a range of internal architectures, Table 9.2 (see pages 204 and 205) summarizes the main parameters from a selection of commercial multipliers. Most multiplier ics employ either the variable transconductance technique or the log–antilog technique. There are also a few so-called multi-function converter ics, which can also be used as multipliers. These are described in the following chapter.

10

Function generating circuits

Various function circuits are described in this chapter, including square, hyperbolic, square root, sine, cosine and arctan function circuits. Some of these circuits are useful for linearizing the response of devices such as thermo-couple transducers which behave in a predictably non-linear fashion. The circuits are grouped by technique rather than by function since there are a few distinct techniques which are commonly employed to produce a function, as compared with the infinite number of functions which can be created.

10.1 Function generating circuits using analog multipliers

You can implement the following functions (square, hyperbola and square root) quite easily using a multiplier in one of the configurations shown in Fig. 10.1. Note that these circuits are shown configured with just a basic multiplier. Commercial multipliers usually have additional circuitry, such as an extra divide input, an output op amp and differential inputs which allow them to be connected in a variety of ways. This means that these circuits can usually be configured directly with a single multiplier ic and the extra op amp will not then be required.

More complicated functions can be realized using multipliers. For instance, every continuous single valued function may be closely approximated using a power series of the type $V_{OUT} = C_0 + C_1 V_{IN} + C_2 V_{IN}^2 + C_3 V_{IN}^3 \cdots C_m V_{IN}^m$. The terms V_{IN}^2, $V_{IN}^3 \cdots V_{IN}^m$ can be obtained using multipliers and then added together using summing circuits to approximate the required function. Clearly, the more terms in the power series (i.e. the higher the value of m) then the more accurately can the desired function be approximated. Further information on multipliers and adders can be found in Chapter 9.

Multipliers are more expensive and more difficult to realize and set up than summing circuits. So, when using this technique, it is necessary to minimize the total number of multipliers:

using 1 multiplier $\cdots m = 2$

using 2 multipliers $\cdots m = 3, 4$

using 3 multipliers $\cdots m = 5, m = 6$ and $m = 8$ (not $m = 7$)

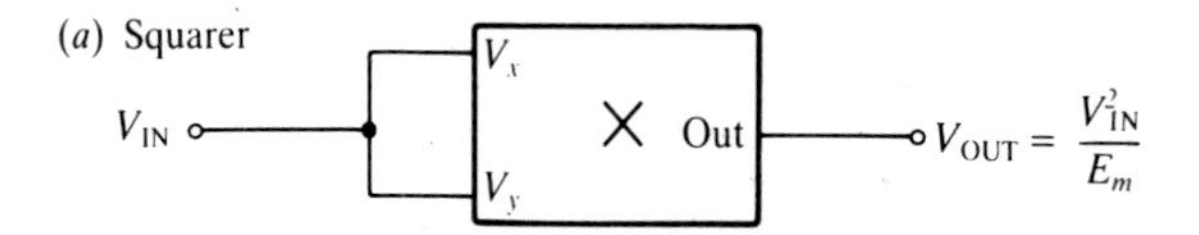

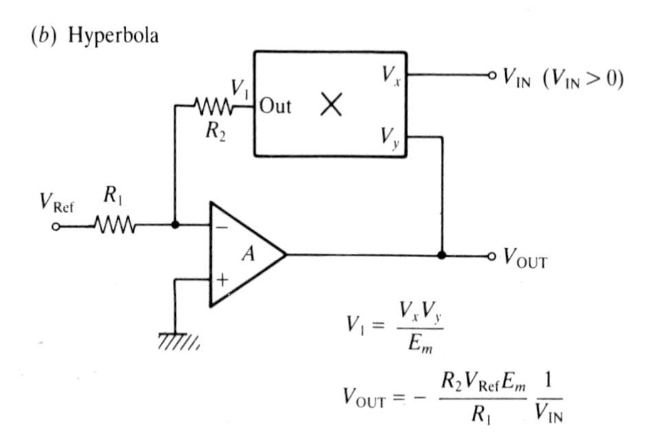

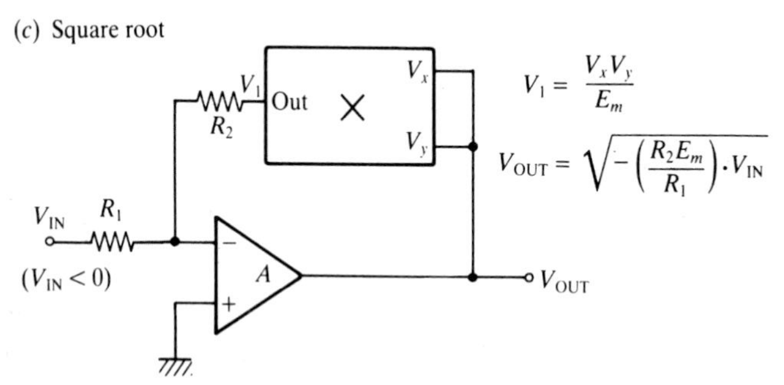

Fig. 10.1. Function generating circuits using multipliers. (*a*) Square. (*b*) Hyperbola. (*c*) Square root.

The actual value you choose for m is a trade-off between the desired accuracy and the real world, i.e. between what you want and what you can get. Ideally, it would be nice to use a very high value of m. This value, however, would need several multipliers and several carefully balanced summing circuits with very accurately matched resistors being used. In the real world, multipliers are not ideal, they can be expensive and they often require trimming. This means the circuit would be very complicated and many adjustments would have to be made. Also, with such a complicated circuit, there would be many sources of error. In other words, the search for greater accuracy results in greater complexity which results in more sources of error, which can reduce the accuracy. For these reasons, it is not very practical to use more than three multipliers for this technique thereby limiting the power series up to the 8th order. This is more than adequate to approximate most functions to within 1%.

The diagram in Fig. 10.2 shows implementations of summers and adders for polynomials up to the 4th order using only one or two multipliers.

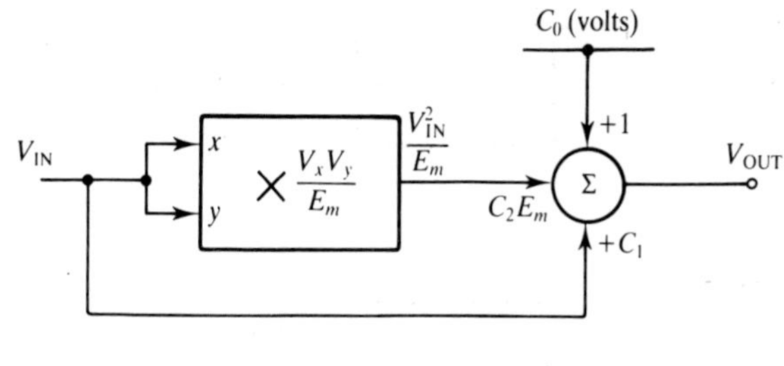

(a) $V_{OUT} = C_0 + C_1 V_{IN} + C_2 V_{IN}^2$

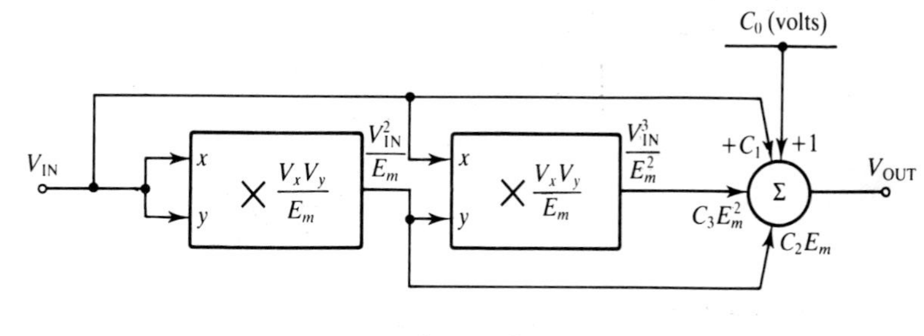

(b) $V_{OUT} = C_0 + C_1 V_{IN} + C_2 V_{IN}^2 + C_3 V_{IN}^3$

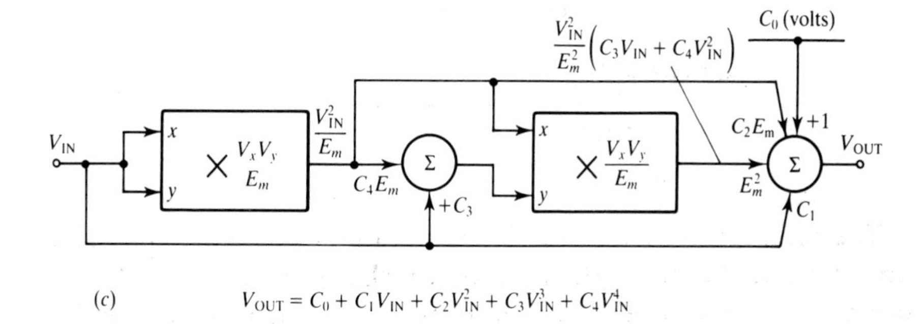

(c) $V_{OUT} = C_0 + C_1 V_{IN} + C_2 V_{IN}^2 + C_3 V_{IN}^3 + C_4 V_{IN}^4$

Fig. 10.2. Realizing 2nd, 3rd and 4th order polynomials. (*a*) Second order. (*b*) Third order. (*c*) Fourth order.

So far, we have only described the implementation of polynomials in circuit form. Before this can be done, however, the polynomial coefficients $C_0, C_1, \ldots, C_m$ must be determined. Two cases can be considered. Firstly, the desired function is a mathematical one, as for example, $\sin(x)$ or $\tan(x)$ etc. Or, secondly, the desired function is a series of empirical values for V_{OUT} and V_{IN} such that a curve needs to be fitted to these values.

In the case of a mathematical function such as $y = \cos(x)$ with values of x equal to $\pm\pi/2$ radians ($\pm 90°$), as shown in Fig. 10.3, the simplest method is to use the terms of the Taylor expansion for $f(x)$, i.e.

$$f(x) = f(x_0) + (x - x_0)f'(x_0) + \frac{(x - x_0)^2}{2!}\cdot f''(x_0) + \cdots + \frac{(x - x_0)^n}{n!}\cdot f^n(x_0)$$

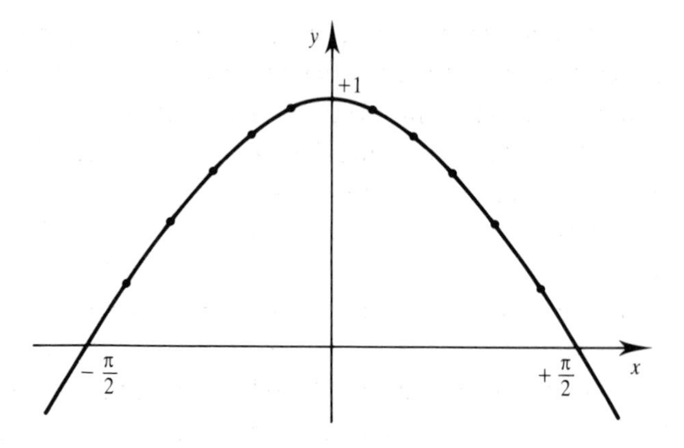

Fig. 10.3. Cosine function.

where $f^n(x)$ is the nth derivative of $f(x)$ with respect to x, or in the case of $\cos(x)$ about $x = 0$

$$\cos(x) = 1 - \frac{x^2}{2!} + \frac{x^4}{4!} - \cdots + (-1)^n \frac{x^{2n}}{(2n)!} + \cdots$$

Selecting the first four terms,

$$\cos(x) = 1 - \frac{x^2}{2} + \frac{x^4}{2 \cdot 4}$$

Now, if V_{IN} is between $\pm \hat{V}_{IN}$ and V_{OUT} is between 0 and $+\hat{V}_{OUT}$ we must scale the terms x and y.

For x and $V_{IN} \cdots$ when $x = \pi/2$ $\quad V_{IN} = \hat{V}_{INMAX}$ $\quad$ i.e. $x = \frac{\pi}{2} \cdot \frac{V_{IN}}{\hat{V}_{IN}}$

For y and $V_{OUT} \cdots$ when $y = 1$ $\quad V_{OUT} = V_{OUTMAX}$ $\quad$ i.e. $y = \frac{V_{OUT}}{\hat{V}_{OUT}}$

Inserting these expressions for x and y into the 4th order polynomial for $\cos(x)$ gives the coefficients

$$V_{OUT} = (\hat{V}_{OUT}) - \left(\frac{\pi^2 \hat{V}_{OUT}}{8\hat{V}_{IN}^{\,2}}\right) \cdot V_{IN}^{\,2} + \left(\frac{\pi^4 \hat{V}_{OUT}}{384\hat{V}_{IN}^{\,4}}\right) V_{IN}^{\,4}$$

so,

$$C_0 = \hat{V}_{OUT}, \qquad C_1 = 0, \qquad C_2 = \frac{\pi^2 \hat{V}_{OUT}}{8\hat{V}_{IN}^{\,2}}, \qquad C_3 = 0, \qquad C_4 = \frac{\pi^4 \hat{V}_{OUT}}{384\hat{V}_{IN}^{\,4}}$$

Determining the polynomial coefficients $C_0 \cdots C_n$ from the Taylor series expansion in this simple manner does not, unfortunately, give the most accurate approximation of the nth order polynomial to the function. A more accurate approximation may be obtained by representing the function as a sum of Chebyshev polynomials and then truncating this sum. The design procedure may get a little complicated and require some tedious computation. (The reader is referred to Wong and Ott (loc cit) since their book includes a very useful FORTRAN program to do the necessary number crunching.)

There are many engineering applications in which the function to be synthesized is not defined mathematically and is only known empirically, represented in either graphical form or as a table of points. This situation is commonly found, for example, when linearizing the response from transducers. In order to implement these functions, firstly it is necessary to approximate the required function to an nth degree polynomial, i.e.

$$V_{\mathrm{OUT}} = C_0 + C_1 V_{\mathrm{IN}} + C_2 V_{\mathrm{IN}}^2 + \cdots - C_m V_{\mathrm{IN}}^m$$

To work this out, $(m + 1)$ points on the response curve as shown in Fig. 10.4 must be known to create $m + 1$ independent linear equations of the form shown in Fig. 10.4 where m is typically greater than 10. These equations can then be solved to determine the polynomial coefficients.

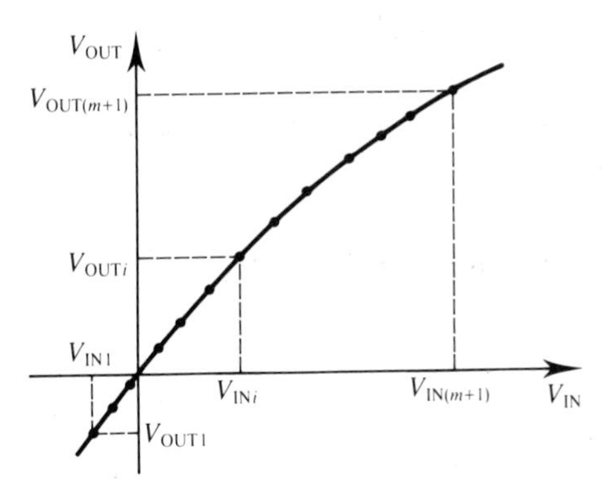

Fig. 10.4. Response curve point analysis.

$$V_{\mathrm{OUT}1} = C_0 + C_1 V_{\mathrm{IN}1} + C_2 V_{\mathrm{IN}1}^2 + C_3 V_{\mathrm{IN}1}^3 + \cdots + C_m V_{\mathrm{IN}1}^m$$

$$V_{\mathrm{OUT}2} = C_0 + C_1 V_{\mathrm{IN}2} + C_2 V_{\mathrm{IN}2}^2 + C_3 V_{\mathrm{IN}2}^3 + \cdots + C_m V_{\mathrm{IN}2}^m$$

$$V_{\mathrm{OUT}m} = C_0 + C_1 V_{\mathrm{IN}m} + C_2 V_{\mathrm{IN}m}^2 + C_3 V_{\mathrm{IN}m}^3 + \cdots + C_m V_{\mathrm{IN}m}^m$$

The more points that are known (i.e. the greater the value of m) the more accurately is the function represented as a polynomial since the order of the polynomial will be higher. Once the function has been approximated to an mth order polynomial, it can then be truncated to a lower order and implemented using the multiplier and summing circuit previously described. Direct truncation is not particularly accurate. Instead the function can be converted into a sum of Chebyshev polynomials which can then be truncated, this allows a more accurate representation. Once again, the reader is referred to Wong and Ott (Chapter 5, Power Series and Function Generators) and the FORTRAN program listings to deal with the large amount of computation which is required.

10.2 Log–antilog function generators (Fig. 10.5)

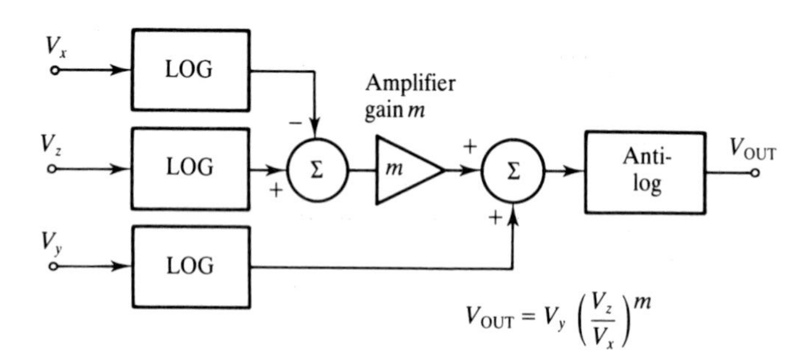

Fig. 10.5. Log–antilog function generators.

Log and antilog circuits are relatively easy to realize. Using the above circuits, it is possible to form the operation

$$V_{OUT} = V_y\left(\frac{V_z}{V_x}\right)^m$$

where m can typically be between 0.1 and 10 (i.e. raising to a non-unity power).

With this configuration as a building block, you can approximate many functions (not just powers) such as sin, cos and $\tan^{-1}$. This configuration is available from several manufacturers in ic form including the AD538 from Analog Devices, the 4301/2 from Burr–Brown and the LH0094 from National Semiconductor. For obvious reasons, these ics are sometimes called multifunction converters. Note, though, that this configuration offers only one quadrant operation, i.e. V_x, V_y and V_z must be unipolar inputs. The configuration can, if required, be customized with relative ease.

The reader is referred to Chapter 8 which deals with log and antilog converters, especially the section on the use of multipliers, to obtain a sounder grasp of this technique. Each of the elements in the schematic diagram, namely log converters, summers, amplifiers and antilog converters, could be realized separately using circuits described in this book. However, the configuration shown in Fig. 10.6 is a particularly efficient

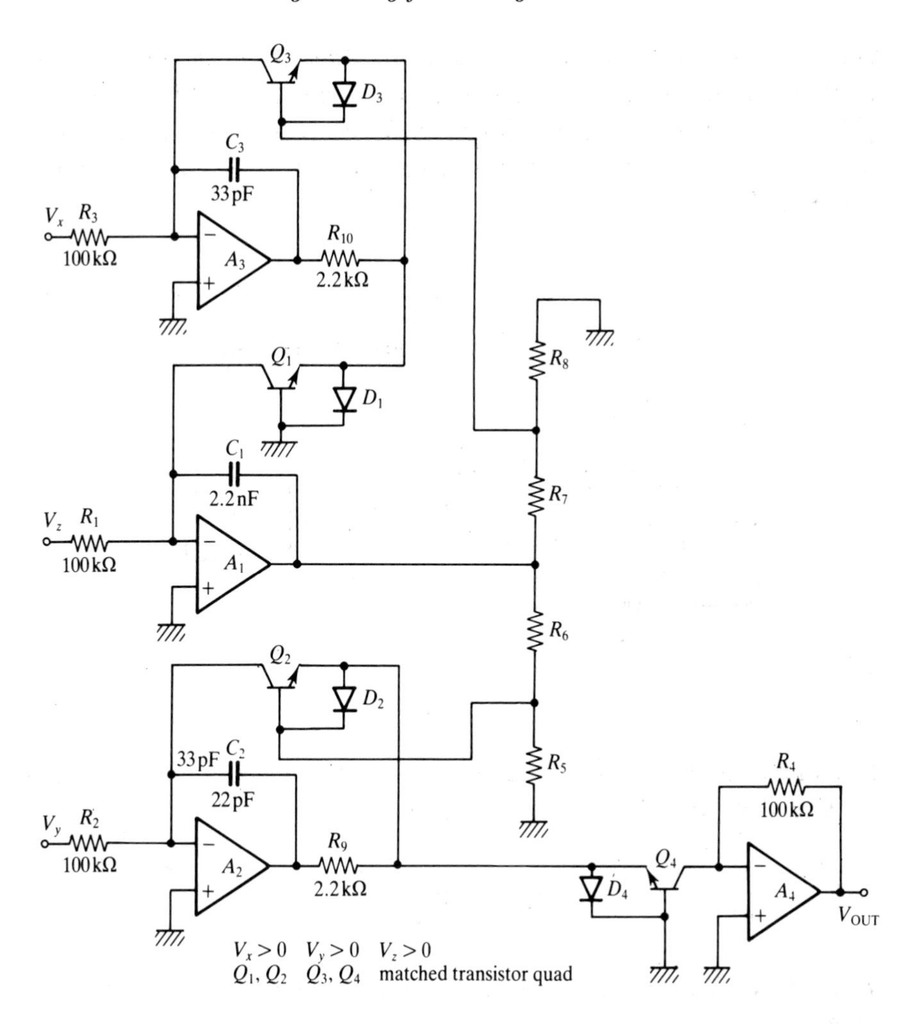

Fig. 10.6. Circuit realization of a multifunction converter.

realization of the design.

$$V_{\mathrm{OUT}} = \frac{R_4}{R_2} \cdot \left(\frac{R_3}{R_1}\right)^m \cdot V_y \left(\frac{V_z}{V_x}\right)^m$$

where

$$m = \frac{R_5(R_7 + R_8)}{R_8(R_5 + R_6)} \qquad \begin{array}{l} V_x > 0 \\ V_y > 0 \\ V_z > 0 \end{array}$$

Usually, the input resistors are chosen equal so that $R_1 = R_2 = R_3 = R_4$ in which case

$$V_{\mathrm{OUT}} = V_y \left(\frac{V_z}{V_x}\right)^m$$

For this circuit to operate accurately, the transistors Q_1 to Q_4 must be very closely matched and integrated on the same ic, otherwise the effects of their temperature dependent parameters will not cancel each other out. In the circuit some typical resistor and capacitor values have been added. For powers (i.e. $m > 1$), then R_6 can be shorted and R_5 omitted. For roots (i.e. $m < 1$), R_7 should be shorted and R_8 omitted. Low resistor values should be used for R_5 to R_8 (i.e. less than a few hundred ohms) to avoid introducing errors caused by the flow of transistor base currents. To improve the accuracy of the circuit and eliminate errors caused by slight mismatches in the transistors, the following resistors can be made trimmable: R_3 (to trim the ratio V_z/V_x); R_2 (to adjust the overall gain); R_6 or R_7 (to trim m). The circuit also suffers from many of the errors common to log converters such as signal-dependent bandwidth. The minimum operating range for input signals is limited mainly by op amp input errors. The maximum operating range for input signals is limited by bulk resistance errors in the transistors.

Some of the commercially available multifunction converter ics have been mentioned earlier. One of the more recent of such ics is the Analog Devices AD538 with a pin-out as shown in Fig. 10.7.

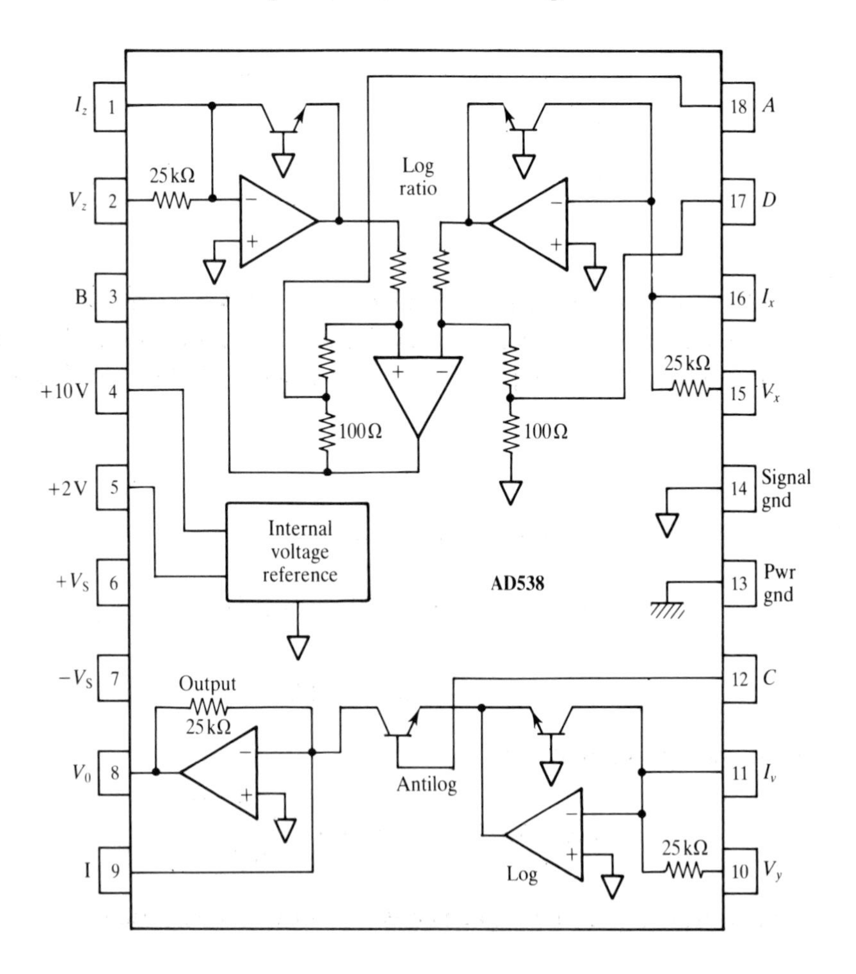

Fig. 10.7. Pin-out of the AD538 multifunction converter.

Some applications of the multifunction converter are given below.

(i) Set $m = 1$ and you have a one quadrant multiplier and divider where $V_{OUT} = (V_z Y_y)/V_x$, $V_x > 0$, $V_y > 0$, $V_z > 0$. To set $m = 1$, omit the resistors R_5 and R_8 and short out R_6 and R_7. Usually, with commercial ics, m is set to 1 by simply shorting pins together with no need for extra resistors.

(ii) Set V_y and V_x to some fixed reference voltage (V_{Ref}) and you have a power or root circuit where $V_{IN} = V_z$ and $V_{OUT} = V_{Ref}\left(\frac{V_z}{V_{Ref}}\right)^m$.

(iii) To make a square root circuit, the configuration shown in Fig. 10.8 is useful since it may avoid using any external resistors.

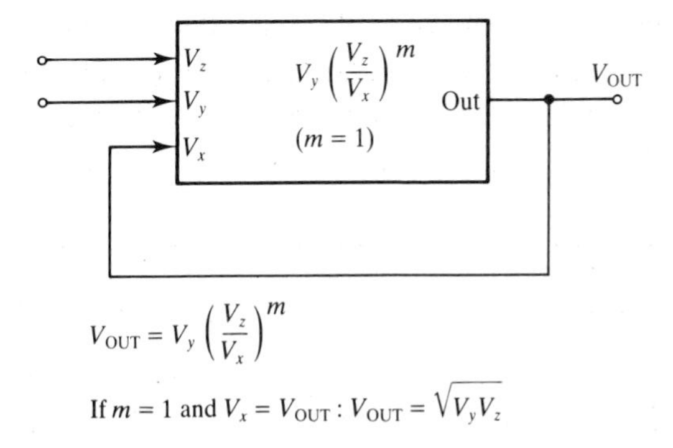

$V_{OUT} = V_y\left(\frac{V_z}{V_x}\right)^m$

If $m = 1$ and $V_x = V_{OUT}$: $V_{OUT} = \sqrt{V_y V_z}$

Fig. 10.8. A square root circuit using a multifunction converter.

Multifunction converters may be used to make a variety of less simple functions. The most popular functions include Sine, Cos and Arctan. These circuits are given below for values of V_{IN} between 0 V and 10 V.

Sine
$$V_{OUT} = 10\sin(9V_{IN})^{\text{DEGREES}}$$

$$0\text{ V} < V_{IN} < 10\text{ V} \quad (1\text{ V} \equiv 10^\circ)$$

so,

$$0\text{ V} < V_{OUT} < 10\text{ V}$$

This can be approximated by

$$V_{OUT} \simeq 1.5708V_{IN} - 1.5924\left(\frac{V_{IN}}{6.366}\right)^{2.827}$$

The function can then be implemented as shown in Fig. 10.9.

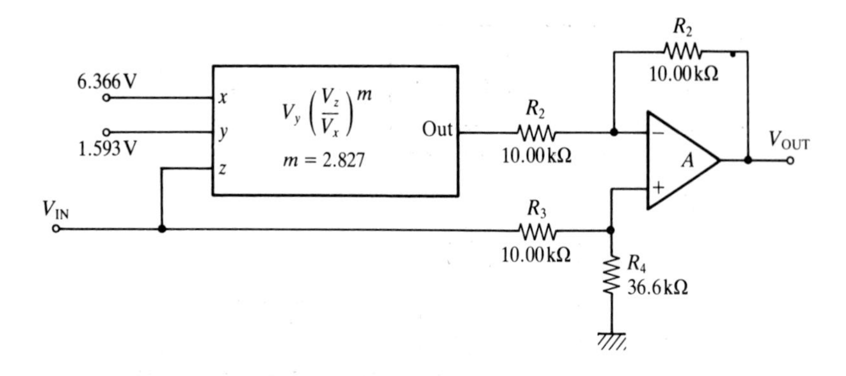

Fig. 10.9. Implementing the sine function.

Cosine $\qquad V_{OUT} = 10\cos(9V_{IN})^{\text{DEGREES}}$

$$0\text{ V} < V_{IN} < 10\ V \quad (1\text{ V} \equiv 10°)$$

so,

$$10\text{ V} > V_{OUT} > 0\text{ V}$$

This can be approximated by $V_{OUT} \simeq 10 + 0.3652V_{IN} - 0.4276V_{IN}{}^{1.504}$. The function is implemented as shown in Fig. 10.10.

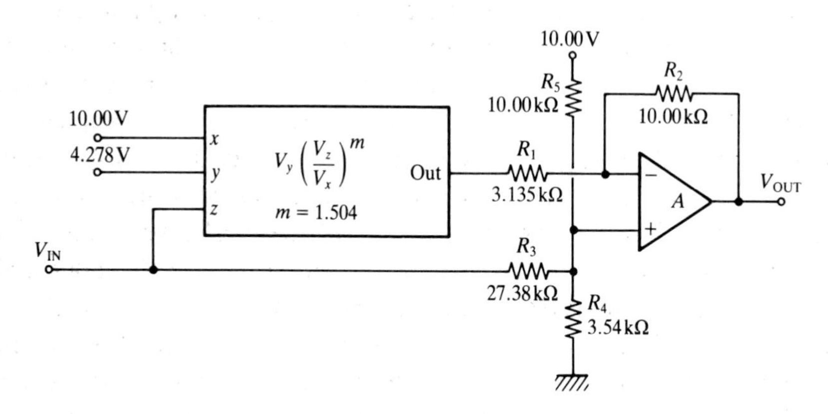

Fig. 10.10. Implementing the cosine function.

Tan^{-1} (arctan) $\qquad V_{OUT} = 9\tan^{-1}(V_x/V_y)$

0 V < V_{OUT} < 9 V corresponding to 0°–90°). This can be approximated to

$$V_{OUT} \simeq \frac{9(V_x/V_z)^{1.2125}}{1 + (V_x/V_z)^{1.2125}} \quad (1\text{ V} \equiv 10°)$$

as implemented in Fig. 10.11.

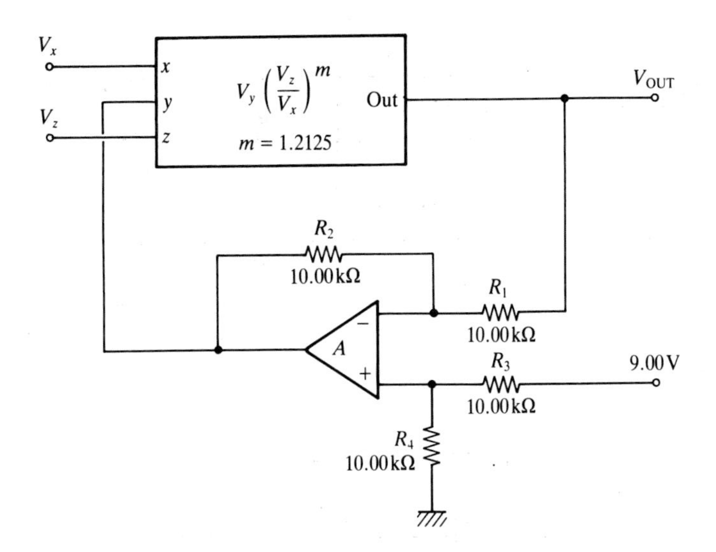

Fig. 10.11. Implementing the arctan function.

10.3 Breakpoint function generators

Most functions can be generated by using this technique. The technique involves a piecewise approximation of the required function by splitting it into a series of line segments as shown in Fig. 10.12. The accuracy of this method depends upon the number of straight line segments which are used and on whether the slope of the function changes rapidly. Further straight line segments could be used over regions where the gradient changes rapidly. The technique is relatively easy to implement using diodes and op amps. The circuit shown in Fig. 10.12(*b*), however, could not be used in critical applications due to temperature variations in the Zener voltages. Zener diodes are not the ideal type of diode in this case since they are more expensive than conventional diodes and only the discrete voltage drops specified by the manufacturer (e.g. 2.7 V, 3.0 V, 3.3 V, 3.6 V, 3.9 V, . . . , etc.) can be used for the breakpoint voltages. Also, only functions which increase in magnitude as the magnitude of V_{IN} increases (see Fig. 10.12(*a*)) can be realized. The technique is losing popularity nowadays due to the availability of function circuits such as multipliers, multifunction converters, A/D and D/A. It remains, though, as a useful approach since its components are relatively cheap.

An alternative approach to that shown in Fig. 10.12 is shown in Fig. 10.13 where an op amp is used to generate each breakpoint. The circuit in this example operates with op amp A_1 as a differential input stage and A_2 as an inverter. Consequently, with switch SW_1 open, the signal is simply passed through linearly from input to output. Closing switch SW_1 enables the function generating circuitry, which consists of

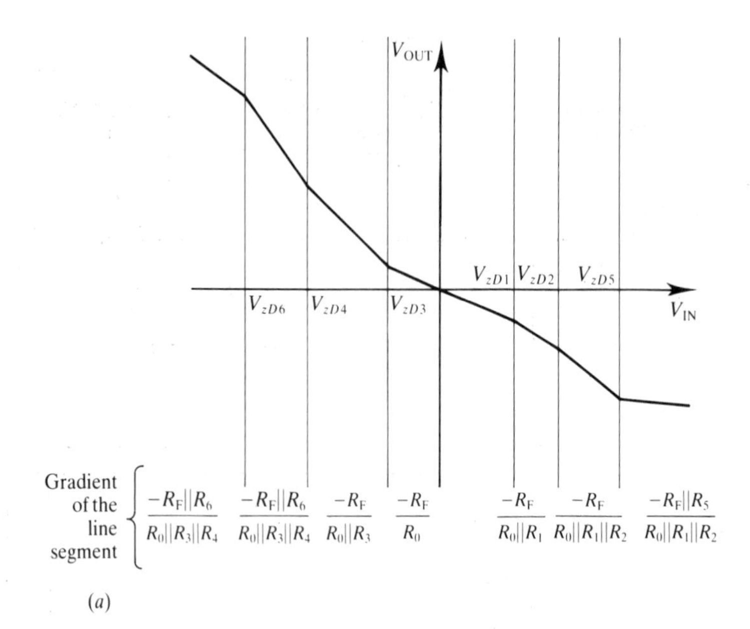

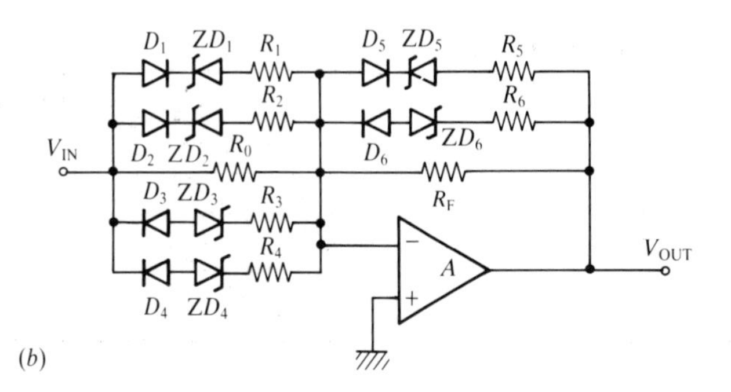

Fig. 10.12. Linear piecewise approximation. (*a*) Transfer characteristic. (*b*) Simple piecewise approximation circuit.

a number of breakpoint stages connected in parallel. A single breakpoint stage is shown inside the dashed box. For input voltages below $+V_{\mathrm{Ref}} \cdot R_A/R_B$, the output contribution of this stage is zero. For input voltages above this value, however, the stage can increase or decrease the gradient of the transfer characteristic up to $\pm(R_2 \cdot R_C \cdot R_E \cdot R_7)/(R_1 \cdot R_A \cdot R_D \cdot R_F)$. Each stage can be modified as follows.

(i) Changing the direction of the two diodes reverses the operation of the stage (i.e. the output contribution is zero when V_{IN} is above the breakpoint rather than below it).

(ii) Using a negative voltage for V_{Ref} introduces negative breakpoints.

(iii) When R_5 is omitted, the output is generated entirely from the breakpoint stages.

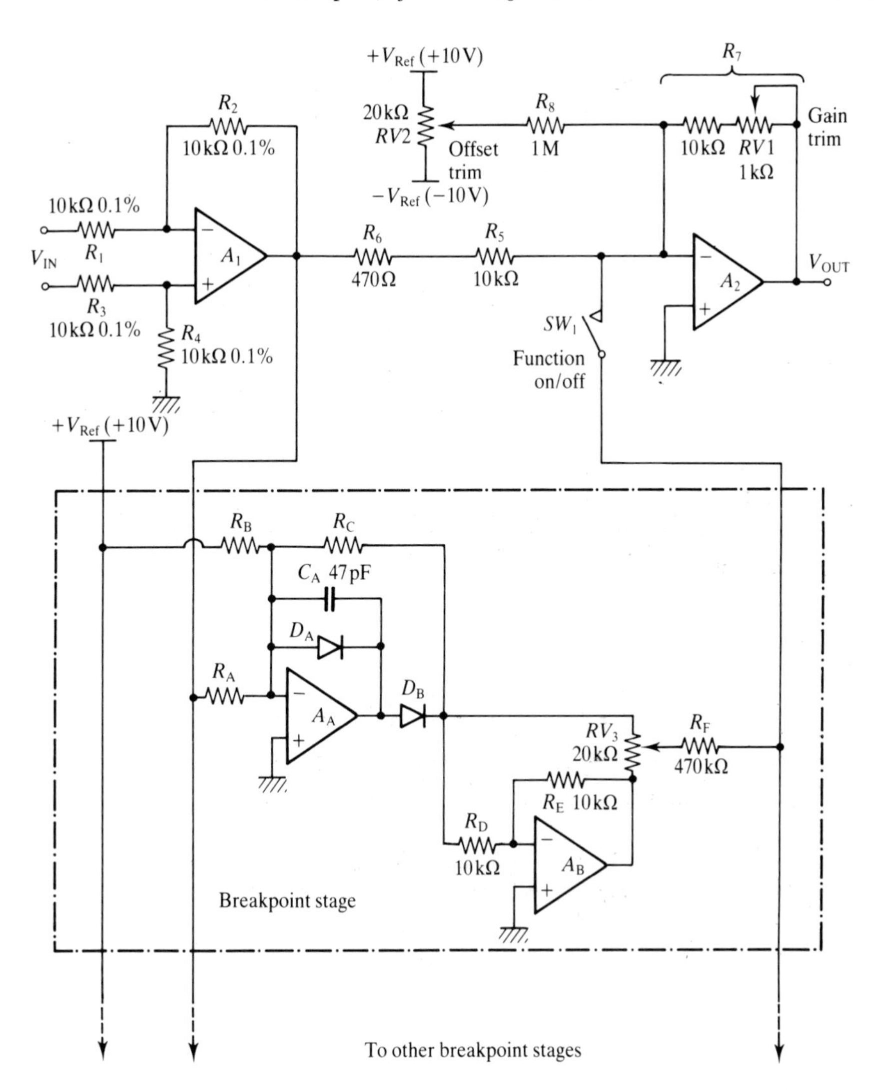

Fig. 10.13. A flexible diode function generator.

(iv) The circuit can be made more 'user-friendly' by adding a comparator to each stage to monitor the output of A_A and detect when the stage is in its operating range. Each comparator can illuminate an LED for example, which helps for the required function to be trimmed. It is a good idea to add a switch so that the LEDs can be turned off during normal operation.

Another example of the diode breakpoint method is used in the circuitry of the popular Intersil 8038 ic which is a waveform generator for generating square, triangular and sine waves. Square and triangular waves are relatively easy to generate. The 8038 converts a triangular wave into a sine wave using a diode breakpoint sine converter block which offers 16 breakpoints, 8 positive and 8 negative.

10.4 Function generating circuits using A/D and D/A converters

This technique shown in Fig. 10.14 is something of a sledgehammer approach in terms of circuit complexity, but with the decreasing costs of A/D and D/A converters this can be a cost-effective, efficient and accurate method of realizing a particular function which is becoming increasingly popular. The A/D converts an input signal into a digital code, which code is used as an address for the ROM. The ROM acts as a look-up table for outputting an appropriate code to the D/A converter which converts this code back into an analog signal. Some suppliers are now offering off-the-shell ROMs so clearly there is an increasing demand for this type of product.

Advantages	Disadvantages
Functions are not subject to drift and offset errors from op amps since the function is stored in digital form.	Accuracy and resolution are limited by the number of bits for the converters.
Any arbitrary function can be implemented using this approach.	Development tools are needed to program the PROM (or EPROM).
Since all the circuitry is within the ics, very few discrete components (such as diodes, resistors, etc.) are needed.	

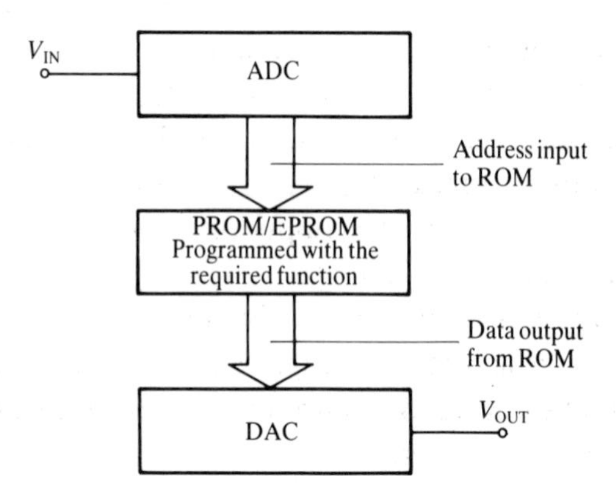

Fig. 10.14. Function generators using A/D and D/A converters.

11

Limiters, peak detectors and rectifiers

This chapter describes a variety of circuits which mainly employ the gross rectifying properties of diodes to perform various functions. Limiters, peak detectors and rectifiers are the main circuits described.

11.1 Limiters

These are circuits which prevent a voltage exceeding a certain range of values. They are sometimes called clamps and are often used for protection purposes. The input and output waveforms for a clamp are shown in Fig. 11.1.

This very simple method involves clamping the output of an op amp or any other similar devices by using either a Zener diode or an ordinary diode. This method needs to be used with caution since the output is clamped by forcing the op amp to current limit its output. This means that the op amp must be able to tolerate an output short circuit for an indefinite length of time at the maximum operating temperature. Also the clamping diodes must be sufficiently rated so that they will not be damaged by the maximum source current. Very often, the input source is an op amp which will usually increase in temperature if the clamp is on for more than a few seconds. So, this technique should be used with caution in sensitive applications where offset drift is important.

The circuit in Fig. 11.1(*a*) uses two back-to-back Zener diodes to give a positive and a negative clamping action. The circuit in Fig. 11.1(*b*) replaces one of the Zeners with an ordinary diode to give only the negative clamping action. The circuit in Fig. 11.1(*c*) is an alternative to using Zener diodes. In circuit (*c*), the clamping levels V_1 and V_2 must be rigid, i.e. they must be able to supply the current limit of the op amp which, typically, is tens of milliamps.

To limit the input current, it might be necessary to add a resistor in series with the source as shown in Fig. 11.2. R should be chosen to force the source to saturate instead of current limiting, but watch out for the thermal rating of this resistor.

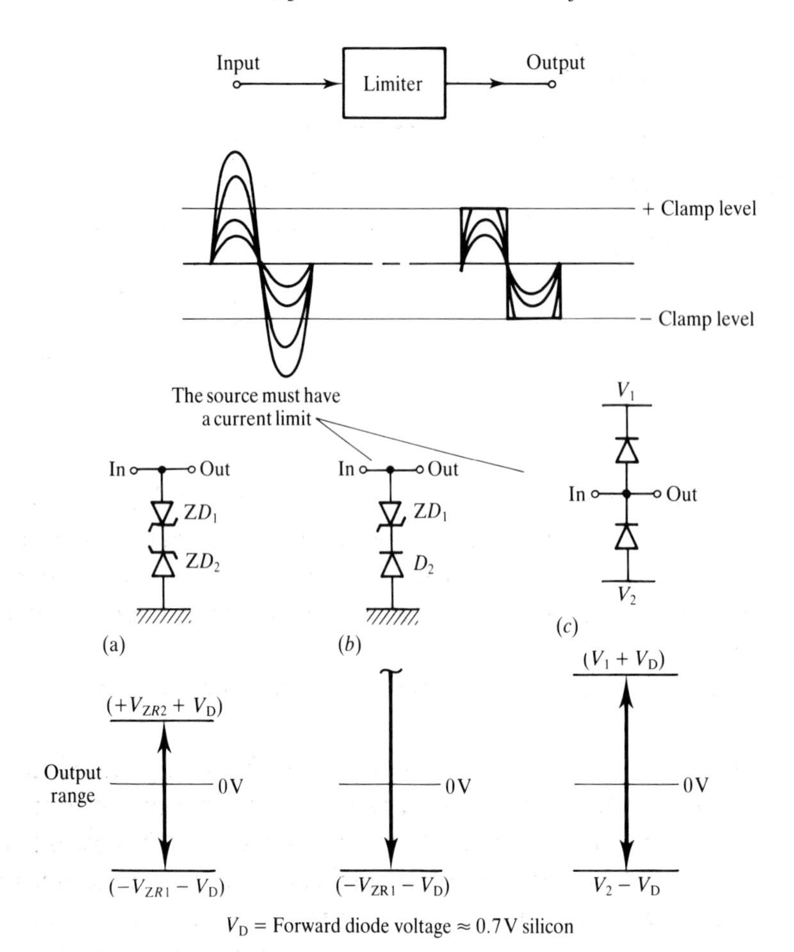

Fig. 11.1. Basic limiter circuits. (*a*) Back-to-back Zeners. (*b*) Zener and a diode. (*c*) Diode clamps to the supply rails.

The hard clamping approach is widely used as a simple circuit protection measure. However, the clamp voltage is not easy to control and power is wasted as unwanted heat. Also, the clamping action may not be very precise because Zener diodes often do not have a very sharp 'knee' in their reverse breakdown characteristic and often do not have a very low reverse breakdown slope impedance. The following pages describe more sophisticated methods for limiting a voltage to within a desired range. Circuits using the inverting op amp configuration can be

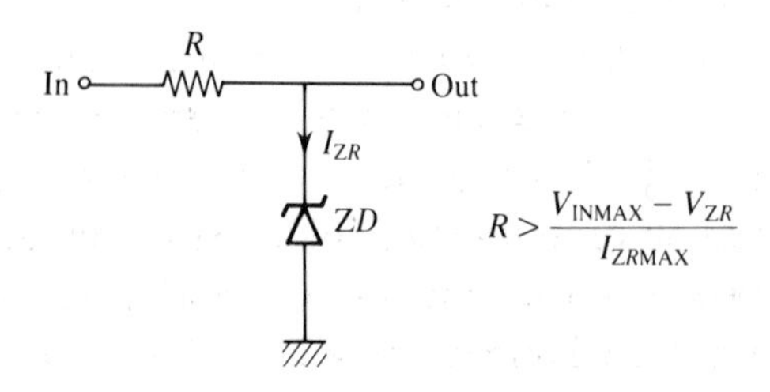

Fig. 11.2. Adding a resistor to limit the input current.

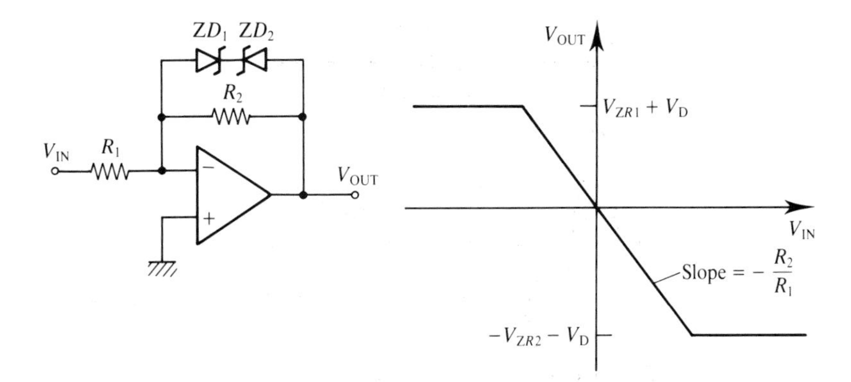

Fig. 11.3. Clamping an inverting amplifier using a Zener diode.

very effectively limited by adding Zener diodes between the output and the virtual earth of the inverting input. This is particularly useful since the output is clamped without forcing the op amp to current limit. The circuit shown in Fig. 11.3 is clamping a simple inverting amplifier.

There are two potential disadvantages with this approach.

(i) Junction capacitance: the speed of the circuit is reduced in two ways due to the junction capacitance of the Zener diodes, this value can be as high as 10s of pF. Firstly, the bandwidth will be reduced due to the shunting effect of the parasitic capacitance across R_2. Secondly, the time taken for the clamp to switch on and off will be increased since the parasitic capacitance must be charged and discharged which will present problems in clamping fast transients.

(ii) Leakage current: there is a relatively large reverse leakage current through the Zener diodes which may cause problems in some sensitive circuits such as integrators.

In the circuit shown in Fig. 11.4, the Zener diode is permanently biased in reverse breakdown and is switched into the rest of the circuit to carry out clamping using the diode bridge. Since the Zener diode is permanently biased ON with a constant DC voltage, the effects of its parasitic capacitances are considerably reduced.

The speed at which the clamp can be applied is limited mainly by the speed of switching of the diodes in the bridge, which is considerably faster than a Zener diode since their junction capacitances and charge storage are at least an order of magnitude less for fast signal diodes. The junction capacitances of the diodes in the bridge have a much reduced effect during normal, unclamped operation. This is because points *A* and

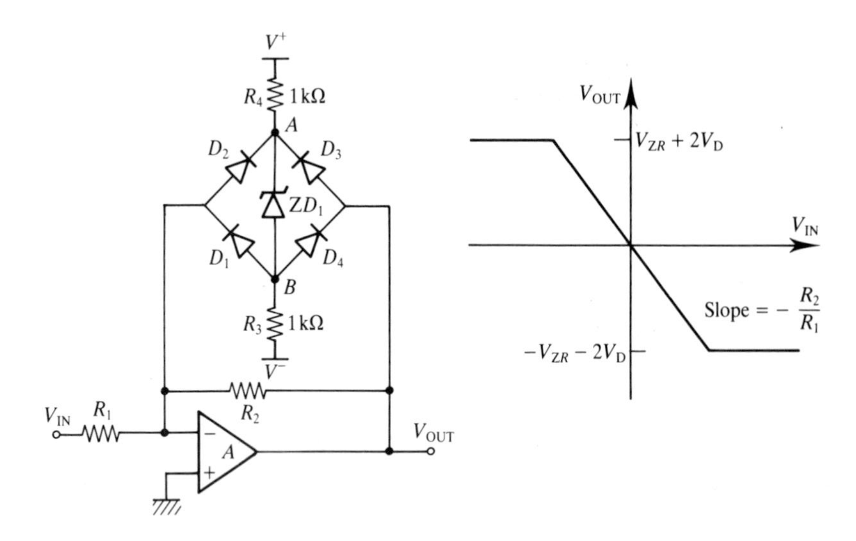

Fig. 11.4. Using a bridge diode arrangement for limiting.

B in the circuit are biased at a fixed dc level. So far as ac signals are concerned, points A and B are a reasonably good ground since $(R_3 + R_4)$ is relatively small. Consequently, the capacitances of diodes D_1 and D_2 are effectively between the op amp inverting input and ground with the parasitic capacitances of D_3 and D_4 between the op amp output and ground. Parasitic capacitances in these positions in the circuit rarely cause the response of the circuit to be greatly slowed down as happens when there is a capacitance across R_2. In fact, these capacitances are more likely to cause instability, particularly with fast op amps, rather than slow down the response of the circuit. A further advantage of this circuit is that it gives sharper clamping because the reverse leakage current and breakdown characteristics of the Zener diode are avoided since the Zener diode is no longer in the feedback loop of the op amp. The clamp is, consequently, switched on using ordinary diodes which have superior leakage current and switching characteristics.

For applications where the leakage current must be very small, and the leakage current through the Zener diode or other diodes cannot be tolerated, the circuit in Fig. 11.5 may be used. This circuit operates by causing the leakage current to flow harmlessly through the Zener diode and R_3 to ground instead of flowing, as before, into the current summing node at the inverting input of the op amp. The resistor R_3 is isolated from the inverting input of the op amp by the back-to-back diodes D_1 and D_2. This resistor must be chosen small enough so that the Zener leakage current flows through it to ground in preference to the alternative path through the diodes D_1 and D_2. If R_3 is made too small, however, the output current from the op amp will be large when the clamp is applied. A typical value for R_3 would be around 100 Ω to 1 kΩ. Low

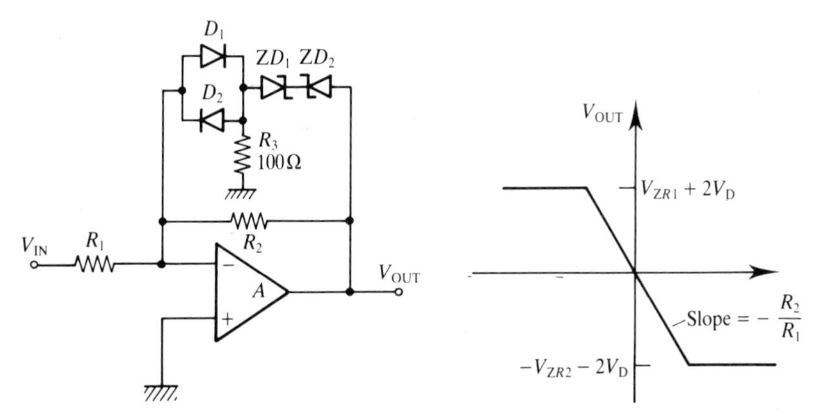

Fig. 11.5. Reducing the effects of the Zener diode leakage current.

leakage diodes such as the PAD series from Siliconix could be used for D_1 and D_2. Note that the addition of diodes D_1 and D_2 adds an extra diode drop to the clamping levels.

Diode bridge limiter

Diode bridge arrangements can be used for limiting purposes as shown in Fig. 11.6. Note that V_1 and V_2 can be positive or negative. V_1 is shown positive in the graph and V_2 negative. During normal operation, with no clamping required, with V_{IN} safely inside the clamping levels set by V_{C1} and V_{C2}, all four diodes are conducting. Point A is one diode drop above V_{IN} and point B is one diode drop below V_{IN}. Since all the four diodes are conducting with approximately equal forward voltages, and since points A and B are common to the input diodes D_1 and D_2 and the output diodes D_3 and D_4, then the output voltage is equal to the input voltage. An alternative way of stating the same point is by describing the output diodes D_3 and D_4 as twins of the input diodes D_1 and D_2 so that V_{OUT} is equal to V_{IN}. For input voltages greater than the upper clamp voltage ($V_1 - V_D$), diode D_1 switches OFF. Diode D_2 pulls

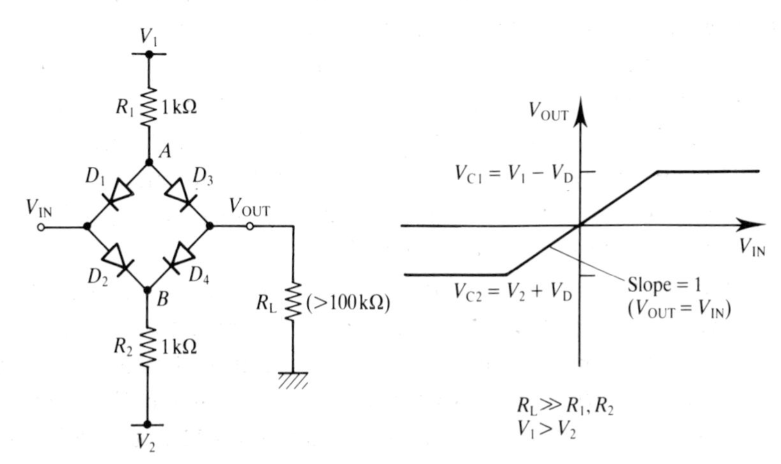

Fig. 11.6. Basic diode bridge limiter.

point B greater than this clamp voltage and diode D_4 also switches OFF. Hence, the output voltage rests below the upper clamp voltage by the forward drop of D_3. Similarly, for inputs less than the lower clamp voltage V_{C2}, diodes D_2 and D_3 are OFF, diodes D_1 and D_4 are ON and the output voltage rests above the lower clamp voltage by the forward drop of V_4.

This circuit is useful since the clamp levels are controlled by voltages V_1 and V_2 and so can be freely varied to an appropriate level. This is not the case with most clamping circuits using Zener diodes, since the clamp levels are fixed by the Zener voltages. The circuit is sometimes used for non-critical applications. There are a number of problems, however, which need to be modified if this approach is to be used in more critical applications:

(i) The clamp voltage and the switch-on voltage are dependent upon the characteristics of the diode. The clamp voltage is therefore prone to small temperature-induced drifts. In addition, the switch-on is not a sharp transition between normal linear operation and the operation of the output clamp.

(ii) For the clamping to be independent of the load, the load resistance must be quite high. The circuit also has a relatively low input impedance.

To achieve a sharper switch-on the diode bridge can be placed inside the feedback network of an op amp as shown in Fig. 11.7. Note that the clamp voltages are still dependent upon the forward voltages of the

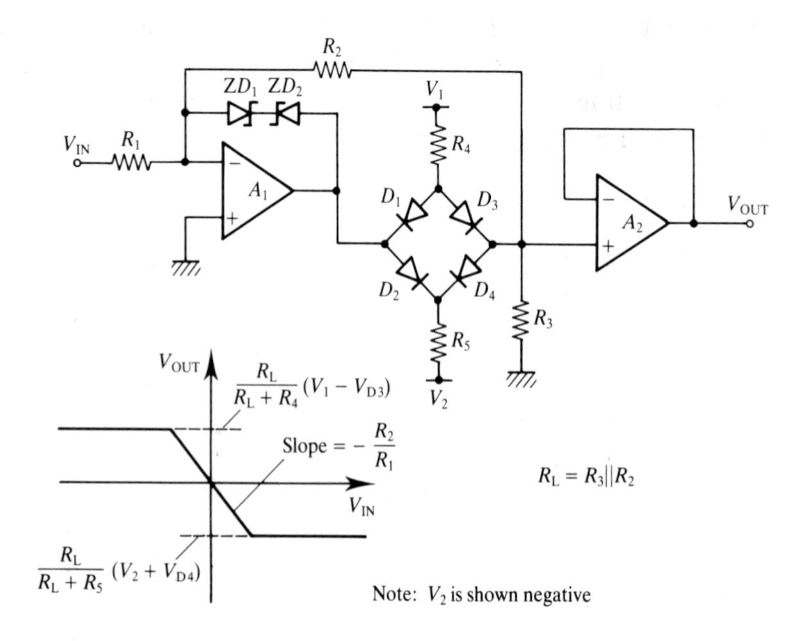

Fig. 11.7. Obtaining a sharper clamping action with a diode bridge limiter.

diodes D_3 and D_4 which will cause a small temperature drift in the clamp level. Op amp A_2 is added as an output buffer. The two Zener diodes around A_1 are included so that feedback is maintained around A_1 when the clamp is in operation. The Zener voltages should be chosen so that they are just outside the clamping range, i.e.

$$V_{ZR2} > V_1 \qquad \text{and} \qquad V_{ZR2} < V_2 \qquad \text{(with } V_2 \text{ negative)}$$

Voltage controlled op amp limiter

A voltage controlled op amp limiter is shown in Fig. 11.8. In this circuit, consider the case where positive clamping is required, as shown in Fig. 11.8(*a*). For input voltages which are less than V_1, the op amp is in positive saturation and so the diode is reverse biased and does not affect the output. The input, consequently, is coupled directly to the output via resistor R. For inputs greater than the clamp voltage V_1, the op amp conducts and forward biases the diode so forcing the output to equal V_1. For negative clamping, as in Fig. 11.8(*b*), the operation is exactly the same but with the polarities reversed. Where positive and negative clamping are both required, you simply cascade the two circuits together. Note that the op amp must be able to tolerate large differential input voltages.

This approach gives sharp clamping since the switching characteristics of the diode are divided down by the gain of the op amp. The clamp input has a high input impedance since the input is connected directly to the non-inverting input of op amp A. One of the main drawbacks of this simple approach is the slowness with which the clamp responds to changes on the input signal since the output of the op amp is forced to slew from saturation to forward biasing the diode. The notes given below provide further improvements on the basic circuit.

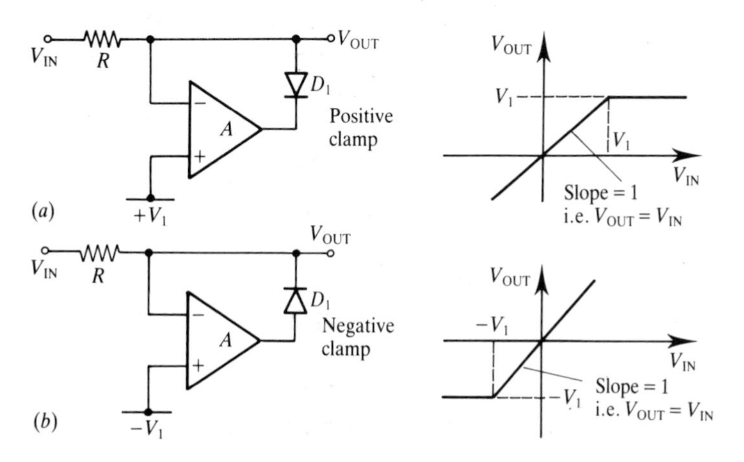

Fig. 11.8. Voltage controlled clamping. (*a*) Positive clamp. (*b*) Negative clamp.

The value of resistor must try and meet two conflicting requirements. On the one hand it represents the output resistance of the circuit during unclamped operation, and should therefore be as low as possible. On the other hand, it is the input resistance of the circuit when the clamp is applied and so should be as high as possible. A value of R of 1 kΩ is typical.

One technique which may be used to speed up the clamp switching rate is to use an uncompensated op amp and only switch in the compensating capacitor when it is needed, i.e. when the the clamp is active. Then, the op amp will have a maximum slew rate since it is not limited by the compensating capacitor when switching on the clamp and will be closed loop stable when the clamp is active. The circuit shown in Fig. 11.9 must be used with an uncompensated op amp, where the op amp may be compensated by connecting a capacitor between the compensation pin and the op amp output.

A voltage controlled clamp which has both upper and lower clamps, high input impedance for both of the clamp inputs and a reasonably low output impedance is shown in Fig. 11.10. A_1 is the usual voltage follower configuration which is clamped by the upper clamp with op amp A_2 and the lower clamp by A_3 where A_2 or A_3 switches in to override the output of A_1. Resistor R_1 is used to ensure that the clamps

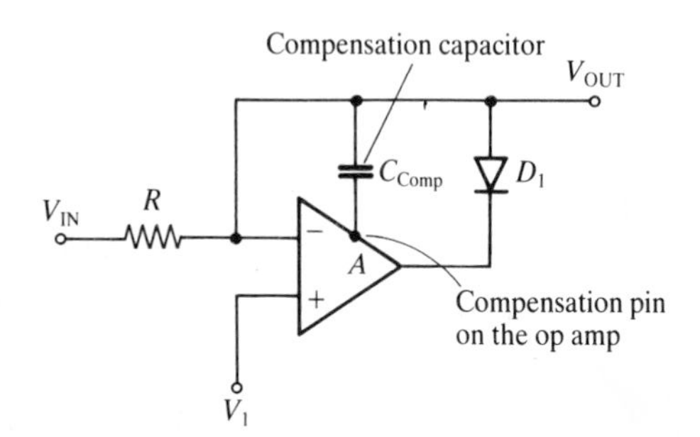

Fig. 11.9. Using an uncompensated op amp for limiting.

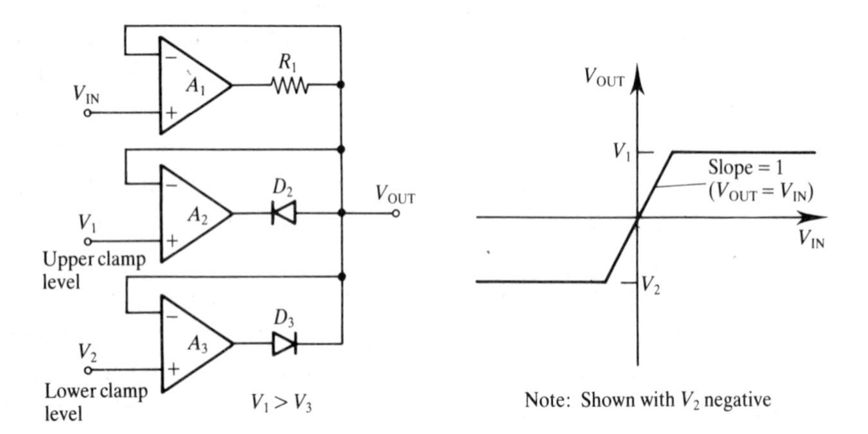

Fig. 11.10. General purpose upper and lower clamping circuit.

can override the output of A_1. This resistor should be chosen so that it is high enough to force op amp A_1 to saturate when either clamp is applied. Typically, a value of around 1 kΩ is chosen. The op amps must again be able to tolerate large differential input voltages.

11.2 Peak detectors

Basic peak detectors

The purpose of this circuit is to detect the maximum value of a signal over a period of time. The operation of a peak detector can be illustrated using the simple diode–capacitor circuit shown in Fig. 11.11 where the components are assumed ideal.

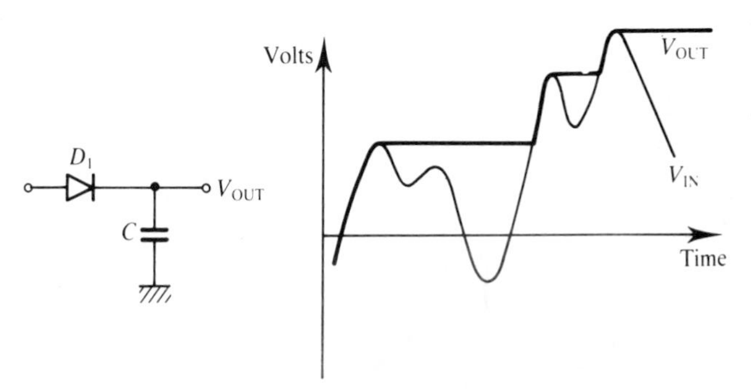

Fig. 11.11. Basic peak detector.

Peak detectors have two distinct modes of operation, either a peak tracking mode (Track) or a peak storage mode (Hold). During the peak tracking mode, the input signal is greater than the previously stored peak value and the output of the peak detector is tracking the input towards a new peak value. In the peak storage mode, the input to the peak detector is less than the previously stored peak value and the output is held constant at a value corresponding to the previous peak. Peak detectors are very similar to sample and hold circuits, both in circuit configurations used and in performance limitations. It should be noted at this point that although this chapter discusses peak detectors, the approach could equally well be used to detect valleys or minimum values since the basic circuit only needs the direction of the diode to be reversed as shown in Fig. 11.12.

Whichever approach is used, this simple detector suffers from major problems. Firstly, there is a 'droop' or slow discharge of the previously stored value through a discharge path provided by the subsequent stage. Generally, droop acts to discharge the capacitor. With more

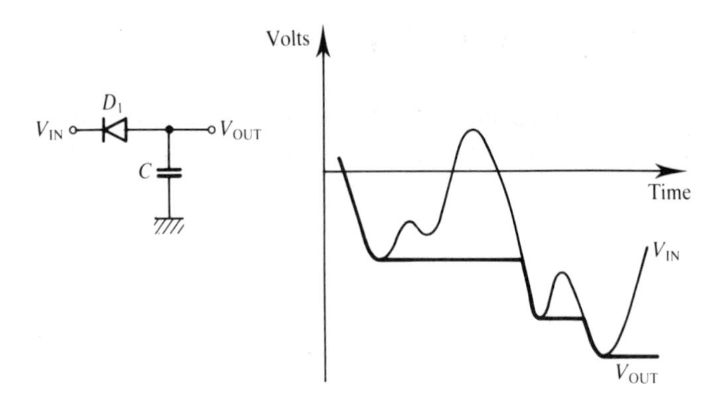

Fig. 11.12. Basic valley detector.

practical peak detector circuits, using op amps, there may be additional unwanted charging of the capacitor due to such things as input bias currents and so on which causes the peak value to increase during the peak storage mode. To confuse you, this movement is also called droop. The second major problem with the simple circuit is the limitation on charging time derived from the capacitor itself, which may determine the shortest duration peak which can be detected, and the maximum slew rate and bandwidth of the peak detector output. The two requirements of reducing droop and increasing slew rate are in conflict with each other when the capacitor value is to be chosen. A large value capacitor, for example, would provide a lower droop rate since it would discharge relatively slowly. The slew rate, though, would also be reduced for the same reason and would not be able to respond adequately to short duration peaks. Once again, the classic engineering situation, where a compromise is required, in this case between slew rate (speed) and droop (dc accuracy).

Usually, the value of the stored peak will be held for only a short period and therefore a resetting mechanism is needed to allow the circuit to respond to a series of peaks. The circuit is modified by using either a reset switch, as shown in Fig. 11.13, or by allowing the droop rate itself to act as a resetting mechanism.

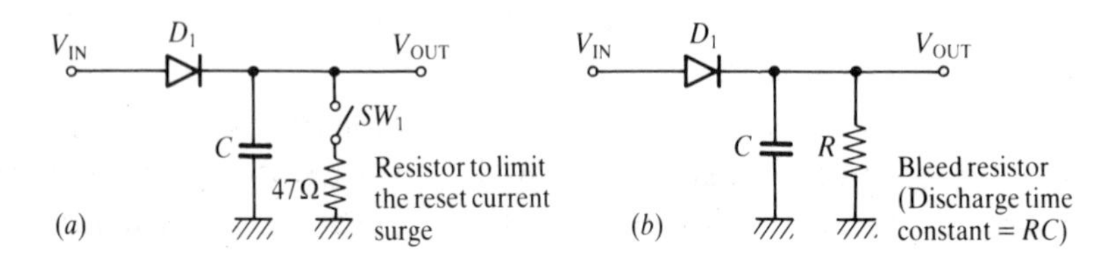

Fig. 11. 13. Resetting a peak detector.

Two stage peak detectors

A basic two stage peak detector is shown in Fig. 11.14. In this circuit, op amp A_1 charges the capacitor up to the peak value and op amp A_2

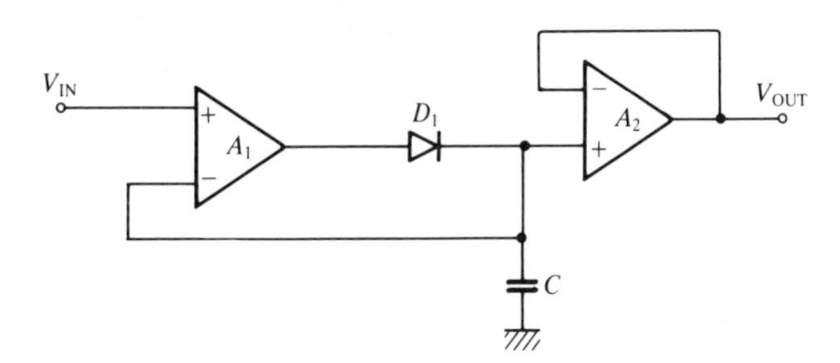

Fig. 11.14. Basic two stage peak detector.

acts as an output buffer. When the input voltage exceeds the voltage stored on C, the output of A_1 swings positive and charges up the capacitor through D_1. So, while the peak value of V_{IN} is increasing, the feedback loop around A_1 is closed through D_1 and the voltage across C follows V_{IN}. Then, when the input voltage reduces from the peak value, the output of A_1 swings into negative saturation. A_1 is now open loop and capacitor C is isolated from the input and stores the peak value.

One of the main causes of unwanted movement of the stored charge is the input bias currents of the op amp flowing into the storage capacitor so increasing its stored charge. With this circuit, C is connected to the inputs of both op amps and so has two lots of input bias currents charging it up. FET input op amps with very low input bias currents can be consequently chosen for many realizations of this design. Also, the input signal has first to go through op amp A_1 before going into A_2, two lots of input offset voltages are thereby added to the signal. So, for most applications, ensure the op amps are chosen to have a low input offset voltage. Since op amp A_1 is the one which actually responds to the peak value, a very fast op amp could be used for A_1 in applications which require only fast captures of peaks and a general purpose op amp can be used for buffer A_2 which may not need to have the fast response. The closed loop response of A_1 must be well damped, otherwise ringing may occur in response to fast peaks so causing overshoot and a false, higher, peak to be stored.

The basic circuit shown in Fig. 11.14 can be modified in a number of ways. In the modification shown in Fig. 11.15, an inverting peak detector with gain is shown. With this modified circuit, the capacitor will

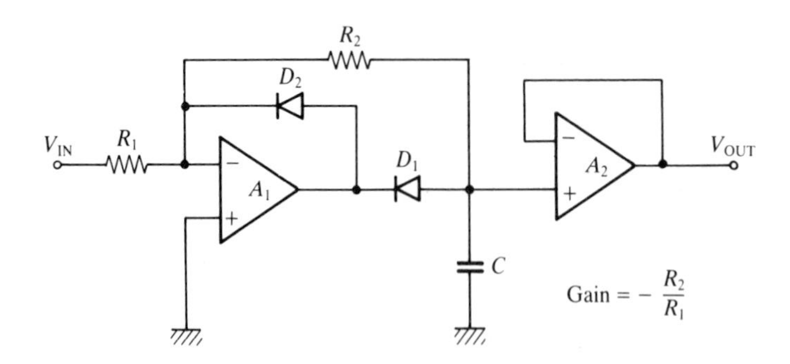

Fig. 11.15. Inverting peak detector.

discharge through R_2 into the virtual earth of the inverting input of op amp A_1. Consequently, the circuit has a high droop with the storage capacitor discharging exponentially with a time constant CR_2 during peak hold mode. The circuit around A_1 is an inverting half wave rectifier. Several of the rectifier circuits described in the later parts of this chapter can be converted into magnitude peak detectors simply by adding a capacitor and a buffer in their output. Note that diode D_2 provides local feedback around A_1 once a peak has been detected. This prevents A_1 from saturating during the peak hold mode and consequently decreases the peak acquisition time. You can omit D_2, but the circuit will be slower at detecting peaks.

In the modified circuit shown in Fig. 11.16, a magnitude peak detector has been created by combining a simple non-inverting peak detector around A_2 with the inverting peak detector around A_1. Droop will again be a problem in some cases since C discharges through R_2 into the virtual earth of the inverting input to A_1 and so this circuit should only be used for applications where a specific amount of droop is required, or the droop can be tolerated.

Notice that the peak value of several signals can be stored by modifying Fig. 11.16 such that each input has its own input stage consisting of A_2, D_3 and D_4 (shown as the dotted box), with each input stage connected to the storage capacitor C and the output buffer A_3.

Overall feedback peak detector

The circuit in Fig. 11.17 shows an overall feedback peak detector. This circuit is very similar to those described earlier and has been treated separately so that the performance differences between the circuits can be emphasized. With peak detector design, one of the most important decisions to be made concerns the configuration, either an individual

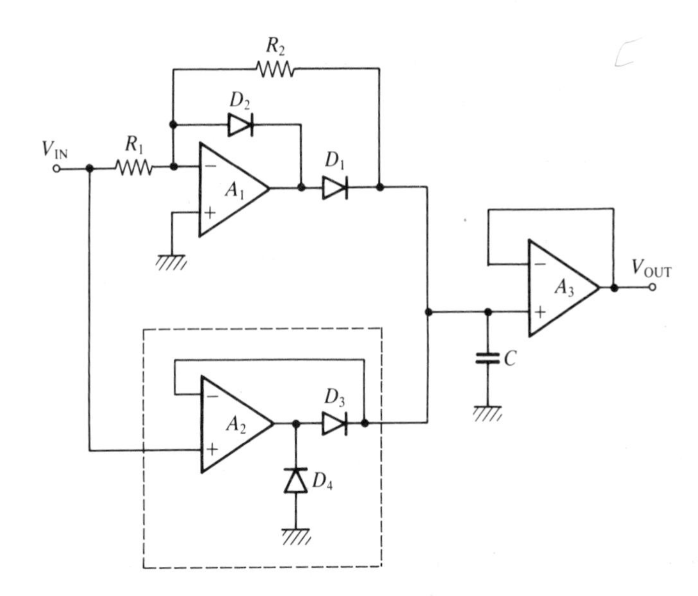

Fig. 11.16. A magnitude peak detector.

loop as described earlier or the overall loop as shown in Fig. 11.17. With the circuit operating in the tracking mode, D_1 is on and A_1 charges up C equal to V_{IN}. Feedback to A_1 is achieved with the buffer A_2. Once a peak has occurred, the input voltage is less than V_{OUT}. As a result, the output of A_1 slews negative, D_1 switches off and the circuit converts into the peak storage mode. The output of A_1 is then clamped at one diode drop below the stored peak voltage by the action of resistor R_1 and D_2, with R_1 typically around 10 kΩ. This circuit is not as fast as the two stage circuit but it does provide better offset and droop performance as explained below.

In this circuit there is only one op amp which is connected to the storage capacitor and so the effects of the total input bias currents of the amplifiers are reduced and therefore less droop occurs. Also, as the circuit consists of an overall feedback around op amp A_1, the output offset is only due to this one op amp, the output offset from A_2 will

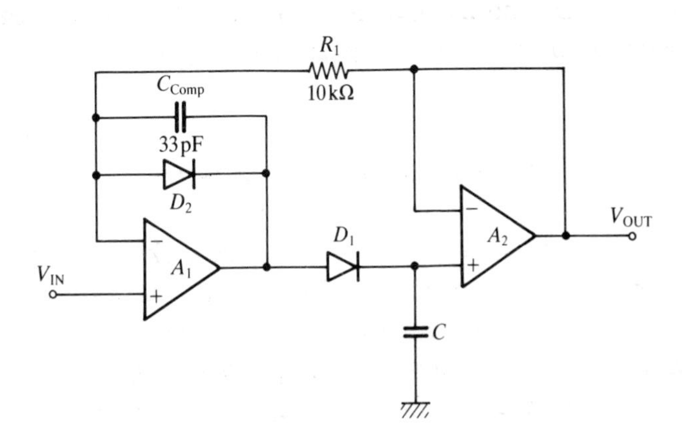

Fig. 11.17. Overall feedback peak detector.

have negligible effects since it is divided down by the gain of A_1. Droop and offset can be minimized by a careful choice of op amps. A_2, for example, could be a FET input amplifier to minimize droop due to bias currents. Op amp A_1, on the other hand, can be a bipolar type for low offset and offset drift.

The dynamics of the overall feedback loop are complicated. Consequently, this circuit is not as fast as the two stage circuit described in the previous section. Peak detectors must have a well-damped response to avoid overshooting, since overshooting will cause an unwanted larger peak to be detected. The circuit in Fig. 11.17 is likely to have an underdamped response which rings slightly due to the added phase shift introduced by A_2 connected as a unity gain buffer. You can damp the circuit by adding capacitor C_{Comp} as shown, with C_{Comp} chosen in the range 10–100 of picofarads depending upon the speed of the op amps. C_{Comp} is omitted from many of the following circuits for the sake of simplicity.

Several modifications of the basic peak detector are shown in Fig. 11.18 including a non-inverting peak detector (*a*), an inverting peak detector (*b*) and a differential input peak detector (*c*). Each of these stages can provide gain.

Several functions can be incorporated into the one peak detector by adding extra op amps as shown in Fig. 11.19.

Improving peak detector performance

The performance limitations of these peak detector circuits are very similar to those of sample and hold amplifiers and the peak detector suffers from similar errors such as droop, feedthrough, pedestal, offset and gain errors, acquisition, aperture and slew rate delays. Some techniques to overcome the droop and speed limitation problems are described below in greater detail.

As discussed earlier, droop occurs mainly due to the flow of leakage and bias currents into or out of the storage capacitor, in the peak storage mode.

$$\text{Droop} = \frac{\mathrm{d}V_{OUT}}{\mathrm{d}t} = \frac{I_{LTOT}}{C}$$

$$I_{LTOT} = I_{LD} + I_{LC} + I_B + I_{LSW} + I_{LSTR}$$

where I_{LTOT} is the total leakage current.

Clearly, as C increases, droop decreases. The total leakage current, I_{LTOT}, is the result of five main factors.

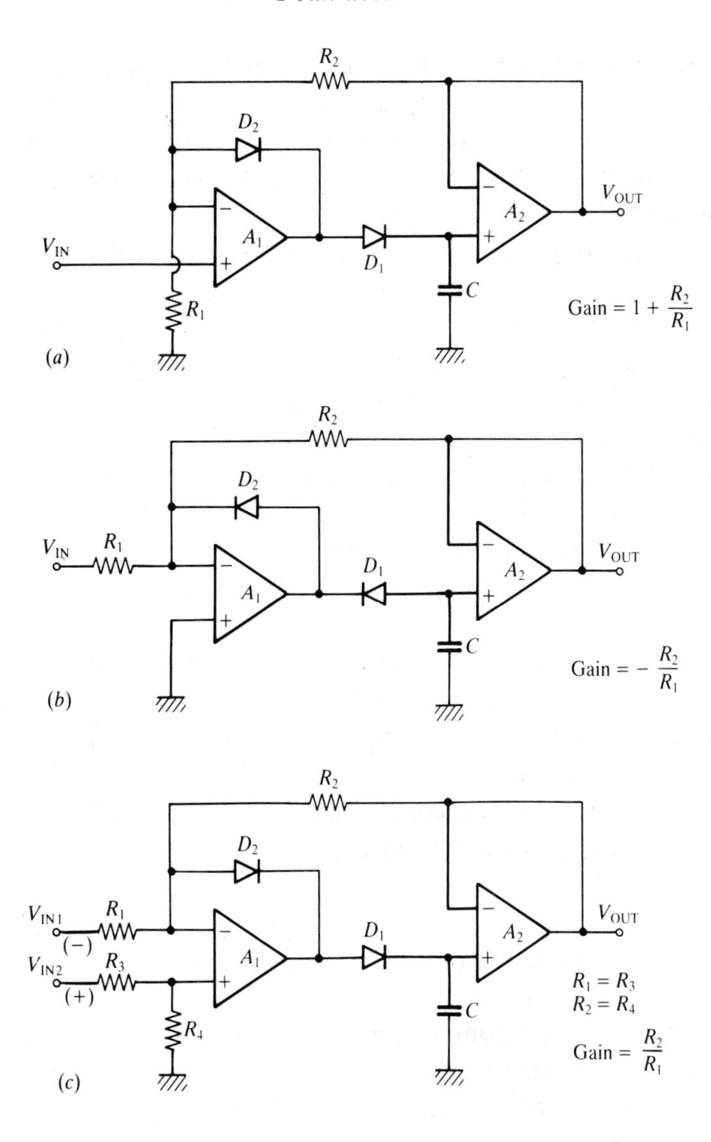

Fig. 11.18. Various overall feedback peak detector circuits. (*a*) Non-inverting peak detector. (*b*) Inverting peak detector. (*c*) Differential input peak detector.

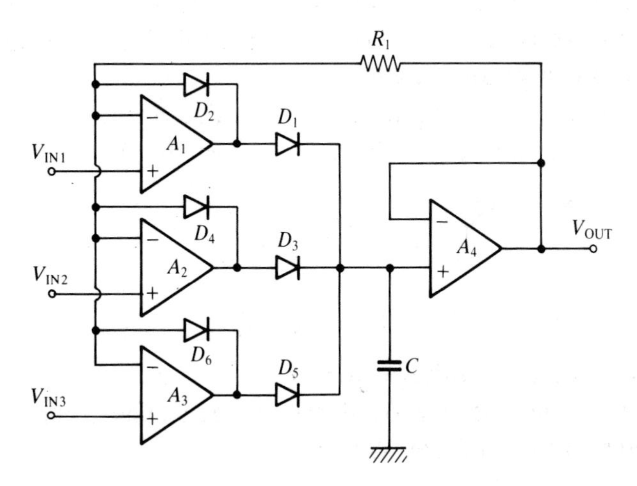

Fig. 11.19. Multiple input peak detection.

(i) I_{LD}, the diode leakage current
(ii) I_{LC}, the internal leakage current of the capacitor ($= V_{OUT}/R_{LC}$ where R_{LC} is the insulation resistance of the capacitor. Polystyrene or PTFE capacitors have a very high insulation resistance of up to 10^6 MΩ/μF)
(iii) I_B, the total input bias currents of the relevant op amps
(iv) I_{LSW}, the switch leakage current
(v) I_{LSTR}, the leakage current through stray paths, such as across the circuit board.

A popular method of substantially reducing droop due to leakage current flowing through the diodes is shown below in Fig. 11.20 and involves the simple addition of an extra resistor and diode. This circuit involves making point A the same voltage as that across the capacitor. As a result, the voltage across D_3 will be zero and so very little leakage current will flow through it. R_2 is typically 100 kΩ and is required to hold point P at a voltage equal to V_{OUT} during the hold mode but without loading the output of A_1 during the peak tracking mode. Very few solutions are completely without their disadvantages. In this case, the cost of this improvement is a decrease in speed since op amp A_1 must slew across three diode drops to change from holding the previous peak to tracking a new peak.

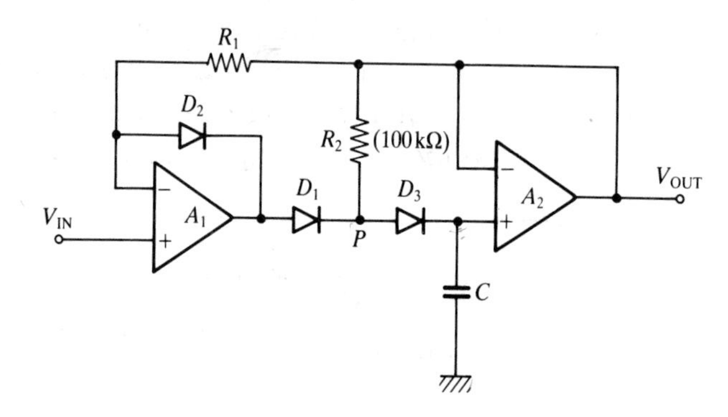

Fig. 11.20. Reducing droop in the detector (1).

Droop due to the flow of op amp bias currents can be reduced by an order of magnitude to the level of the input offset bias current by adding an extra capacitor as shown in Fig. 11.21. Note that a reset switch is needed across both of the capacitors. A further method for reducing droop is to inject a current into the capacitor which will cancel the discharging current as shown in Fig. 11.22.

The circuit shown in Fig. 11.22 is an overall feedback type of detector with A_2 as an integrator. This circuit is intended for long storage times

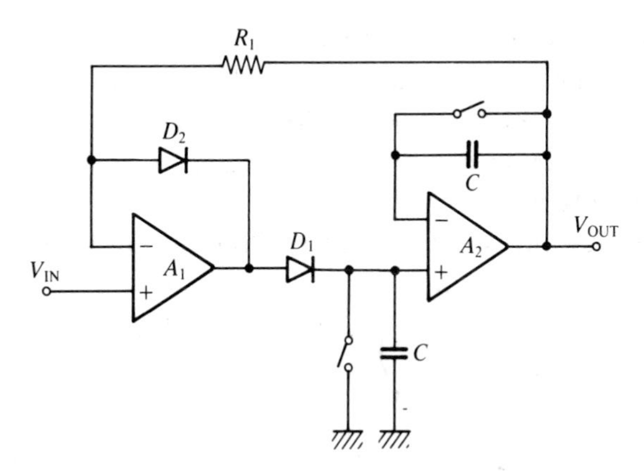

Fig. 11.21. Reducing droop in the detector (2).

with ultra low droop. Consequently, C is as large as practically possible for a good quality capacitor, i.e. up to 10 μF, and op amp A_2 is an ultra low bias current, electrometer grade amplifier with an input bias current of the order of 0.1 pA. The droop due to bias current alone $= I_B/C =$ 0.1 pA/10 μF = 0.01 μV/s. Such a high droop value is not, however, practically achievable since the leakage resistance of the capacitor leaks more current than the input bias current of A_2. The leakage resistance of C is 10^5 MΩ for a very high quality capacitor, so the leakage for a 1 V peak is $1/10^5$ MΩ = 10 pA and there is a corresponding droop of 1.0 μV/s.

To decrease droop even further, op amp A_3 is added which injects current into C to cancel dielectric leakage. Note that A_3 is a simple inverter configuration to ensure that the injected charge is the opposite polarity to the leakage. This circuit can achieve an order of magnitude reduction in droop caused by dielectric leakage. Temperature variations

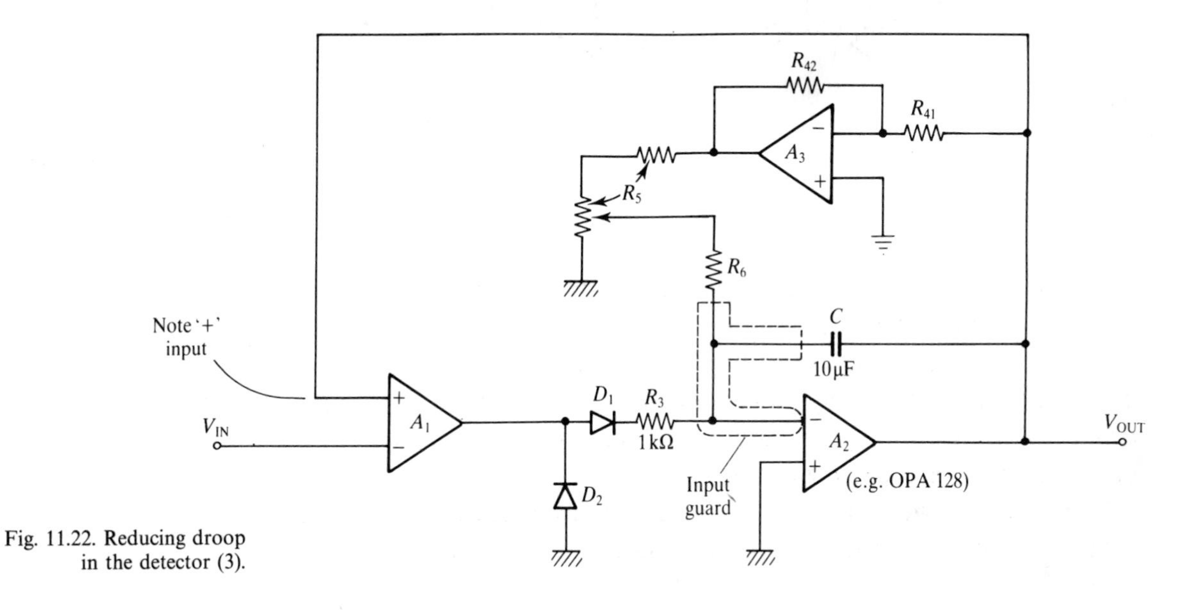

Fig. 11.22. Reducing droop in the detector (3).

in the leakage resistance of C, however, eventually limit its effectiveness. Note that the same technique should not be used to cancel droop due to bias currents since bias currents are much more sensitive to temperature (double every 10°C rise) variations than capacitor leakage resistance.

Typical values:

$R_{41} = R_{42} \simeq 10\ \text{k}\Omega$

R_6 is as high as possible, use a high ohmic resistor ($>100\ \text{M}\Omega$)

R_5 will have to divide the output of A_3 down by up to 1000:1.

To realize low droop, a guard ring is essential to prevent parasitic leakage currents flowing across the surface of the circuit board and into the storage capacitor. It is also worth using special mounting arrangements such as a PTFE stand-off to avoid leakage through the circuit board. Careful cleaning of the circuit board is essential also to remove contaminations which might increase leakage current flows such as solder flux.

The speed at which the peak detector can follow a new peak value is generally limited by either the slew rate of the amplifiers or by how quickly the storage capacitor can be charged by the maximum output drive current, I_{OUTMAX}, as given by I_{OUTMAX}/C.

Clearly, as the value of C increases, the speed reduces.

The slew rate of the peak detector can often be increased by current boosting using a single transistor as shown in Fig. 11.23. In this circuit, D_1 protects the base–emitter junction of the transistor against reverse breakdown while the circuit is storing the previous peak. R_2 is also reducing droop by ensuring a near-zero voltage across the base–emitter junction of the transistor as before. R_2 is typically 100 kΩ. R_3 limits the charging current through Q_1 to prevent damage to Q_1. The maximum operating base current of the transistor must be greater than

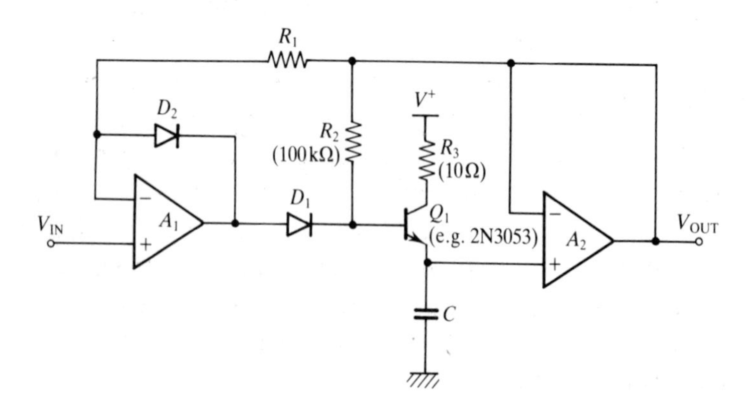

Fig. 11.23. Using a discrete transistor for improving slew rate.

the output current drive of the op amp, otherwise the op amp will destroy the transistor.

The main factor governing the speed at which fast changes in the peak value of the input may be followed is the finite slew rate of the op amp A_1. For a step change in the peak value of the input, the output of A_1 must slew from the input level before the step change The new peak can then be tracked. There are two ways to speed up this transition. Firstly, the slew rate of A_1 can be maximized by choosing a very fast op amp. Secondly, the voltage range over which A_1 must slew can be limited, for example by using Schottky diodes with a slow forward voltage and using the technique shown in Fig. 11.24.

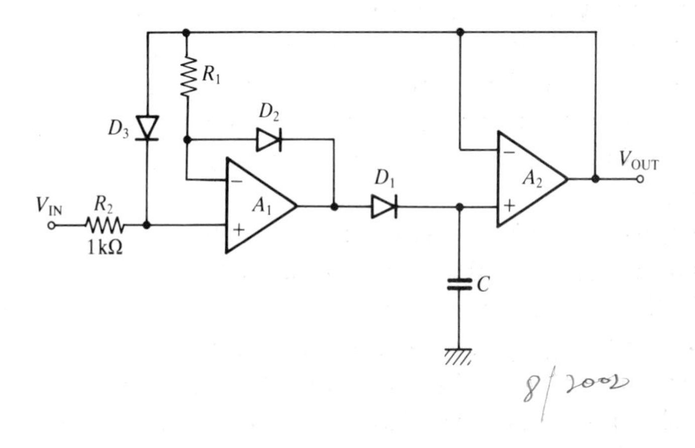

Fig. 11.24. Increasing the acquisition speed of a peak detector.

In this circuit, diode D_3 and resistor R_2 prevent the non-inverting input of A_1 from dropping more than one diode drop below the output voltage. Consequently, the output of A_1 will not drop more than two diode drops below the output. Resistor R_2 isolates the D_3 clamp from the input. R_2 is the input resistance of the circuit during the storage mode. The result of this arrangement is that the voltage range which A_1 must slew across is limited to a maximum of two diode drops, before a new peak can be tracked.

To reset the peak detector, an analog switch is commonly used, such as a FET or reed relay. A simple method of switching is to use a MOSFET as shown in Fig. 11.25. In general, choose a FET with a leakage current at least an order of magnitude smaller than the input bias current of the op amp. However, this will be difficult with electrometer grade op amps since their input bias currents are so low. Note that with these switches and particularly with reed relays, a small value resistor may be needed in series with the switch to limit the maximum reset current.

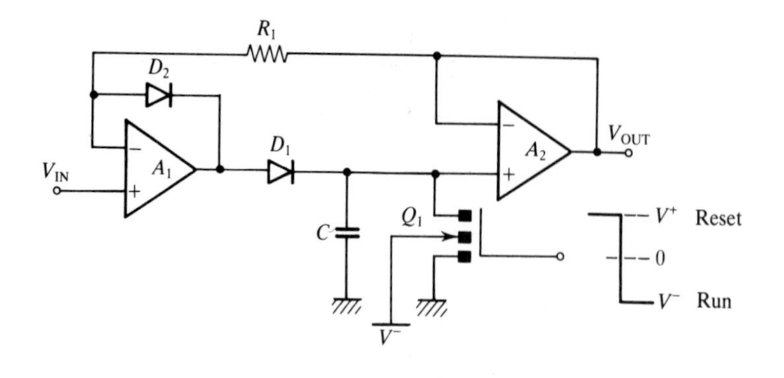

Fig. 11.25. Resetting the detector.

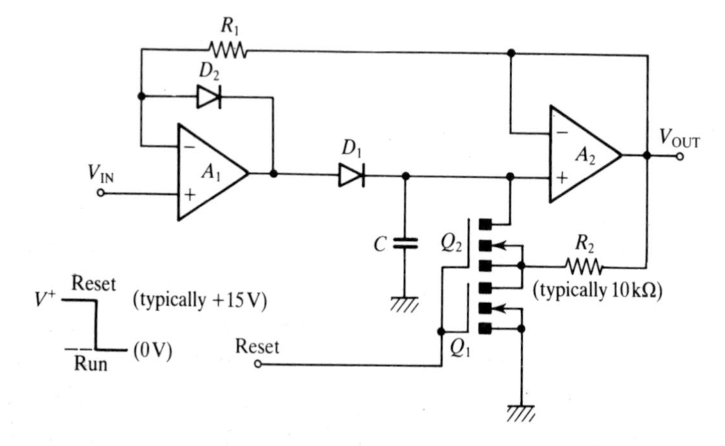

Fig. 11.26. Reducing reset switch leakage current.

A technique for reducing the gate–source leakage current through the FET is shown in Fig. 11.26 where an extra switch is employed with a bootstrap resistor R_2 connected across the switch to maintain zero voltage across it.

As a final point on this theme, you can obtain a fast peak detector with low droop by cascading a low droop peak detector and a fast peak detector as shown in Fig. 11.27(*a*). Lastly, a peak-to-peak detector can also be configured using a peak detector, a valley detector and a differential amplifier as shown in Fig. 11.27(*b*).

It is worth noting, lastly, that the negative-going output of op amp A1, in some of the previous peak detector circuits, which occurs as the circuit switches from peak tracking to peak holding, can be used with a comparator to indicate the detection of a new peak.

Notes on component selection

The capacitor value There are two conflicting requirements concerning the value of capacitor to be used. A high value capacitor gives low

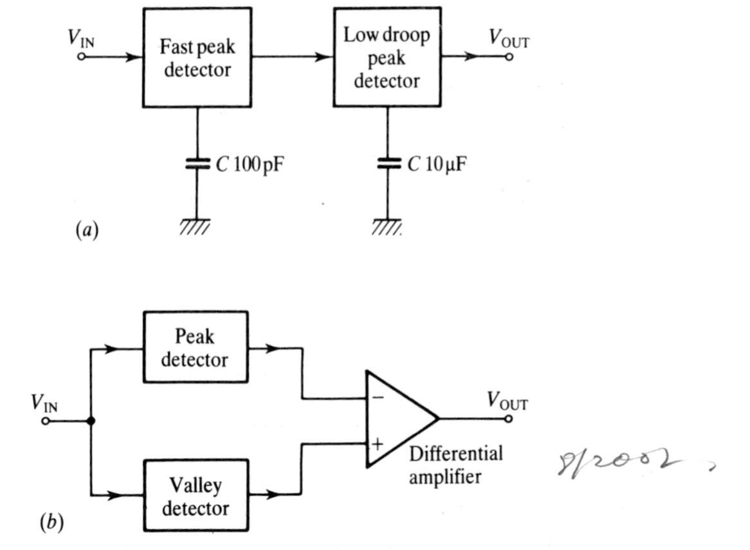

Fig. 11.27. Peak detector variations. (*a*) Fast composite peak detector. (*b*) Peak-to-peak detector.

droop, i.e. it is good at storing the peak value. This high value of capacitance also, however, reduces the slew rate at the output of the charging op amp(s) which thereby reduces the speed at which the peak can be captured and limits the fastest peaks to be detected. Obviously, some kind of compromise between the two requirements is needed. As an example, consider a FET input op amp with 50 pA input bias current and output current drive of 20 mA and slew rate of 10 V/μs, all of which are typical values. A typical range of values for C is from 100 pF to 10 μF. Various droop values and slew rates are shown here for capacitors within that range.

	C	$S_r = I_{OUTMAX}/C$ (V/μs)	Droop $= I_B/C$
High speed			
High droop	100 pF	10	0.5 V/s
	1000 pF	10	0.05 V/s
	0.01 μF	2	5 mV/s
	0.1 μF	0.2	500 μV/s
	1 μF	0.02	50 μV/s
Low speed			
Low droop	10 μF	0.002	5 μV/s

The capacitor type Generally, it is best to use good quality capacitors with a high insulation resistance, some values for different types of capacitor are shown in the Table.

Type	Range	Insulation resistance
polystyrene	up to 0.01 μF	10^6 MΩ
polycarbonate	up to 10 μF	10^5 MΩ–μF typical at 25°C
polypropylene	up to 0.047 μF	10^5 MΩ–μF
Teflon	up to 0.01 μF	10^6 MΩ–μF

General purpose polyester or ceramic capacitors are a bad choice due to their lower insulation resistance (polyester 10^4 MΩ, ceramic <5000 MΩ). If the peak is fast and stored for times approaching a second or more, then dielectric absorption may be important for high accuracy applications. The amount of charge absorbed, (i.e. lost) for four capacitor types are as follows

polystyrene	0.05%
polypropylene	0.05%
polycarbonate	0.5%
Teflon	0.01%

The op amp The op amps connected to the storage capacitor can be FET input types to minimize droop due to the input bias currents (FET input types ≃ 50 pA, bipolar input types ≃ 10 nA). Bipolar input type op amps should only be considered for these op amps in very short peak storage times. For extremely low droop rates, use an electrometer grade op amp such as the AD515 or the OPA128. The input op amp of the overall feedback configuration is best as a bipolar type to minimize output offset voltages, where FET input op amps compare unfavourably with bipolar types. For ultra low offsets, use a chopper stabilized op amp.

The diodes For most applications which do not require an ultra low droop value, general purpose low power signal diodes are quite adequate, e.g. the IN4148. For very low droop applications, however, it is possible to use a low leakage JFET connected as a diode as shown in Fig. 11.28.

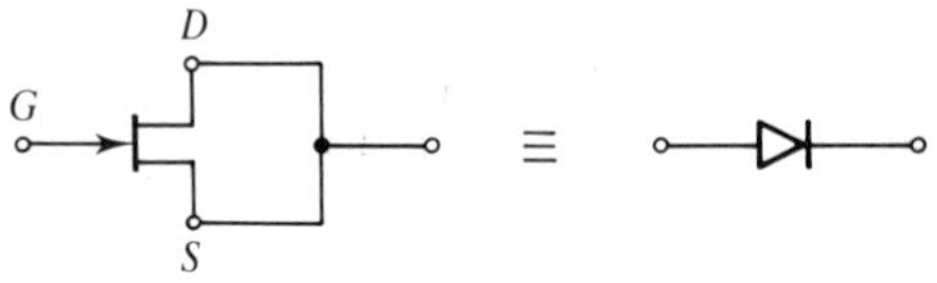

D
G
S
≡

Fig. 11.28. Connecting a JFET as a diode.

A popular low leakage JFET is the 2N4117A. Alternatively, use ultra low leakage diodes such as the ID101 from Intersil. For high speed applications, silicon Schottky barrier diodes are worth considering since they have a low forward 'on' voltage of only 0.4 V (silicon p–n diodes $\simeq$ 0.7 V). Consequently, the op amp outputs have less voltage to slew across when detecting fast peaks. Be careful, though, since some silicon Schottky barrier diodes have low reverse voltages (i.e. less than 10 V).

Peak detectors using ic building blocks

As an alternative to realizing the peak detector circuits described earlier out of individual op amps, diodes, resistors and capacitors, it is possible to construct these circuits using dedicated analog functional blocks such as peak detector ics, sample and hold circuits and comparators. In addition, this section also describes a different peak detector circuit using a D/A converter: a peak detector which has zero droop. One ic, for example, which contains all of the necessary op amps, diodes and reset switches is the PKD-01 from Precision Monolithics.

A peak detector can be realized very conveniently with two ics as shown in Fig. 11.29, where you can see a sample and hold being used alongside a comparator. For inputs less than the previously stored peak value, the comparator output is low and so the sample and hold is in the hold mode. If the input goes greater than the previously stored peak, then the comparator output goes high and the sample and hold is switched into sample mode. As a result, the input is tracked by the sample and hold until a new peak value is obtained. You can add a small amount of hysteresis (a few mV) to the comparator for increased noise immunity.

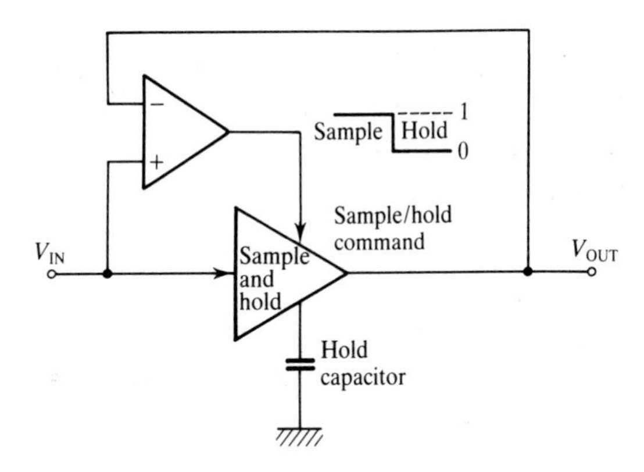

Fig. 11.29. Peak detection using a sample and hold.

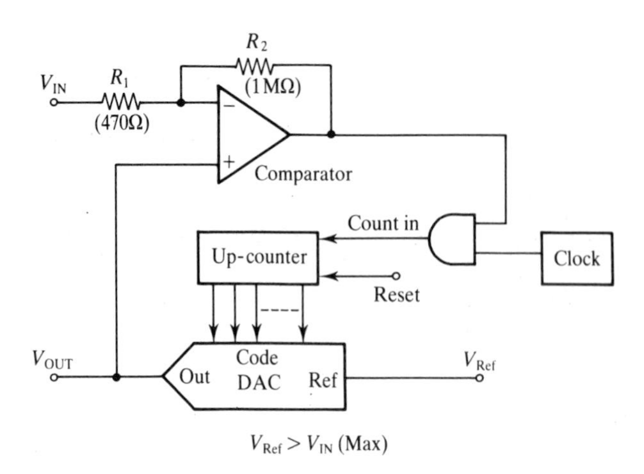

Fig. 11.30. Peak detection using a D/A converter.

The ever-versatile DAC can be used as shown in Fig. 11.30 to provide peak detection circuitry. The output from the comparator gates the clock into the up-counter until V_{IN} is equal to V_{OUT}. At this stage, the comparator blocks the clock until the next time $V_{IN} > V_{OUT}$. The great advantage of this approach is that the peak value of V_{IN} is stored digitally, rather than as a charge on a capacitor, and so there is zero droop of the stored peak. A small amount of hysteresis may be added to the comparator to increase noise immunity by using feedback around the positive input of the op amp shown in Fig. 11.30.

11.3 Half wave precision rectifiers

The four half wave rectifiers shown in Fig. 11.31 have been classified according to whether they are inverting or non-inverting types and also according to whether they pass the positive portion of the signal or the negative portion. Non-inverting half wave rectifiers have a much higher input impedance than inverting rectifiers. In the inverting type, D_2 conducts to allow the signal to be amplified while D_1 is reverse biased. In the non-inverting type, D_2 conducts to allow the appropriate polarity of the signal to be amplified via R_1 and R_2 while D_1 is reverse biased to block the opposite polarity signal.

Diode D_1 is added to the non-inverting rectifier to speed up its operation. If this diode is not used, the op amp would saturate when the circuit is at zero output. Consequently, the transition between positive and negative signals would take longer because the op amp must first recover from saturation and then slew over a large voltage range before rectification takes place. Adding diode D_1 prevents the op amp from saturating and limits the voltage range over which the

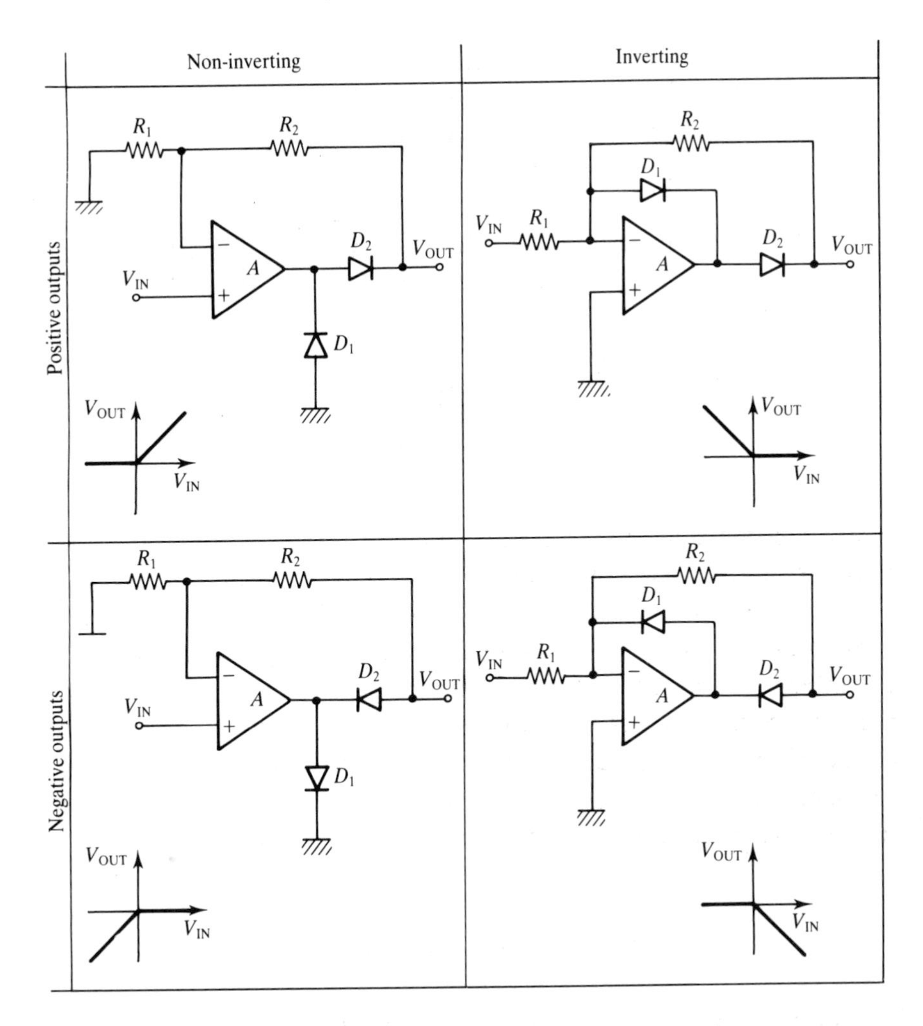

Fig. 11.31. Half wave rectification.

op amp output must slew for the circuit to switch between input signal polarities. Diode D_1 presents a hard clamp to the output of the op amp in the non-inverting circuit and so the op amp must be able to tolerate an output short circuit indefinitely at the maximum operating temperature. In addition, with the non-inverting circuit, the op amp must be able to tolerate a large differential input voltage and display a fast recovery from an input current limit.

Loading the output with a capacitor will change the circuit into a simple peak detector. After a peak has occurred, the output will droop exponentially with a time constant of:

inverting $\quad C_L(R_2//R_L) = C_L \cdot (R_2R_L)/(R_2 + R_L)$

non-inverting $\quad C_L(R_L//(R_1 + R_2)) = C_L \cdot (R_1 + R_2)R_L/(R_1 + R_2 + R_L)$

where C_L is the capacitor value and R_L is the load resistance.

11.4 Full wave rectifiers

Single op amp full wave rectifier (Fig. 11.32)

In this circuit, positive input signals forward bias D_1 and reverse bias D_2; D_1 holds the inverting input at virtual earth. Output is given by the divider network formed partly by R_3 and partly by R_2 in parallel with R_L. For negative signals, D_2 is ON and D_1 is OFF. The circuit now acts as a non-inverting amplifier. For each polarity the circuit acts alternatively as a dividing network and a non-inverting amplifier.

Advantages	Disadvantages
Uses only one op amp	Low input resistance which differs for +ve and −ve signals
	Source and load resistance affect the rectifying action
	Cannot tolerate reactive loads
	Requires three matched resistors
	Attenuates the input signal

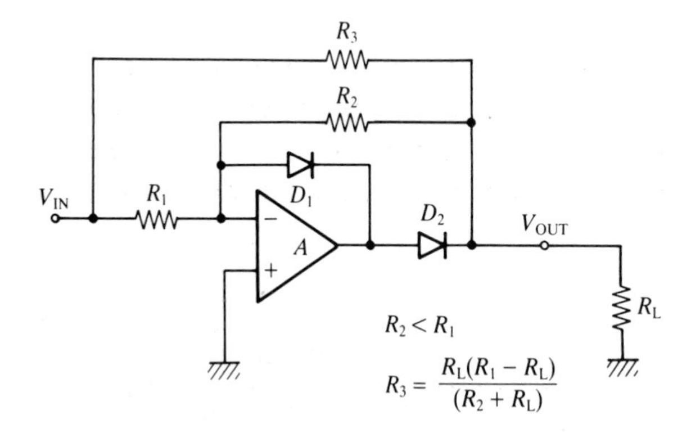

Fig. 11.32. Single op amp full wave rectification.

Gain (A_V)

$$\text{Positive input } A_V = (R_2//R_L)/(R_3 + (R_2//R_L)$$

$$= (R_2R_L)/(R_3R_2 + R_3R_L + R_2R_L)$$

$$\text{negative input } A_V = -R_2/R_1$$

resistor values for equal positive and negative gain $\Rightarrow$ $R_3 = R_L(R_1 - R_2)/(R_2 + R_L)$

$R_1 = R_3 + R_2 \qquad (\text{if } R_L = \infty)$

Input resistance (R_{IN})

$$\text{positive input } R_{IN} = R_1//(R_2 + R_3)$$
$$= R_1(R_2 + R_3)/(R_1 + R_2 + R_3)$$
$$\text{negative input } R_{IN} = R_1//(1 + R_2/R_1)R_3$$
$$= R_1R_3(R_1 + R_2)/(R_1{}^2 + R_3(R_1 + R_2))$$

Transition time from inverting to non-inverting operation

$$T = 2V_D S_r$$

(assuming fast diodes and small stray capacitances) where V is the forward diode voltage and S_r is the slew rate of the op amp.

This circuit is best used between a low impedance source and a high impedance load. High source and low load impedances will cause gain errors and gain mismatch errors. If $R_3 = R_2 = R_1/2$ then the circuit will have a gain of $\frac{1}{2}$. A matched resistor network of equal resistors (R) can then be used where $R_3 = R_2 = R$ and R_1 consists of two Rs in parallel. Putting a capacitor C on the output will change the circuit into an absolute value peak detector.

Current output full wave rectifier (Fig. 11.33)

This circuit consists of an op amp with a full wave rectifier in its feedback path. The bridge rectifier rectifies both polarities of the input signal and delivers a current output into its floating load Z_L. Resistor R_1 ensures a current output. Also, resistor matching is not needed and R_1 can be varied to alter the gain.

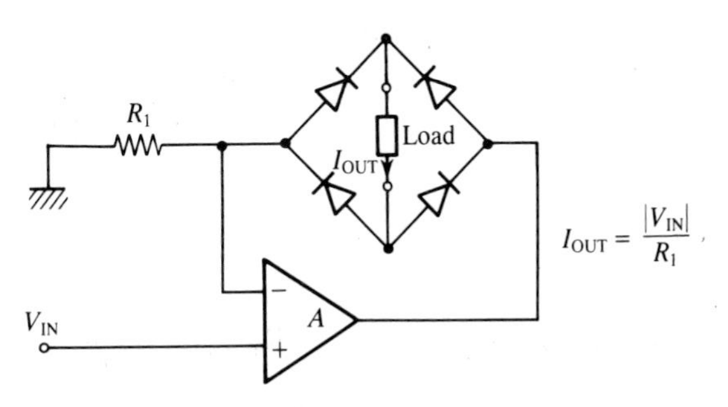

Fig. 11.33. Current output full wave rectification.

Gain $\qquad I_L = |V_{IN}|/R_1$

Input impedance = common mode input impedance of the op amp

Changeover speed $= 4V_D/S_r$ where S_r is the slew rate and V_D is the forward diode voltage.

The maximum load current must not be high enough to saturate the op amp, i.e. $I_L < (V_{SAT} - V_{IN} - 2V_D)/Z_L$ where Z_L is the load resistance and V_{SAT} is the op amp saturation voltage.

Current input full wave rectifier (Fig. 11.34)

With this approach, for negative input currents, D_2 is ON and D_1 is OFF. For positive input currents, D_1 is ON and D_2 is OFF with the op amp now operating as a voltage follower. For equal gain for both positive and negative input currents, choose R_1 to equal R_2.

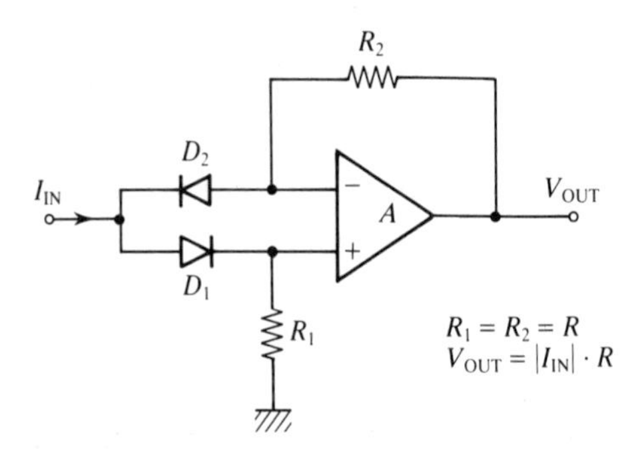

Fig. 11.34. Current input full wave rectifier.

Two op amp full wave rectifier with minimum components

This circuit (Fig. 11.35) combines an inverter/buffer based on op amp A_1 and an active clamp based on A_2. For positive inputs, the clamp A_2 does not affect the circuit since diode D_1 is reverse biased. A_1 acts, under this condition, as a unity gain buffer. The output of A_2 is clamped at a negative voltage by D_2. For negative inputs, diode D_1 is forward biased which causes the non-inverting input of A_1 to be clamped to ground. Under this condition, A_1 acts as an inverting amplifier.

Advantages	Disadvantages
Small number of external components	Fixed gain of unity
Low output impedance from A_1	Low input resistance which changes for +ve and −ve inputs
Requires only one pair of matched resistors	Slow changeover from +ve to −ve inputs

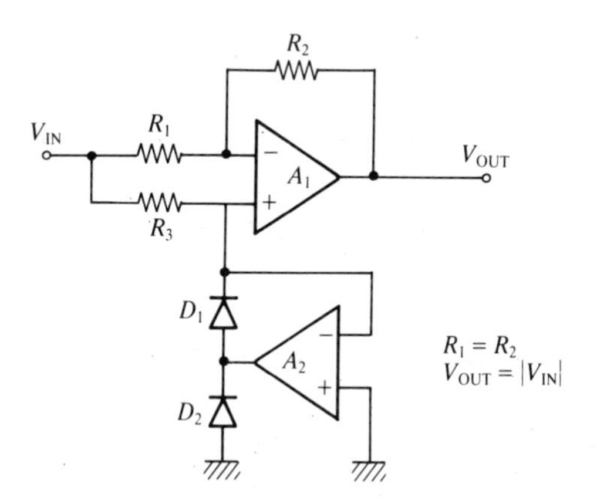

Fig. 11.35. Two op amp full wave rectifier.

Gain (A_0)

$$\text{positive input} \quad A_V = 1$$

$$\text{negative input} \quad A_V = -R_2/R_1$$

Resistor ratio for equal positive and negative gains

$$R_1 = R_2$$

Input impedance

$$\text{positive input} = \text{common mode input resistance of } A_1$$

$$\text{negative input} = R_1 /\!/ R_3 = (R_1 R_3)/(R_1 + R_3)$$

The input resistance of the circuit is low and varies according to the polarity of the input signal. High source resistances, consequently, will not only cause gain errors but also errors in matching the positive and negative signal gains. Resistor R_3 is generally made equal to $\frac{1}{2}R_1$ but it need not be matched with the same accuracy.

General purpose full wave rectifier (Fig. 11.36)

This circuit is a good, general purpose precision rectifier with high input impedance, low output impedance and requires only one pair of matched resistors for unity gain. With positive input signals, D_1 is OFF and D_2 is ON. Feedback is applied around both A_1 and A_2 via the three resistors R_1, R_2 and R_3. Feedback forces the inverting input of A_1 to V_{IN} thus setting the output voltage. With negative input signals, D_1 is ON and D_2 is OFF. A_1 now acts as a unity gain buffer and A_2 as an inverter with a gain of R_2/R_1.

Advantages	Disadvantages
High input impedance	Closed loop gain may require frequency compensation with capacitor C
Low output impedance from A_2 will drive most loads	
Circuit requires only one pair of matched resistors for gain of unity ($R_3 = \infty$)	

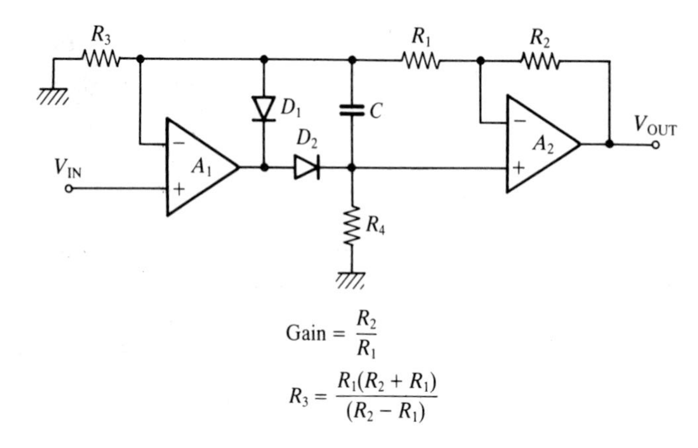

Fig. 11.36. General purpose full wave rectifier.

Gain (A_V)

$$\text{positive input} \quad A_V = (R_1 + R_2 + R_3)/R_3$$

$$\text{negative input} \quad A_V = R_2/R_1$$

Resistor ratio

$$R_3 = R_1(R_2 + R_1)/(R_2 - R_1)$$

With positive inputs, both op amps are in the feedback loop. This causes problems with frequency stability and so capacitor C may be required to provide frequency compensation. Values of C around 100 pF are usual.

If $R_3 = \infty$ (i.e. not in the circuit), and $R_1 = R_2$, then the circuit will have a gain of unity and only requires matching of R_1 and R_2.

Current summing full wave rectifier (Fig. 11.37)

In this approach, for positive inputs, D_1 is ON and D_2 is OFF. This configures op amp A_1 as an inverter. A_2 is also acting as an inverter with

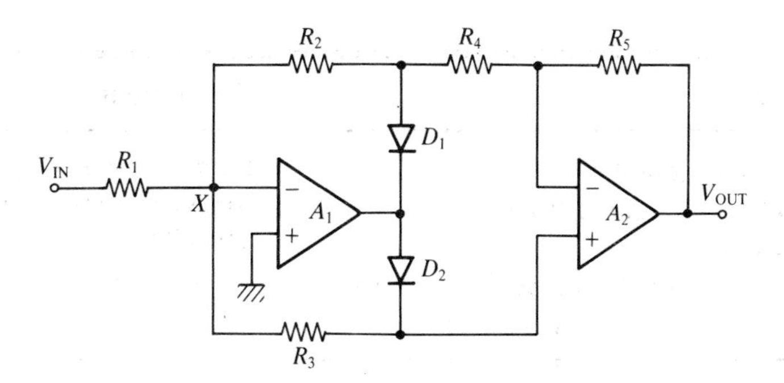

Fig. 11.37. A current summing full wave rectifier.

its non-inverting input at virtual ground via resistor R_3 (point X). For negative inputs, D_1 is OFF and D_2 is ON. This configures A_1 as an inverter driving the non-inverting input of A_2 which results in an overall negative gain.

Advantages	Disadvantages
Current summing at point X allows several signals to be added	Requires several matched resistors
Low output impedance from A_2 can drive most loads	

Gain (A_0)

positive inputs $A_V = (R_2/R_1) \cdot (R_5/R_4)$

negative inputs $A_V = R_3(R_2 + R_4 + R_5)/R_1(R_2 + R_3 + R_4)$

Resistor ratios

$$R_2R_5/R_3R_4 = (R_2 + R_4 + R_5)/(R_2 + R_3 + R_4)$$

This circuit allows several signals to be added since it will sum input currents to point X. Usually, $R_2 = R_3 = R_4 = R_5 = R$ which gives a circuit gain of R/R_1. A resistor network in the same package can then be used for the four equal-value resistors with the value of R_1 being used to control the gain.

General notes on rectifier circuits

You can remove dc offset errors by using the offset nulls of the op amps. Reducing the op amp offsets to zero, however, will not give the minimum overall circuit dc offset error since overall dc offset error is dependent upon the point at which the diodes switch and not on any zero op amp offset. Diodes switch when the current through them is zero and the diode current includes dc leakage currents and the op amp bias current. Op amp offsets, consequently, should be adjusted for zero dc error at the output when the input signal crosses zero. With two op amps, the adjustments should be done at the same time since the offset adjustments of one will affect the offset errors of the other op amp.

For fast switching, low resistor values should be chosen since they allow the diode capacitances and stray capacitances to be quickly discharged. Low resistor values also help to reduce offsets due to bias and leakage currents.

You should also choose an op amp with a high slew rate if speed is important. A high slew rate op amp is required since the output of the op amp has to change from forward biasing one diode or set of diodes to forward biasing the other diode or diodes.

When the op amp of a rectifier is slewing during changeover, the diodes are OFF and the op amp is operating, effectively, in its open loop mode. Since the op amp is in open loop mode, compensation can be removed to speed up the op amp so increasing its slew rate and providing a faster changeover. One technique for improving switching times involves using an op amp in the rectifier circuit which is compensated by a capacitor, connected externally, between the output of the op amp and its compensation pin. In this technique, the diodes are used to remove the compensation capacitor C during changeover so that the capacitor is only connected when it is needed for stability and not during transitions of the op amp. Fig. 11.38 shows the configuration of the diodes and the compensating capacitor for an inverting half wave rectifier.

If a negative output is required from a rectifier, then one method is to change the direction of all of the diodes in a circuit so that they

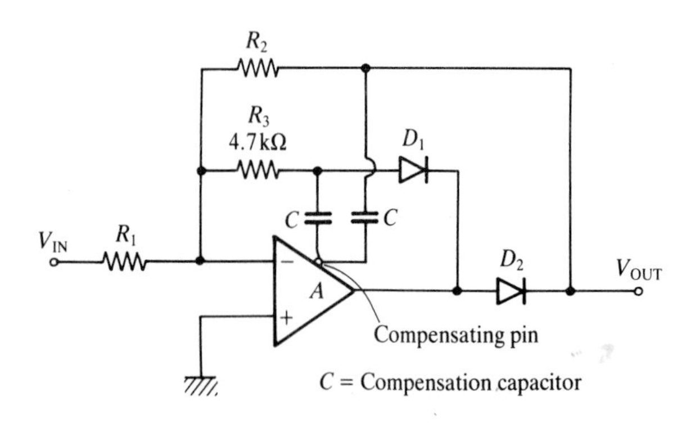

Fig. 11.38. Frequency compensation with improved slew rate.

conduct in an opposite direction to that shown in the circuit diagram. This method will work with all of the previously described configurations.

For most applications, general purpose, low power signal diodes such as the IN4148 are adequate. For high accuracy applications, however, where offset errors are critical and reverse diode leakage current must be very small, then either an ultra-low leakage diode such as the ID101, or alternatively a low leakage FET connected as a diode such as the 2N4117A, may be used. In applications where speed is important, silicon Schottky barrier diodes are worth considering since they have a low forward ON voltage of only 0.4 V (silicon p–n diodes $\simeq$ 0.7 V). Consequently, the outputs from the op amps have less voltage to slew across when detecting peaks. Note, though, that some silicon Schottky barrier diodes have comparatively low reverse voltages (i.e. less than 10 V).

12
Peak, average and RMS circuits

With DC signals, only two quantities are required to specify the signal exactly, the magnitude of the signal and its polarity. AC signals are much more difficult to measure. To measure all the properties of an AC signal with maximum effectiveness, a series of many samples of the waveform must be taken at a rate of sampling (by Shannon's sampling theorem) which is at least twice the maximum signal frequency content to counter aliasing distortion and produce a series of values which are a faithful representation of the sampled signal. In many applications, however, all that is required for a measurement of the signal is a single value. The three parameters used for measuring the size of a signal are the peak value, the mean absolute value (MAV) and the root mean square (RMS). The peak value, $E(\text{peak})$, is variously described as the maximum value of amplitude or the peak or peak-to-peak size of the signal.

The mean absolute value, MAV, is the average value of the rectified AC signal,

i.e.
$$E(\text{MAV}) = \frac{1}{T}\int_{t=0}^{T} |e(t)|\, dt$$

where $e(t)$ is the AC signal and T is the time interval over which the average value is of interest; usually T is very much longer than the period of the waveform.

The root mean square value (RMS) is, for the reasons outlined below, the most important parameter for measuring the size of a signal. As the name suggests, this parameter consists of the square root of the average value of the square of the signal and is defined as

$$E(\text{RMS}) = \sqrt{\frac{1}{T}\int_{t=0}^{T} e(t)^2\, dt}$$

root mean square

The RMS value is such an important measurement for the following reasons:

(i) It is the most common way of describing the size of an AC signal; in most cases, the RMS value is assumed where it is not explicitly stated.

(ii) The RMS value is a measure of the energy content in the signal since the RMS value of an AC voltage or signal is defined as being equivalent to that DC voltage or current which has a similar heating effect to the AC signal when applied to an identical resistor.

(iii) The RMS value is useful in some statistical operations; for example, with any stationary, zero mean random signal, such as white noise, the RMS value is equal to the standard deviation of the signal.

(iv) When two orthogonal or uncorrelated signals are summed, for example where noise sources are added together, the RMS value of their sum is equal to the square root of the sum of the square of their individual RMS values.

Two parameters are commonly used to relate the peak value, the RMS value and the mean absolute value: the form factor and the crest factor. The form factor is the ratio of $E(\text{RMS})$ divided by $E(\text{MAV})$, i.e. $E(\text{RMS})/E(\text{MAV})$. The crest factor is the ratio of $E(\text{peak})$ divided by $E(\text{RMS})$, i.e. $E(\text{peak})/E(\text{MAV})$. Note that the crest factor is a measure of the 'spikiness' of a waveform and is equal to 1 for a square wave and a large value (>10) for a narrow pulse train.

Table 12.1 shows the wide variations in peak, MAV and RMS values for different AC waveform shapes. These variations can cause significant problems when older or low cost types of digital voltmeters are used for measuring the RMS value of a signal. These problems arise due to the cost and complexity of the circuitry for RMS to DC conversion. The technique which has evolved with these voltmeters is to measure the MAV of the input waveform since the MAV is, from the circuit viewpoint, a relatively easy value to measure as will be shown in the following sections. These instruments are then calibrated to read RMS for a sine wave input and consequently need scaling up by 1.11 to give the correct value. With any other waveform input, as can be seen from inspecting the various values of form factor in Table 12.1, an error is introduced into the measuring instrument which is unavoidable. Consequently, significant measuring errors may be present on the cheaper RMS measuring DVMs when non-sine wave AC signals, or a sine wave is combined with a DC component, are being measured. The advent of low cost, high performance RMS to DC converter ics has led to the effective elimination of these types of errors in good quality DVMs.

12.1 Peak responding circuits

The two circuits shown in Fig. 12.1 are intended merely to highlight the nature of peak responding circuits. For fuller details on peak detector circuits, the reader is referred to Chapter 11. In the circuit of Fig. 12.1(*a*), the simplest form of peak detector is shown with D_1 half wave rectifying the ac signal and capacitor C holding the peak charge with R and C determining the time constant of the circuit. The circuit shown in Fig. 12.1(*b*) is a more sophisticated peak detector but, once again, there is a rectifier and a reservoir capacitor with the time constant determined by the values of R and C. One of the most important design decisions with this type of circuit is the choice made for the time constant RC. If you choose too large a value, the output will be very smooth but the circuit will not respond quickly to reductions in input signal size. If you make the time constant too small, however, the circuit will be able to respond quickly to reductions in signal size but the output voltage will now contain a large ripple content. Alternatively, resistor R can be replaced with an analog switch, which will enable these circuits to sample and hold the peak value of the input signal over a specific time interval. The analog switch is closed to allow the circuits to be reset. In some applications, a precision full wave rectifier may be needed before the input to the peak detector circuit so that the maximum peak, positive or negative deviation, can be assessed.

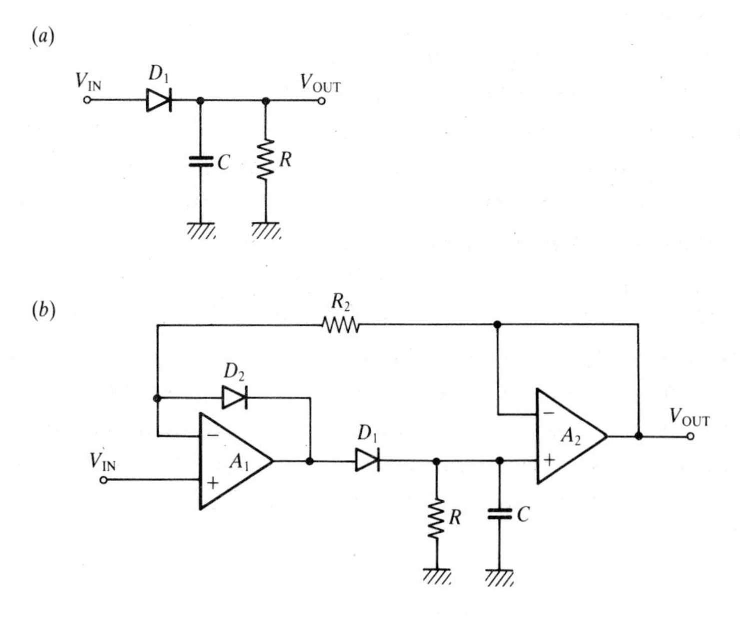

Fig. 12.1. Peak responding circuits. (*a*) Simple peak detector. (*b*) Improved peak detector.

Table 12.1. *AC signal measurement parameters*

Waveform	Peak	RMS	MAV	F.F.	C.F.
DC 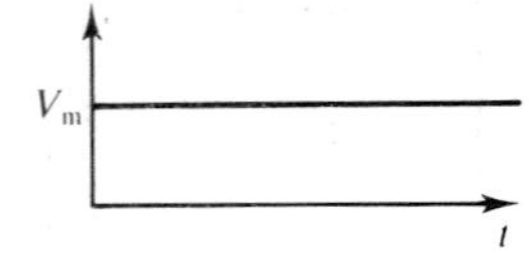	V_m	V_m	V_m	1	1
Sine wave or full wave (rectified wine wave) 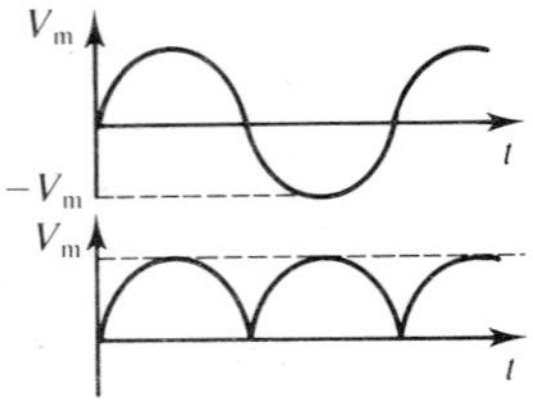	V_m	$\frac{V_m}{\sqrt{2}} = 0.707V_m$	$\frac{2}{\pi}V_m = 0.637V_m$	$\frac{\pi}{2\sqrt{2}} = 1.111$	$\sqrt{2} = 1.414$
Half wave rectified sine wave 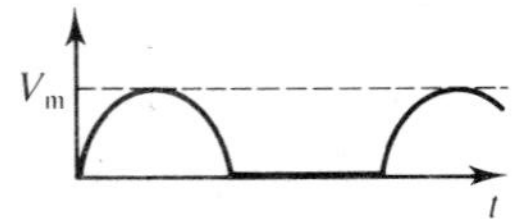	V_m	$\frac{1}{2}V_m$	$\frac{V_m}{\pi} = 0.318V_m$	$\frac{\pi}{2} = 1.571$	2
Square wave (symmetrical or nonsymmetrical) 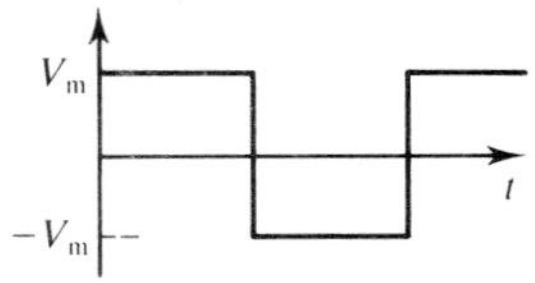	V_m	V_m	V_m	1	1
Saw tooth/tri-angular wave 	V_m	$\frac{V_m}{\sqrt{3}} = 0.577$	$\frac{V_m}{2}$	$\frac{2}{\sqrt{3}} = 1.155$	$\sqrt{3} = 1.732$

Gaussian noise	Peak is theoretically unlimited. The amount of the time peak is under a certain C.F. – see opposite	V_{RMS}	$\sqrt{\frac{2}{\pi}}\cdot V_{RMS} = 0.798\,V_{RMS}$	$\sqrt{\frac{\pi}{2}} = 1.253$	C.F. / % of time less than C 1 / 32% 2 / 4.6% 3 / 0.37% 3.3 / 1000 ppm 3.9 / 100 ppm 4.0 / 63 ppm 4.4 / 10 ppm 4.9 / 1 ppm
Rectangular pulse train $K = \frac{t_1}{T}$ $[0 < K < 1]$	V_m	$V_m\sqrt{K}$	KV_m	$\frac{1}{\sqrt{K}}$	$\frac{1}{\sqrt{K}}$
Exponential pulse train Time constant τ	V_m	$V_m\left[\frac{\tau}{2T}\left(1-e^{-\frac{2T}{\tau}}\right)\right]^{\frac{1}{2}}$ $\approx V_m\sqrt{\frac{\tau}{2T}}$ if $T \gg \tau$	$\frac{\tau V_m}{T}\left(1-e^{-\frac{T}{\tau}}\right)$ $\approx \cdot\frac{\tau}{T}$ if $T \gg \tau$	$\frac{\left[\frac{T}{2\tau}\left(1-e^{-\frac{2T}{\tau}}\right)\right]^{\frac{1}{2}}}{\left(1-e^{-\frac{T}{\tau}}\right)}$ $\approx\sqrt{\frac{T}{2\tau}}$ if $T \gg \tau$	$\left[\frac{\tau}{2T}\left(1-e^{-\frac{2T}{\tau}}\right)\right.$ $\approx\sqrt{\frac{2T}{\tau}}$ if $T \gg \tau$
Rectangular pulse train average = 0	$V_m(1-K)$	$V_m\sqrt{1-K)K}$	$2V_mK(1-K)$	$\frac{1}{2\sqrt{K(1-K)}}$	$\sqrt{\left(\frac{1}{K}-1\right)}$

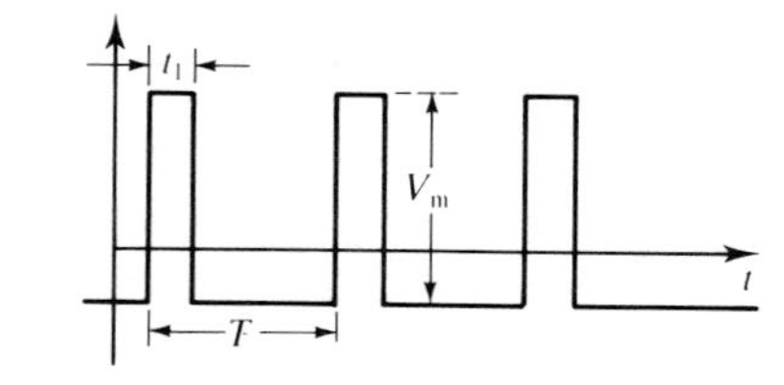

$K = \frac{t_1}{T}$ $(0 < K < \frac{1}{2})$

12.2 Mean absolute value circuits

A mean absolute value block diagram is shown in Fig. 12.2. There are two parts to these circuits, the rectifier and the low pass filter with each part being very easy to realize in terms of electronic components. The rectifier can be either half wave or full wave (see Chapter 11 for further information on rectifier circuits). The low pass filter is not usually a critical design and in most cases a simple resistor/capacitor/buffer arrangement will suffice. The filter is designed to provide the averaging part of the operation. The time constant of the averager is usually made much greater than the period of the input waveform and, as with peak detectors, the choice of the smoothing time constant is a trade-off between output ripple error and speed of response. Full wave rectification will often be needed to minimize ripple in low frequency applications.

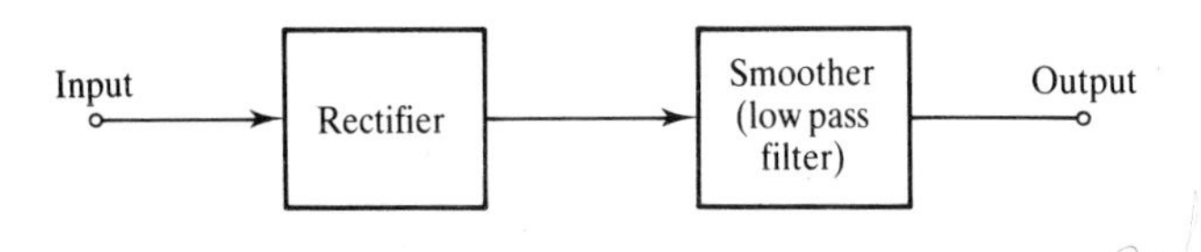

Fig. 12.2. Mean absolute value circuit block diagram.

12.3 Root mean square circuits

You can determine the RMS value of a waveform electronically with multiplier circuits by using either of the two techniques shown in Fig. 12.3. One technique is known as direct computation which directly computes the square, the mean and the square root of the analog value. There are several major drawbacks with using this direct approach. Firstly, two multiplier/divider circuits are needed. Secondly, there is a restricted dynamic range because the squaring is followed by the square rooting operation. An input signal with a dynamic range of 1000:1 (for example, 10 mV:10 V), would require a signal at the output of the squarer with a dynamic range of 1 000 000:1 (i.e. 10 μV:10 V). Due to the problems with dynamic range and the need for two multiplier/divider circuits, the RMS value is seldom implemented by a direct computation circuit and this approach has only been described here to highlight the advantages of the implicit computation approach described below.

The implicit computation approach requires only one multiplier/divider block and has a good dynamic range. The operation of this circuit is as follows:

$$V_{OUT} = \overline{V}_1 \qquad \text{where } \overline{V}_1 \text{ denotes the mean of } V_1$$

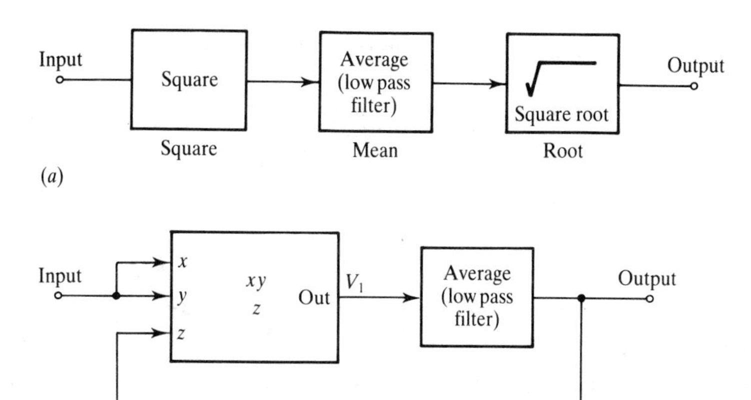

Fig. 12.3. RMS converter block diagrams using multipliers. (*a*) Direct computation. (*b*) Implicit computation.

but

$$V_1 = V_{IN}^2/V_{OUT}$$

so

$$V_{OUT} = \overline{\left(\frac{V_{IN}^2}{V_{OUT}}\right)} = \frac{\overline{V_{IN}^2}}{V_{OUT}}$$

(since V_{OUT} is effectively constant with respect to the the input waveform) therefore

$$V_{OUT} = \sqrt{\overline{V_{IN}^2}} \qquad \text{i.e. the RMS.}$$

There are several ways of realizing this RMS circuit. One obvious way is to buy commercial RMS to DC converters (by far the most popular and convenient approach). Alternatively you could use a multiplier and divider ic (a semi-customized approach). Lastly, you could custom design the entire circuit using super-matched transistors (for the DIY enthusiast).

A 'semi-custom' designed RMS to DC converter can be realized, as shown in Fig. 12.4, by using a multiplier and divider ic (or a multi-function converter). The transfer function of this circuit is

$$V_{OUT}^2 = \frac{V_{IN}^2}{(1 + \frac{1}{2}RC)}$$

and so if the input frequency is $\gg \dfrac{1}{\pi RC}$ Hz

then

$$V_{OUT} = \sqrt{\overline{V_{IN}^2}}, \qquad \text{i.e. the RMS}$$

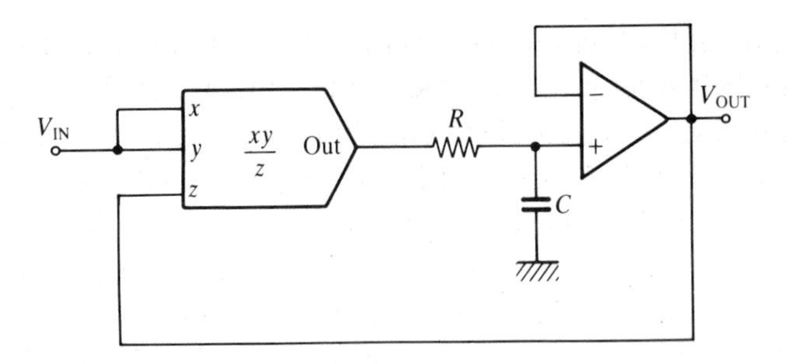

Fig. 12.4. RMS to DC converter using a multiplier/divider block.

For those independent, dyed-in-the-wool customizers, the entire circuit can be designed using discrete components as shown in Fig. 12.5. The circuit is described here mainly because many of the commercial ics are based upon this configuration. Assuming that all the transistors and resistors are perfectly matched, and that the input frequency

$$\gg \frac{1}{\pi R_3 C_3} \text{ Hz},$$

then

$$V_{OUT} = \sqrt{\frac{R_3 R_4}{R_1{}^2}} \cdot \sqrt{\overline{V_{IN}{}^2}}$$

The circuit is built up from two main parts. Firstly, a full wave rectifier based around op amp A_1 (see Chapter 11 for further details on this part of the circuit). Secondly, a multiplier/divider circuit employing the log–antilog principle (see Chapter 8 for further details on this part of the circuit).

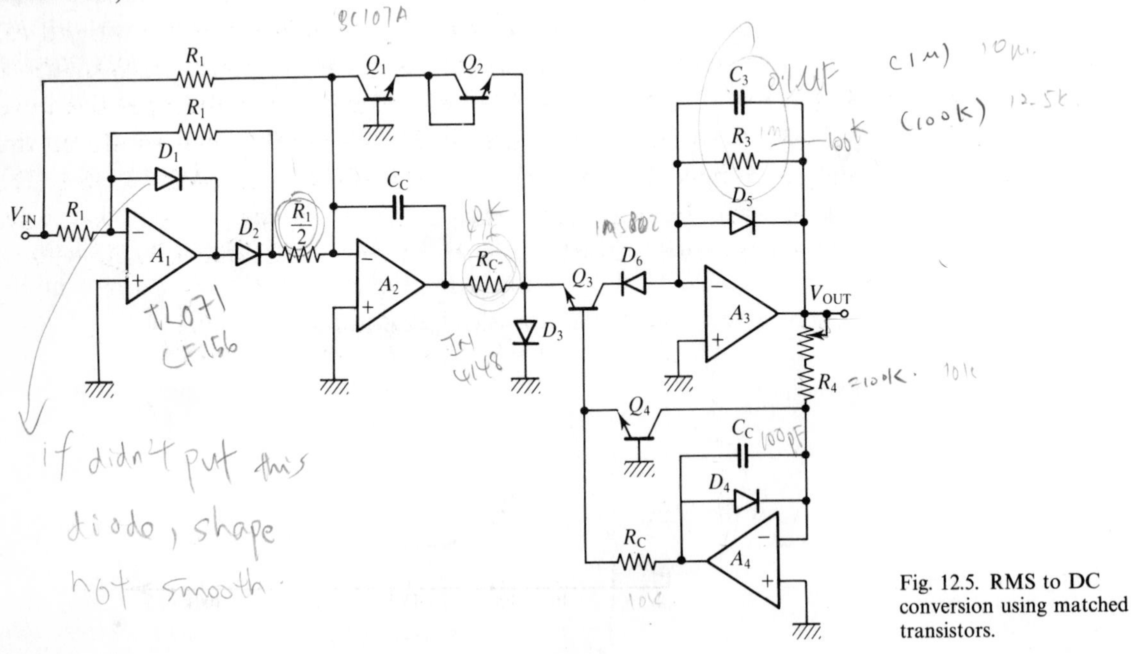

Fig. 12.5. RMS to DC conversion using matched transistors.

Index

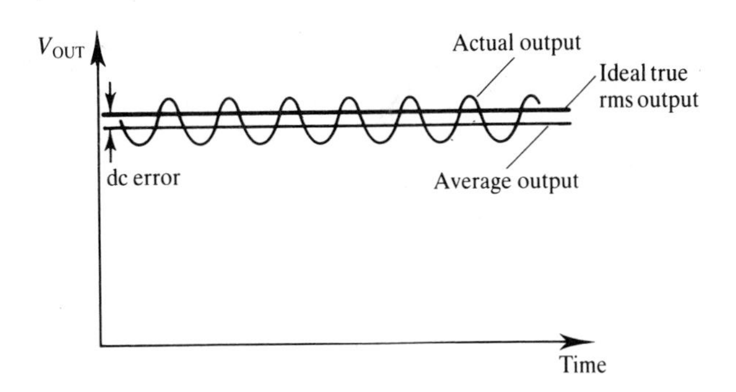

Fig. 12.7. Low frequency output of an RMS to DC converter.

error less than 1% for most commercial RMS to DC ics. This frequency range is much greater than $1/T$ where T is the time constant of the smoothing circuit being used. As the frequency of the input signal is reduced, the smoothing filter is not able to average as effectively and a greater ripple content appears on the output together with a DC output error as shown in Fig. 12.7.

Clearly, the ripple and DC error can be reduced by increasing the time constant of the filter, however this will slow down the response of the circuit. So, for low input frequencies, around 10 Hz, a compromise needs to be made between having a large time constant for smoothing which provides a slow response time with good accuracy or having a small time constant which provides a faster response time but poorer accuracy. The following expressions can be used for determining the minimum time constant for a particular DC error or ripple.

The transfer function for the RMS to DC converter has earlier been given as

$$V_{OUT}^2(s) = \frac{V_{IN}^2(s)}{(1 + sT)}$$

where T is $RC/2$ for the implicit computational circuit.

For an input sine wave,

$$V_{IN} = \sqrt{2} \cdot V_{RMS} \sin(\omega t) \qquad \text{Frequency} = \frac{\omega}{2\pi} \text{ Hz}$$

Then

$$V_{OUT}(t) = V_{RMS} \sqrt{\left(1 + \frac{\sin 2\omega t}{\sqrt{(1 + 4\omega^2 T_2)}}\right)}$$

By expanding the square root using the Taylor series approximation and using some simple trigonometric algebra, this yields:

Briefly, the following points should be borne in mind when designing. Resistors labelled R_1 must be closely matched, preferably 0.1% or better, for accurate full wave rectification of the input signal. Transistors Q_1, Q_2, Q_3 and Q_4 must be derived from a super-matched quad-transistor ic. Resistor R_3 and capacitor C_3 provide the averaging function. Diodes D_3, D_4 and D_5 protect the four transistors against damage from reverse base–emitter voltages. Resistors labelled R_c and capacitors labelled C_c provide frequency compensation in the circuit. Diode D_6 may be included to maintain the base–collector voltage of Q_3 near zero and so improve accuracy. Precision metal film or wirewound resistors should be used for R_1, R_3 and R_4.

For the DIY enthusiast who is not enthusiastic about deriving the values of components, you could try the following values.

$R_3 = 100\ \text{k}\Omega$ $\quad R_4 = 10\ \text{k}\Omega$ (with trimmer) $\quad R_3 = 1\ \text{M}\Omega$

$R_c = 10\ \text{k}\Omega$

$C_c = 100\ \text{pF}$ $\quad C_3 = 0.1\ \mu\text{F}$

Q_1, Q_2, Q_3, Q_4 a super-matched transistor array, e.g. MAT-04, CA3086 etc.

A_1, A_2, A_3, A_4 FET input op amps, e.g. LF156, TL071 etc.

$D_1, D_2, D_3, D_4, D_5, D_6$ general purpose diodes, e.g. IN4148

The frequency response figure of an RMS to DC converter provides a measure of how well the circuit can measure the RMS of sine wave input signals of varying frequencies. This frequency response is different from that used when describing filters or amplifiers which are essentially linear devices. The RMS to DC converter is a non-linear device with an AC input and a DC output and so has a frequency response which is defined as the % error on the output against the frequency of the input sine wave. A typical response is illustrated in Fig. 12.6. The values shown on this graph are typical for many presently available commercial RMS to DC converters. The 'mid frequencies' refer to the normal operating range of the converter and are typically between 10 Hz and 100 kHz with an

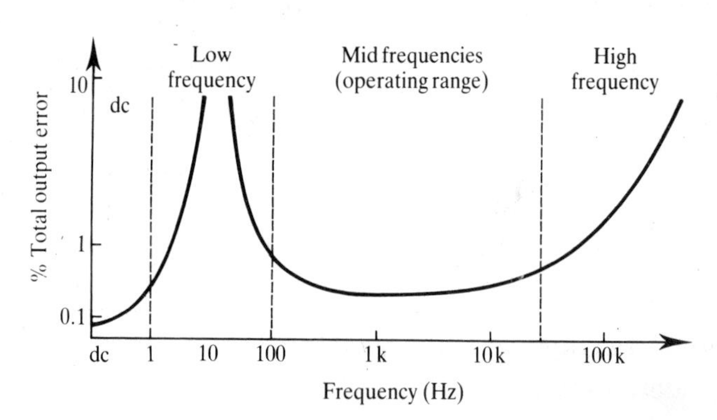

Fig. 12.6. Frequency response of an RMS to DC converter.

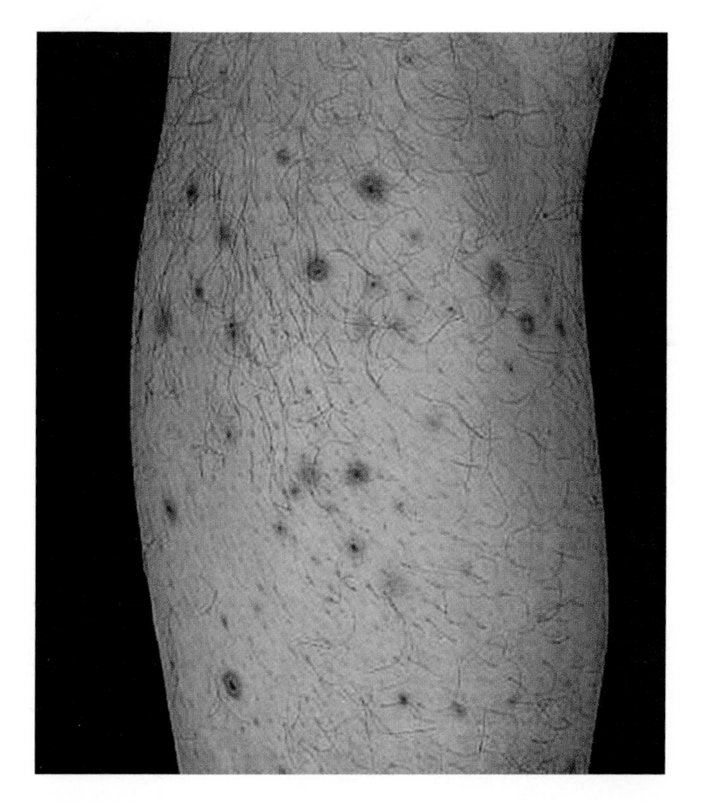

Figure 22.7 Vasculitis.

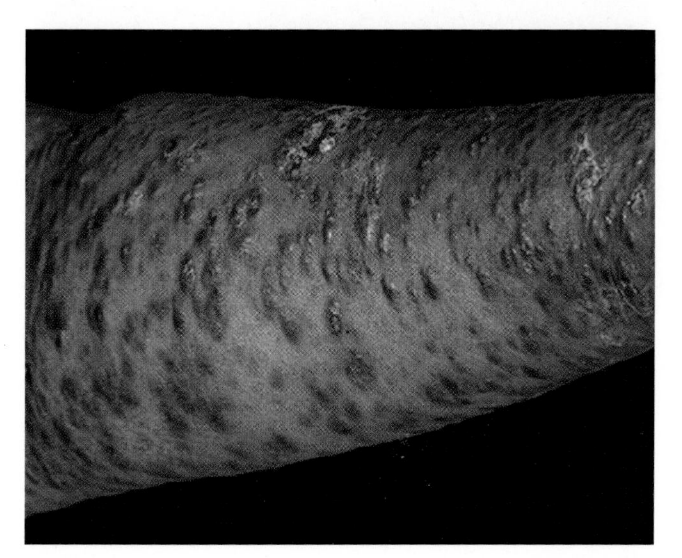

Figure 22.8 Lichenoid drug eruption.

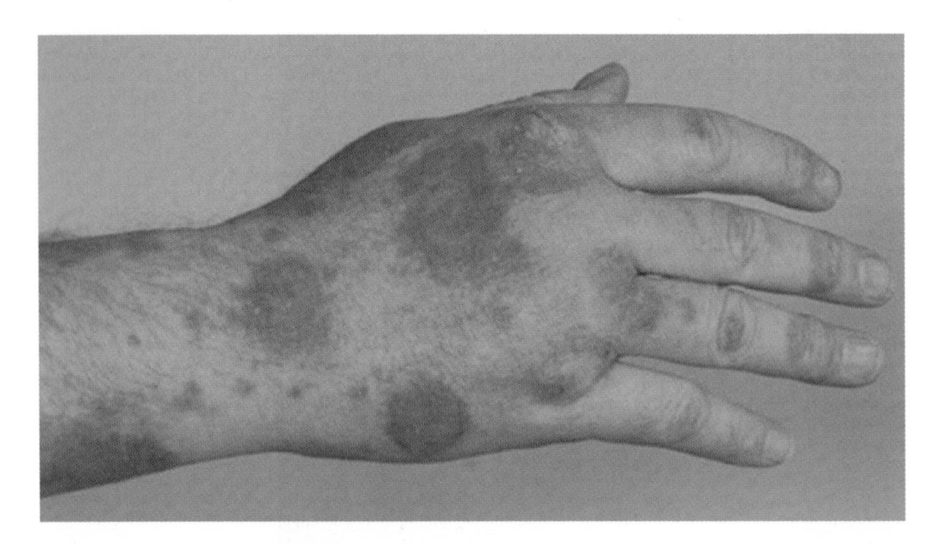

Figure 22.5 Erythema multiforme

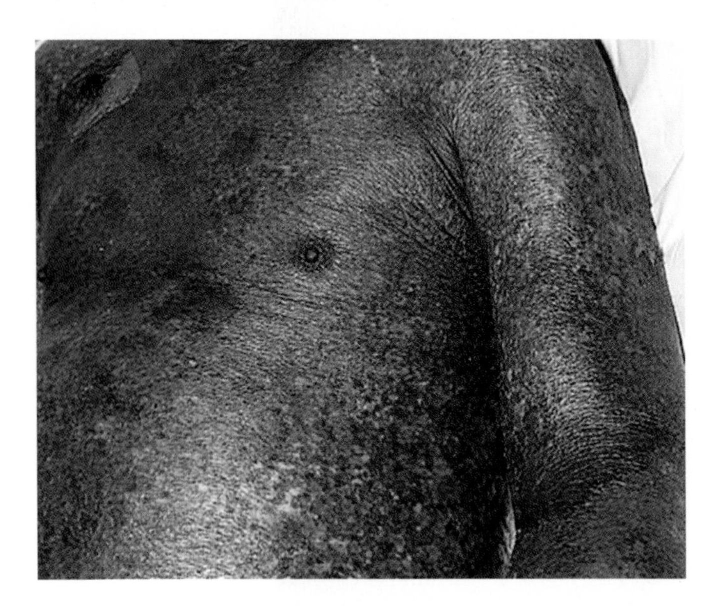

Figure 22.6 Exfoliative dermatitis.